CLIMATE CHANGE, FORCED MIGRATION, AND INTERNATIONAL LAW

Climate Change, Forced Migration, and International Law

JANE McADAM

OXFORD
UNIVERSITY PRESS

OXFORD
UNIVERSITY PRESS

Great Clarendon Street, Oxford OX2 6DP

Oxford University Press is a department of the University of Oxford.
It furthers the University's objective of excellence in research, scholarship,
and education by publishing worldwide in

Oxford New York

Auckland Cape Town Dar es Salaam Hong Kong Karachi
Kuala Lumpur Madrid Melbourne Mexico City Nairobi
New Delhi Shanghai Taipei Toronto

With offices in

Argentina Austria Brazil Chile Czech Republic France Greece
Guatemala Hungary Italy Japan Poland Portugal Singapore
South Korea Switzerland Thailand Turkey Ukraine Vietnam

Published in the United States
by Oxford University Press Inc., New York

First published 2012

British Library Cataloguing in Publication Data
Data available

Library of Congress Cataloging in Publication Data
Data available

Typeset by Newgen Imaging Systems (P) Ltd, Chennai, India
Printed and bound by
CPI Group (UK) Ltd, Croydon, CR0 4YY

ISBN 978–0–19–958708–7

3 5 7 9 10 8 6 4

Acknowledgements

It is commonly assumed that academia is an isolating occupation, where one's primary companions are books, databases, and computers. However, while this book has of course necessitated the companionship of these three things, it has also been greatly enriched by the insights of many people around the world who have given generously of their time and expertise.

In particular, the project has benefited from fieldwork undertaken in the Pacific island countries of Kiribati and Tuvalu, and in Bangladesh and India. I am very grateful to all those who agreed to be interviewed for the project, whose insights helped me to understand the impacts of climate change on their homes.

I am indebted to colleagues who were willing to bounce around ideas with me, share literature, and give feedback on evolving arguments. I thank them for their support and encouragement, and particularly for pushing me to look at new angles, approaches, and sources. While there are too many people to list individually, special thanks go to Jon Barnett, John Campbell, Jean-François Durieux, Alice Edwards, Elizabeth Ferris, Rolf Einar Fife, Laurie Fransman, François Gemenne, Matthew Gibney, Guy Goodwin-Gill, Nina Hall, Agnès Hurwitz, Walter Kälin, Vikram Kolmannskog, Rosemary Rayfuse, Volker Türk, Koko Warner, and Roger Zetter. In particular, I would like to thank Maryanne Loughry for being a wonderful travel companion during our fieldwork in the Pacific, and for the rich discussions we had about the research, especially during its formative stages.

I have been fortunate to have the invaluable support of a number of highly talented research assistants during the course of this project. I would particularly like to thank Emily Crawford, my Research Associate for the first year, and Trina Ng, my research assistant for the following 18 months, for their excellent work. Trina's eye for detail and capacity to turn around work in record time was a great asset. Emma Dunlop and Naomi Hart scored the unenviable tasks of assisting me with proofreading, cross-referencing, and indexing, carrying them out with precision and goodwill. Matthew Albert, Kate Purcell, and Emma Truswell also provided ad hoc assistance at various points of the project, for which I am grateful.

This research would not have been possible without the support of a three-year Australian Research Council Discovery Grant. It enabled me to undertake fieldwork, teaching relief, and to employ research assistants, and gave me the space to think about the big picture. I am very grateful for this opportunity.

I have benefited from an exceptionally supportive research environment at the Law Faculty of the University of New South Wales. I thank the Dean, David Dixon, and the former and current Associate Deans (Research), Kathy Bowrey and Janet Chan, for fostering this research culture and for their personal encouragement. I also thank the current and former Directors of the Gilbert + Tobin Centre of Public Law, Andrew Lynch and George Williams, for their support and generosity.

Thank you also to Lincoln College at the University of Oxford for hosting me as a Visiting Fellow in 2009. In particular, I thank Paul Langford, Simon Gardner, Christopher McCrudden, and Louise Durning.

I would also like to thank John Louth, Merel Alstein, Anthony Hinton, Zoë Organ, and Matthew Humphrys at Oxford University Press for their professionalism and patience throughout the writing and editing process, as well as the copy-editors and proofreaders who worked so carefully on the manuscript.

As always, my family has been exceptionally supportive, interested, and tolerant. I thank them for their understanding when the books, databases, computers, and travel beckoned. Finally, I thank Ben Saul, my intrepid fieldwork companion in Bangladesh and India, for his partnership on the homefront.

The material in this book is current as at October 2011.

Jane McAdam
Sydney
December 2011

Contents

Table of Treaties, Other International Instruments, and Legislation

TREATIES AND OTHER INTERNATIONAL INSTRUMENTS

LEGISLATION

List of Abbreviations

AIT	Asylum and Immigration Tribunal (UK)
ALP	Australian Labor Party
APF	Asia Pacific Forum of National Human Rights Institutions
ASK	Ain o Salish Kendra (Bangladesh)
AusAID	Australian Agency for International Development
BELA	Bangladesh Environmental Lawyers Association
BMET	Bureau of Manpower Employment and Training (Bangladesh)
CAT	Convention against Torture and Other Cruel, Inhuman or Degrading Treatment or Punishment
CCEMA	Climate Change, Environment and Migration Alliance
CDMP	Comprehensive Disaster Management Programme (Bangladesh)
CRC	Convention on the Rights of the Child
DED	Deferred Enforced Departure (US)
DFID	Department for International Development (UK)
EACH-FOR	Environmental Change and Forced Migration Scenarios
ECHR	European Convention on Human Rights
ECOSOC	United Nations Economic and Social Council
ECOWAS	Economic Community of West African States
ECRE	European Council on Refugees and Exiles
ECtHR	European Court of Human Rights
EEZ	Exclusive Economic Zone
EU	European Union
EVD	Extended Voluntary Departure (US)
FAO	Food and Agriculture Organization of the United Nations
GHG	Greenhouse gas
H&C	Humanitarian and Compassionate (Canada)
HRC	United Nations Human Rights Committee
IASC	Inter-Agency Standing Committee
ICCPR	International Covenant on Civil and Political Rights
ICESCR	International Covenant on Economic, Social and Cultural Rights
ICJ	International Court of Justice
ICRC	International Committee of the Red Cross
IDMC	Internal Displacement Monitoring Centre
IDPs	Internally Displaced Persons
IFRC	International Federation of Red Cross and Red Cresent Societies

IISD	International Institute for Sustainable Development
ILC	International Law Commission
ILO	International Labour Organization
IOM	International Organization for Migration
IPCC	Intergovernmental Panel on Climate Change
IRB	Immigration and Refugee Board (Canada)
KANI	Kiribati–Australia Nursing Initiative
LDC	Least Developed Country
NAPA	National Adaptation Programme of Action
NGO	Non-governmental organization
OAU	Organization of African Unity
OCHA	United Nations Office for the Coordination of Humanitarian Affairs
OHCHR	United Nations Office of the High Commissioner for Human Rights
PAC	Pacific Access Category (New Zealand)
PICs	Pacific Island Countries
PRRA	Pre-Removal Risk Assessment (Canada)
RRT	Refugee Review Tribunal (Australia)
RSAA	Refugee Status Appeals Authority (New Zealand)
RSE	Recognised Seasonal Employer Scheme (New Zealand)
SIDS	Small-Island Developing States
TPS	Temporary Protected Status (US)
UDHR	Universal Declaration of Human Rights
UK	United Kingdom
UN	United Nations
UNDP	United Nations Development Programme
UNEP	United Nations Environment Programme
UNFCCC	United Nations Framework Convention on Climate Change
UNFPA	United Nations Population Fund
UNGA	United Nations General Assembly
UNGAOR	United Nations General Assembly Official Records
UNHCR	United Nations High Commissioner for Refugees
UNICEF	United Nations Children's Fund
UNSC	United Nations Security Council
UNSCOR	United Nations Security Council Official Records
UNTS	United Nations Treaty Series
US	United States
WFP	United Nations World Food Programme
WHO	World Health Organization

Introduction

> We want to begin that [the migration process] now, and do it over the next twenty, thirty or forty years, rather than merely, in fifty to sixty years time, simply come looking for somewhere to settle our one hundred thousand people because they can no longer live in Kiribati, because they will either be dead or drown. We begin the process now, it's a win–win for all and very painless, but I think if we come as refugees, in fifty to sixty years time, I think they would become a football to be kicked around.[1]

Movement in response to environmental and climate change is a normal human adaptation strategy. It can provide a means of escaping danger and increasing resilience, especially when it is planned. The difficulty is that people cannot simply migrate as and when they choose: national immigration laws restrict the entry of non-citizens into other countries. International law recognizes only a very small class of forced migrants as people whom other countries have an obligation to protect: 'refugees', 'stateless persons', and those eligible for complementary protection. This means that unless people fall within one of those categories, or can migrate lawfully for reasons such as employment, family, or education, they run the risk of interdiction, detention, and expulsion if they attempt to cross an international border without a legal entitlement to enter that other country, and they may face a precarious existence there if they do manage to evade detection. Cross-border displacement stemming from natural disasters and the effects of climate change has thus been identified as a normative gap in the international legal protection regime.[2]

The impacts of climate change will be felt differently in different communities, since people's ability to cope with them will be affected by underlying political, economic, and social conditions. The economic, social, and political capacity of a country—including its infrastructure, economic resilience, and ability to assist people to rebuild their lives—will necessarily affect mobility decisions. This is the case in relation both to people's initial resilience to a disaster (linked to the adaptation and prevention strategies that have been put in place), as well as

[1] President Anote Tong of Kiribati, cited in Duncan Wilson, 'Climate Change: Nobody is Immune' (*Islands Business*, 2008) <http://www.islandsbusiness.com/islands_business/index_dynamic/containerNameToReplace=MiddleMiddle/focusModuleID=18087/overideSkinName=issueArticle-full.tpl> accessed 13 August 2011.

[2] UNHCR, 'High Commissioner's Dialogue on Protection Challenges: Breakout Session 1: Gaps in the International Protection Framework and its Implementation: Report by the Co-Chairs' (8–9 December 2010) 3; 'Statement by António Guterres, United Nations High Commissioner for Refugees' (Nansen Conference on Climate Change and Displacement in the 21st Century, Oslo, 6 June 2011) 4; 'Chairperson's Summary' (Nansen Conference on Climate Change and Displacement in the 21st Century, Oslo, 6–7 June 2011) para 23.

the duration of its impacts on their lives (linked to their capacity to rebuild and rehabilitate).

While there remains uncertainty about how severe global warming will be and its precise impacts on human society,[3] there is consensus among 97 per cent of scientific experts that a rapid build-up of greenhouse gas is a result of human activities and this will worsen without mitigation.[4] Most climate change projections are based on a two degrees Celsius global mean temperature rise above the temperature in 1850, which has been agreed by States to be the threshold for 'dangerous' climate change.[5] Since two degrees Celsius warming is the benchmark that underscores most climate modelling and emissions targets, and is implicit in studies of the social impacts of climate change, it is adopted in this book. While global warming had generally matched the predictions by the Intergovernmental Panel on Climate Change (IPCC),[6] some scientists now argue that the IPCC's estimates are conservative and that change may be occurring faster than anticipated.[7] There has been little analysis in the social sciences of the impacts of temperature rise beyond two degrees Celsius, but it seems clear that if warming were to exceed four degrees Selsius, the impacts could be significantly different since such temperature increases could threaten people's very survival in certain parts of the world.[8]

During the 1980s and 1990s, climate change was predominantly conceived of as a scientific and environmental issue. However, in 1990, the potential impacts of climate change on human migration were identified by the IPCC. It noted that millions of people would likely be uprooted by shoreline erosion, coastal flooding, and agricultural disruption,[9] and that climate change might necessitate consideration of 'migration and resettlement outside of national boundaries'.[10]

[3] Australian Academy of Science, *The Science of Climate Change: Questions and Answers* (Australian Academy of Science, 2010) 16.

[4] Peter T Doran and Maggie Kendall Zimmerman, 'Examining the Scientific Consensus on Climate Change' (2009) 90 Eos, Transactions, American Geophysical Union 22; William RL Anderegg and others, 'Expert Credibility in Climate Change' (2010) 107 Proceedings of the National Academy of Sciences 12107; Naomi Oreskes, 'Beyond the Ivory Tower: The Scientific Consensus on Climate Change' (2004) 306 Science 1686.

[5] This is as much a political decision as a scientific one, as there is no particular scientific rationale for selecting this figure: Ben Saul and others, *Climate Change and Australia: Warming to the Global Challenge* (Federation Press, forthcoming, 2012).

[6] Stefan Rahmstorf and others, 'Recent Climate Observations Compared to Projections' (2007) 316 Science 709; Australian Academy of Science (n 3) 12.

[7] Australian Academy of Science (n 3), 12 suggests that 3°C may be more accurate; see also AP Sokolov and others, 'Probabilistic Forecast for Twenty-First-Century Climate Based on Uncertainties in Emissions (Without Policy) and Climate Parameters' (2009) 22 Journal of Climate 5175.

[8] See, eg, François Gemenne, 'Climate-Induced Population Displacements in a 4°C+ World' (2011) 369 Philosophical Transactions of the Royal Society A 182.

[9] Intergovernmental Panel on Climate Change (IPCC), *Climate Change: The IPCC Scientific Assessment: Final Report of Working Group I* (Cambridge University Press, 1990). This was underscored in its most recent report: see, eg, Gabriele C Hegerl and others, 'Understanding and Attributing Climate Change' in Susan Solomon and others (eds), *Climate Change 2007: The Physical Science Basis: Contribution of Working Group I to the Fourth Assessment Report of the Intergovernmental Panel on Climate Change* (Cambridge University Press, 2007); IPCC, *Climate Change 2007: Synthesis Report:Summary for Policymakers* (Cambridge University Press, 2007).

[10] Robert T Watson and others (eds), *IPCC Special Report: The Regional Impacts of Climate Change: An Assessment of Vulnerability: Summary for Policymakers* (IPCC, November 1997) Part 6.8.

Five years earlier, the term 'environmental refugee' had been used in international bureaucratic discourse for the first time. In a 1985 report for the United Nations (UN) Environment Programme, El-Hinnawi had employed the term as a means of describing people 'forced to leave their traditional habitat, temporarily or permanently, because of a marked environmental disruption (natural and/or triggered by people) that jeopardized their existence and/or seriously affected the quality of their life'.[11] Although he used the language of refugee protection, his aim was to draw attention to the damaging effects of anthropogenic climate change on human settlement, rather than to advocate for the extension of the international protection regime to people displaced by it.

In legal terms, there is no such thing as a 'climate change refugee'. There is, however, evidence that people are moving in response to the effects of climate change—both in terms of its direct physical impacts (eg flight from a flood or cyclone), and its attendant socio-economic effects (on infrastructure, shelter, food and water supplies, livelihoods, and so on). On the one hand, this is nothing new. Movement away from disasters and their effects has always been a rational, adaptive response. Similarly, movement away from the impacts of longer-term processes, such as drought, is a well-documented survival strategy. In some respects, discussions about climate-related movement simply represent a repackaging of the broader debate in the 1990s about environmental displacement.[12]

On the other hand, the 'newness' of displacement triggered (at least in part) by climate change is its underlying anthropogenic basis,[13] the large number of people thought to be susceptible to it,[14] and the speed with which climate change will magnify the risk of extreme weather events and rises in temperature, sea levels, and so on, which means that people's traditional coping strategies are likely to be overwhelmed at some point. As the IPCC has observed, '[w]hile physical exposure can significantly influence vulnerability for both human populations and natural systems, a lack of adaptive capacity is often the most important factor that creates a

[11] Essam El-Hinnawi, *Environmental Refugees* (United Nations Environment Programme, 1985) 4.

[12] See, eg, Jodi L Jacobsen, 'Environmental Refugees: A Yardstick of Habitability' (1988) Worldwatch Paper 86; Astri Suhrke and Annamaria Visentin, 'The Environmental Refugee: A New Approach' [1991] Ecodecision 73; Norman Myers, 'Environmental Refugees in a Globally Warmed World' (1993) 43 BioScience 752; Gregory S McCue, 'Environmental Refugees: Applying International Environmental Law to Involuntary Migration' (1993) 6 Georgetown International Environmental Law Review 151; Astri Suhrke, 'Environmental Degradation and Population Flows' (1994) 47 Journal of International Affairs 473; Graeme Hugo, 'Environmental Concerns and International Migration' (1996) 30 International Migration Review 105; Gaim Kibreab, 'Environmental Causes and Impact of Refugee Movements: A Critique of the Current Debate' (1997) 21 Disasters 20; Betsy Hartmann, 'Population, Environment and Security: A New Trinity' (1998) 10 Environment and Urbanization 113; Jessica B Cooper, 'Environmental Refugees: Meeting the Requirements of the Refugee Definition' (1998) 6 New York University Environmental Law Journal 480; Steve Lonergan, 'The Role of Environmental Degradation in Population Displacement' (1998) 4 Environmental Change and Security Project Report 5.

[13] That is not to say that 'natural' disasters are without anthropogenic bases: see, eg, Ben Wisner and others, *At Risk: Natural Hazards, People's Vulnerability and Disasters* (2nd edn, Routledge, 2004), who argue that few disasters are ever 'natural'; they are a combination of environmental *and* socio-economic and political factors.

[14] Debates about numbers remain highly contentious: see Ch 1.

hotspot of human vulnerability'.[15] A country's level of development is central to its adaptive capacity, since resources and technology increase capacity, while poverty limits it.[16] Thus, although the effects of climate change are indiscriminate, they will be felt more acutely in some parts of the world than in others.[17]

Since the IPCC first observed in 1990 that '[m]igration and resettlement may be the most threatening short-term effects of climate change on human settlements',[18] the relationship between climate change and forced migration has emerged as an increasingly studied—but contested—field of inquiry. Reasoned, empirically grounded analysis has too often been overshadowed by alarmist and ill-informed assumptions about the numbers of people on the move and the nature of that movement.[19] This has fed populist notions of 'sinking nations' and mass influxes of 'climate refugees', which in turn have fuelled the idea that climate change-related displacement is a threat to international security. This is both unfortunate and ironic. For while an underlying objective of those who take an alarmist perspective is typically to highlight the deleterious effects of climate change on human society—in particular, on those who are displaced—their focus on mass movement also leverages an anti-immigration–security agenda.[20] Thus, rather than expanding movement options for the displaced, they may inadvertently close them down. A

[15] Martin L Parry and others (eds), *Climate Change 2007: Impacts, Adaptation and Vulnerability: Contribution of Working Group II to the Fourth Assessment Report of the Intergovernmental Panel on Climate Change* (Cambridge University Press, 2007) 317. 'The two main elements that give rise to risk are the *hazards*—the potential damaging events or phenomenon—and the *vulnerability* of populations to these hazards. Natural hazards by themselves do not cause disasters; it is the combination of an exposed, vulnerable and ill-prepared population or community with a hazard event that results in a disaster. Human activity, such as land use changes, environmental exploitation and unplanned settlement, often exacerbates the level of disaster risk': Inter-Agency Standing Committee (IASC), 'Disaster Risk Reduction Strategies and Risk Management Practices: Critical Elements for Adaptation to Climate Change' (Submission to the UNFCCC Ad Hoc Working Group on Long Term Cooperative Action by the Informal Taskforce on Climate Change of the Inter-Agency Standing Committee and the International Strategy for Disaster Reduction, 11 November 2008) 4.

[16] International Organization for Migration (IOM), *Assessing the Evidence: Environment, Climate Change and Migration in Bangladesh* (IOM, 2010) 8.

[17] Rex V Cruz and others, 'Asia' in Parry and others (n 15) 488; W Neil Adger and others, 'Assessment of Adaptation Practices, Options, Constraints and Capacity' in Parry and others (n 15) 733–4; Robert J Nicholls and Richard SJ Tol, 'Impacts and Responses to Sea-Level Rise: A Global Analysis of the SRES Scenarios over the Twenty-First Century (2006) 364 Philosophical Transactions of the Royal Society A 1073. See also Council of the European Union, 'Climate Change and International Security', (Report from the Commission and the Secretary-General/High-Representative to the European Council) Doc 7249/08 (3 March 2008).

[18] William JM Tegart, Gordon W Sheldon and D Colin Griffiths (eds), *Climate Change: The IPCC Impacts Assessment* (Australian Government Publishing Service, 1990) 5–9.

[19] As Kibreab notes, numbers are based on 'educated guesses and cannot therefore be the basis of evidence-based policy which can help with pre-emption, mitigation or adaptation': Gaim Kibreab, 'Climate Change and Human Migration: A Tenuous Relationship?' (2009) 20 Fordham Environmental Law Review 357, 400.

[20] On which, see Lorraine Elliott, 'Climate Migration and Climate Migrants: What Threat, Whose Security?' in Jane McAdam (ed), *Climate Change and Displacement: Multidisciplinary Perspectives* (Hart Publishing, 2010); Jane McAdam and Ben Saul, 'Displacement with Dignity: International Law and Policy Responses to Climate Change Migration and Security in Bangladesh' (2010) 53 German Yearbook of International Law 233.

number of developed States have already 'flipped' the security discourse away from the human security of the displaced, towards an insular self-protection response.[21]

This book takes a more sober approach in examining how climate change impacts may affect patterns of human mobility, and whether existing legal and policy frameworks are sufficiently well equipped to respond. Through a combination of theoretical analysis and empirically grounded case studies based on fieldwork in Kiribati, Tuvalu, and Bangladesh,[22] it argues that acknowledgement of the multicausality of climate-related movement may in fact open up a wider range of legal and policy 'solutions' to address it. The book is based on the following premises:

1. Climate change is having real impacts on people's lives, but in most cases it is only one of a number of reasons why people decide to move. This is because climate change tends to multiply pre-existing stressors, rather than solely causing movement. In other words, climate change acts as a threat multiplier, which magnifies existing vulnerabilities.
2. The complexity of migration decisions and the interconnectedness of environmental, economic, social, and political factors make it virtually impossible to provide an accurate estimate of people who move 'because of' climate change.
3. Most movement is likely to be internal, rather than across international borders.
4. Existing legal regimes do not provide adequate protection or migration pathways for cross-border movement.
5. Principles of international human rights law, refugee law, and general international law principles of dignity, humanity, and international cooperation provide a useful normative framework to underpin a variety of different legal and policy strategies.
6. Since climate change-related movement manifests in a number of very different ways, a one-size-fits-all response is inappropriate.
7. It is essential that international, regional, and national responses are informed by a bottom-up approach, taking account of the desires of affected communities and responding appropriately to their needs and concerns.

It is likely that the three main conceptual factors that will shape legal and policy responses will be whether movement is perceived as being (a) voluntary or forced; (b) temporary or permanent; and (c) internal or across an international border. The first of these is especially complex when it comes to climate change-related movement. Human movement is theorized as a *continuum*, with forced migration at one end and voluntary movement at the other. It is based on the degree of compulsion or choice that motivates individual and household decisions to

[21] See Ch 9.

[22] This methodology seeks to embed a human rights-sensitive participatory approach, which is recognized as being fundamental to decisions relating to climate-related movement.

move. The international protection regime is predicated on the idea of forced exile, and is premised on the responsibility of other States to extend legal protection (as opposed to humanitarian assistance) if a particular individual engages a State's *non-refoulement* obligations under international law.

'Migration', by contrast, does not enliven international legal duties beyond States' obligations under human rights law generally, which are owed to all people within their territory or jurisdiction.[23] Thus, the nature of movement, and motivations for it, are significant when it comes to legal responses. These neat legal categorizations do not sit easily with sociological explanations for movement, which reveal an enormous degree of complexity in decision-making. Slow-onset impacts of climate change, in particular, pose a challenge to traditional understandings of 'forced' migration: although people may have no prospect of a sustainable livelihood if they remain in their home, they are not—yet—facing imminent harm. While pre-emptive movement in such circumstances may be a rational human response, neither international law nor most national laws facilitate this.

Partly as a result of these conceptual issues, there is no uniform terminology used to describe people who move in response to the impacts of climate change. Other reasons include concerns about the use of the term 'climate refugee'—which is not only erroneous as a matter of law but is a label rejected by many to whom it is attributed[24]—and whether climate change can be singled out as a cause of movement, or is simply a multiplier of pre-existing threats. This book uses the terms 'climate change-related movement' and 'climate-related movement' interchangeably to describe movement that is linked, at least in part, to the impacts of climate change. 'Displacement' is used to denote a response to a sudden extreme weather event, such as a flood, cyclone, or collapse of a riverbank, even though people may only move temporarily or over a short distance. The term 'migration' is used to imply a degree of decision-making in the nature, timing, and location of movement.[25] 'Choice'—which is a feature of the prevailing forced/voluntary, displacement/migration paradigm in the scholarly discourse—is generally avoided as a distinguishing criterion because decisions to move are often compelled by underlying, inter-related conditions of poverty, environmental degradation, socio-political factors, and the adverse impacts of climate change, making it difficult to disentangle precise motivations. Although this book at times refers to people fleeing from the impacts, or moving on account, of climate change, this is for ease of reference only. It is not intended to undermine the complex multicausality of movement discussed above and in Chapter 1.

[23] The poorly ratified Migrant Workers Convention contains particular responsibilities towards migrant workers and their families, but again the focus is on treatment in the host State rather than any obligation to admit people. See International Convention on the Protection of the Rights of All Migrant Workers and Members of Their Families (adopted 18 December 1990, entered into force 1 July 2003) 2220 UNTS 3.

[24] See Ch 2.

[25] This is not intended to downplay the degree of agency that people exercise in determining when to leave and where to go, or, conversely, not to leave in a 'displacement' scenario. Rather, it is employed as a means of distinguishing between the two forms of movement without relying on the language of 'choice'.

Indeed, the fact that there is still no internationally agreed definition of what it means to be an environmental 'migrant', 'refugee', or 'displaced person' makes it difficult to systematically progress deliberations about appropriate multilateral legal and institutional responses. Questions of definition have governance implications, because they inform the appropriate location of climate-related movement institutionally—as an international, regional or local, developed and/or developing State concern/responsibility—as well as normatively—for example, within the existing refugee protection framework or under the UN Framework Convention on Climate Change (UNFCCC).[26] Furthermore, there is a risk that a one-size-fits-all response could downplay the cultural and livelihood needs of displaced communities and local knowledge bases for adaptation.

When the law is faced with a novel challenge, it may be brought to bear in a number of different ways. Existing legal principles might be elongated, adapted, or particularized to respond to new circumstances, whether through creative interpretation or extrapolation by analogy. Thus, for example, norms developed to protect refugees have been transplanted to address the similar situation (but for the fact of crossing an international border) of some internally displaced persons,[27] while women, children, and the disabled are subject to specialist treaty regimes while simultaneously coming within the protective framework of general 'human' rights treaties.[28]

An alternative strategy is to recognize the deficiency of existing legal norms, and to develop new norms in response. This option is normally more difficult because it requires the galvanizing of political will, and even when it does succeed, it carries the risk that agreement will be achieved at the lowest common denominator, and may suffer from lack of implementation and enforcement.[29] It must be recalled that the international legal regime rests on the consent of States, and while their actions sometimes can be constrained by their existing commitments, the creation of new law cannot be imposed by some omnipotent authority.

The purpose of this book is to examine the scope of existing international law to respond to climate change-related movement, and to identify its potential for future development and expansion. Stepping aside from political debates and

[26] United Nations Framework Convention on Climate Change (adopted 9 May 1992, entered into force 21 March 1993) 1771 UNTS 107 ('UNFCCC').

[27] See Guiding Principles on Internal Displacement, UN Doc E/CN.4/1998/53/Add.2 (11 February 1998); Ch 9.

[28] See respectively Convention on the Elimination of All Forms of Discrimination against Women (adopted 18 December 1979, entered into force 3 September 1981) 1249 UNTS 13; Convention on the Rights of the Child (adopted 20 November 1989, entered into force 2 September 1990) 1577 UNTS 3 ('CRC'); Convention on the Rights of Persons with Disabilities, UNGA res 61/106, Annex 1 (adopted 13 December 2006, entered into force 3 May 2008); International Covenant on Civil and Political Rights (adopted 16 December 1966, entered into force 23 March 1976) 999 UNTS 171 ('ICCPR'); International Covenant on Economic, Social and Cultural Rights (adopted 16 December 1966, entered into force 3 January 1976) 993 UNTS 3 ('ICESCR').

[29] This is examined in Jane McAdam and Ben Saul, 'An Insecure Climate for Human Security? Climate-Induced Displacement and International Law' in Alice Edwards and Carla Ferstman (eds), *Human Security and Non-Citizens: Law, Policy and International Affairs* (Cambridge University Press, 2010).

climate change advocacy, it systematically analyses international legal principles, jurisprudence, and State practice to determine where regulatory gaps exist, and how they might best be responded to. The book sounds a note of caution about advocating for the creation of new norms unless such advocacy is informed by empirical evidence and the needs and desires of affected groups.

The book takes a human rights-based approach to climate change-related movement. This requires more than simply identifying which rights are at risk. It also mandates understanding about what drives such movement, what the nature of that movement is likely to be, and what those who are affected by it actually desire in terms of legal and policy solutions. As the International Law Commission (ILC) has explained:

> a rights-based approach deals with situations not simply in terms of human needs, but in terms of society's obligation to respond to the inalienable rights of individuals, empowers them to demand justice as a right, not as a charity, and gives communities a moral basis from which to claim international assistance when needed.[30]

Top-down approaches will be ineffective unless they recognize the complex interrelationships between migration, development, adaptation, and sustainability; the great diversity of mobility responses that are encapsulated by the umbrella term 'climate change displacement/migration'; and the idiosyncratic features of particular contexts, which require flexibility and creativity in crafting solutions. While overarching legal principles of human rights law, dignity, humanity, and international cooperation provide helpful normative guidance, a one-size-fits-all solution is impossible. Certainly, such principles should underpin multiple and complementary interventions at the international, regional, national, and local levels, but they must also be given meaning by grassroots initiatives and participation.

I. The Book's Structure

The term 'climate change refugee' is sometimes used to describe people who will be forced to leave their homes as a result of climate change. Chapter 1 explains why the use of this term is misleading, but also why there are gaps in the existing international legal frameworks for protecting such people. It considers the empirical nature of climate change-related movement; the numbers of people likely to move on account of climate change impacts; whether climate change can be considered 'the' cause of such movement; and the fundamental importance of understanding historical demographic and migration patterns in particular contexts, without which policy interventions are likely to be inappropriate.

International law is sometimes viewed as an abstract form of governance, wedded in technical definitions, mired by political compromise, and removed from the

[30] International Law Commission (ILC), 'Preliminary Report on the Protection of Persons in the Event of Disasters by Mr Eduardo Valencia-Ospina, Special Rapporteur', 60th Session, UN Doc A/CN.4/598 (5 May 2008) para 174.

social context it seeks to regulate. This is a common criticism of international refugee law, which premises protection on a 60 year old definition of a 'refugee', which, some argue, fails to protect the much broader category of 'forced migrants' presently on the move. It is true that international refugee law does not offer protection to all who might need it, and nor does it purport to do so. The refugee definition prescribes a class of person entitled to a particular legal status, and in this respect it serves an instrumental purpose. There is a considerable literature about the moral and legal justifications for this prescribed class.[31]

Nevertheless, State practice and the jurisprudence that has emerged over the decades since the Refugee Convention entered into force affirm that the 'refugee' definition is not fixed in a particular historical moment.[32] Rather, its interpretation has evolved as developments in international human rights law have informed the meaning of 'persecution' and the bases on which human beings inflict severe harm on others. While the text of the refugee definition has remained virtually static over time,[33] its interpretation has not. Chapter 2 therefore explores whether people forced to move from their homes on account of climate change-related impacts could meet the 'refugee' definition. It concludes that there are a number of obstacles to this, including that most people are likely to be displaced internally rather than across an international border (thus not triggering the Refugee Convention); the harm feared is not considered to amount to 'persecution' (including because there is no identifiable agent of persecution); even if the harm is regarded as persecution, it is not for a Convention reason; and/or the country of origin itself is not withholding protection. Furthermore, many of those affected by climate change will not necessarily conceptualize the cause of their displacement as such—even if they can describe its physical impacts on their livelihoods[34]—and therefore may never seek 'protection'. The chapter also recounts the strong rejection of the 'refugee' label by Pacific islanders in Kiribati and Tuvalu, who see it as undermining their sense of agency and human dignity.

In the absence of considerable jurisprudential development, people fleeing the impacts of climate change are also unlikely to benefit (at least in the near future) from human rights-based 'complementary protection', which safeguards against return to arbitrary deprivation of life, torture, or cruel, inhuman or degrading treatment. Courts have carefully limited the meaning of 'inhuman or degrading treatment' so that it cannot be used as a remedy for general poverty, unemployment, or

[31] See, eg, Andrew Shacknove, 'Who is a Refugee?' (1985) 95 Ethics 274; James C Hathaway, 'Is Refugee Status Really Elitist? An Answer to the Ethical Challenge' in Jean-Yves Carlier and Dirk Vanheule (eds), *Europe and Refugees: A Challenge?* (Kluwer Law International, 1997); Matthew E Price, *Rethinking Asylum: History, Purpose, and Limits* (Cambridge University Press, 2009).

[32] Convention relating to the Status of Refugees (adopted 28 July 1951, entered into force 22 April 1954) 189 UNTS 137, Art 1A(2), read in conjunction with Protocol relating to the Status of Refugees (adopted 31 January 1967, entered into force 4 October 1967) 606 UNTS 267.

[33] The 1967 Protocol removed the temporal limitation of Art 1A(2) ('[a]s a result of events occurring before 1 January 1951') and, for new Contracting States, removed the option of entering a geographical reservation ('events occurring *in Europe*') (emphasis added).

[34] Interview with Abul Kalam Azad, Research Director, Bangladesh Institute of International and Strategic Studies (Dhaka, Bangladesh, 15 June 2010).

lack of resources or medical care, except in the most exceptional circumstances. Chapter 3 therefore examines when socio-economic forms of harm may be regarded as triggering the principle of *non-refoulement*, including whether they can do so independently or whether they must be re-characterized as violations of civil and political rights. It also examines whether States may be held responsible for climate change under international environmental law, and if so, whether this might provide another cause of action for people displaced on account of climate change impacts.

Chapter 4 turns to a consideration of legislative and ad hoc schemes that States have developed to assist people fleeing from the impacts of disasters and other forms of serious harm.[35] It provides a relatively descriptive compilation of temporary protection responses, asylum-type mechanisms, and ad hoc humanitarian schemes (group and individual) in order to identify the extent to which existing domestic and regional frameworks might respond to climate change-related movement. Primarily, it suggests that while there is already considerable State practice in this area, it is inconsistent and unpredictable. This means that protection is precarious—both in terms of when it is granted, and the legal status accorded to beneficiaries. In particular, there remains a protection gap for people forced to move on account of the slower-onset impacts of climate change.

There has been much speculation about the potential disappearance of small island States on account of climate change. Grounded in a case study of Kiribati and Tuvalu, Chapter 5 critiques the underlying assumptions about the nature of such disappearance. It argues that the loss of international legal personality by small island States is not inevitable, and even if territory is ultimately rendered uninhabitable, this will not automatically bring about the extinction of the State itself.[36] In particular, the chapter examines empirical evidence about the possible impacts of climate change on Kiribati and Tuvalu, intertwined with pre-existing environmental and socio-economic stressors. While climate change is not the sole contributing factor to island States' vulnerability, a combination of poor socio-economic conditions, natural resource and space limitations (including population growth and associated demographic pressures, ecosystem degradation, and competition for limited resources), and the impact of natural hazards complicates the ability of such States to adapt to climate change.[37] The chapter also shows why

[35] As UNHCR's Assistant High Commissioner (Protection) notes: '[A]sylum is also one of the responses suitable to situations which do not fit the classical refugee paradigm, and which involve a temporary protection need. This has been partially recognised, particularly in Europe, through subsidiary protection arrangements, but also through discretionary provisions of various sorts in the immigration laws of a number of countries outside this region': Erika Feller, Assistant High Commissioner (Protection), 'The Refugee Convention at 60: Still Fit for Its Purpose?' (Workshop on Refugees and the Refugee Convention 60 Years On: Protection and Identity, Prato, 2 May 2011) 8.

[36] See also 'Summary of Deliberations on Climate Change and Displacement' (UNHCR Expert Roundtable on Climate Change and Displacement, Bellagio, 22–25 February 2011) (April 2011) ('Bellagio Deliberations') para 30.

[37] See John Connell, 'Environmental Change, Economic Development, and Emigration in Tuvalu' (1999) 22 Pacific Studies 1; John Connell, 'Losing Ground? Tuvalu, the Greenhouse Effect and the Garbage Can' (2003) 44 Asia Pacific Viewpoint 89. There is also evidence that 'islands which have been subject to substantial human modification are inherently more vulnerable than those that

people who may move from affected small island States are unlikely to be regarded as 'stateless persons' as a matter of international law. It examines why proposals for *en masse* relocation of national groups to other States is problematic from the perspective of human rights law. There is a long history of relocation in the Pacific,[38] much of which occurred (involuntarily) during the colonial period.[39] The cultural and social problems associated with it can continue for generations, and in the absence of careful, culturally-sensitive negotiations with affected populations and potential host countries, it is unlikely to achieve favourable long-term outcomes. Finally, the chapter considers alternative governance structures to maintain nationhood, so that small island States may retain their culture, identity, and community, even if they are forced to move.

Chapter 6 turns to a case study of Bangladesh. It considers the extent to which climate change is likely to impact on displacement and migration in and from Bangladesh, and the kinds of legal and policy responses that might be implemented—at the national and international levels—to respond to it. An analysis of current patterns of movement, which are the most likely indicators of future movement,[40] reveals that most climate-related displacement in Bangladesh is *internal* rather than cross-border in nature. This counters alarmist forecasts that there will be mass international displacement from Bangladesh by the middle of this century, which has important ramifications for policy development.

In particular, the chapter underscores the need for multiple legal and policy strategies. For example, while adaptation must be financed and enhanced, migration and development options also need to be considered. Movement can be a sign of resilience, not just vulnerability, and can be implemented as a climate change adaptation strategy in and of itself. Nevertheless, adaptation cannot come at all costs, and it is essential that policies ensure that human dignity and human rights are respected.

Finally, the chapter considers the way in which internal climate-related displacement may have a 'domino' effect on international migration. As increased internal movement to cities increases pressure on urban infrastructure and settlements, city-dwellers with the requisite skills and resources to obtain existing migration visas may move abroad. It is unlikely that such migration will ever be described as

have not been modified': Nobuo Mimura and others, 'Small Islands' in Parry and others (n 15) 698. For example, the cost of infrastructure and settlement protection is a significant proportion of their GDP, which most small island States cannot afford: 694.

38 See Chs 1, 5, 7; John Campbell, 'Climate-Induced Community Relocation in the Pacific: The Meaning and Importance of Land' in McAdam (n 20); Michael D Lieber (ed), *Exiles and Migrants in Oceania* (The University Press of Hawaii, 1977).

39 Martin G Silverman, 'Introduction: Locating Relocation in Oceania' in Lieber (n 38) 4.

40 See, eg, Graeme Hugo, 'Climate Change-Induced Mobility and the Existing Migration Regime in Asia and the Pacific' in McAdam (n 20) 9; Jon Barnett and Michael Webber, *Accommodating Migration to Promote Adaptation to Climate Change* (Commission on Climate Change and Development 2009) 17.

climate change-related, but it may represent an indirect form of climate change-related movement.[41]

Chapters 7, 8, and 9 consider some possible legal and policy responses to climate change-related movement. Governance is very important because it can itself impact on how people respond to disasters or slow-onset change. Either it may cast movement as a sign of vulnerability, or it may harness it as a form of adaptation that is proactively developed in consultation with affected communities. Furthermore, the efficacy of governance will play a key role in whether the displaced return home, or stay away permanently.[42] Policymakers may in fact exacerbate displacement if they do not carefully plan for responses to it at the local, national, and international levels.

Chapter 7 draws on fieldwork undertaken in Kiribati, Tuvalu, and Bangladesh to argue that calls for a new treaty to address climate change-related movement are presently misplaced. First, the chapter queries the utility—and, importantly, the policy consequences—of pinning 'solutions' on a multilateral refugee-like instrument, in light of the likely nature of movement, the desires of communities affected by it, and the fact that a treaty will not, without wide ratification and implementation, 'solve' the humanitarian issue. The argument is developed by examining certain conceptual and pragmatic difficulties in attempting to construct a refugee-like instrument for people fleeing the effects of climate change, and by critiquing whether there are legal, as opposed to political, benefits to be gained by advocating for such an instrument. The second part of the chapter examines an alternative paradigm for responding to movement: migration. This approach builds on calls by some affected States, such as Kiribati, for the expansion of existing migration opportunities in order to create planned movement pathways for their nationals. Such responses are better aligned with the likely impacts of climate change on the environment (slow and gradual) and patterns of movement (pre-emptive and gradual); the desires of people in those countries; and the history of movement in those regions.

Protection mechanisms are remedial in nature and do not respond adequately to pre-emptive movement. For example, most movement away from Tuvalu and Kiribati is likely to be slow and gradual, even if extreme weather events such as cyclones or king tides trigger more sudden, but probably temporary (and internal) moves.[43] Migration has long been an adaptation strategy to environmental

[41] 'In the face of slow-onset environmental change those who are able to move—those with money, social networks, and alternative livelihoods—may migrate independently. The vulnerable poor, those with no capacity to move, the very young and the elderly may be left behind initially, and forced to resettle later. Gender and demographic structure also play a role in environmentally induced migration patterns': Koko Warner, 'Assessing Institutional and Governance Needs Related to Environmental Change and Human Migration' (Study Team on Climate-Induced Migration, German Marshall Fund of the United States, June 2010) 6.

[42] ibid, 2.

[43] In May 2009, the Red Cross in Tuvalu indicated that only four families had moved in response to flooding from king tides, and this was temporary: interview with Red Cross representative in Tuvalu (Funafuti, Tuvalu, 27 May 2009). Most movement after the Samoan tsunami in 2009 was internal.

variability. But whereas Pacific islanders could once move freely to other islands in times of resource scarcity or climate change,[44] the legal (and sometimes physical) barriers to entry imposed by States today considerably restrict freedom of movement. Accordingly, a key policy objective of the I-Kiribati government is to enhance existing migration options to developed countries in the region, primarily Australia and New Zealand,[45] thereby building up 'pockets' of its community abroad.[46]

If at some future point States do decide to create a treaty responding specifically to climate change-related movement, it is important to recognize that this will be a conscious, political decision to privilege climate change as a driver of movement over other drivers.[47] While some argue that this is appropriate, given the role of industrialized States in contributing to climate change and the need for them to accept responsibility for responding to its impacts,[48] the multicausality of movement, and the conceptual difficulties discussed in this and other chapters, makes this argument less compelling than it might otherwise be.

Chapter 8 examines institutional responses to climate change-related movement. Although the issue cuts across several areas of international governance—migration and asylum, the environment, development, human rights, humanitarian aid and assistance, and security—each of which is represented by a number of different UN and other bodies, no international institution has a particular responsibility for governing it. The chapter examines the responses of the (former) UN Commission on Human Rights, the Human Rights Council, the Office of the High Commissioner for Human Rights, the General Assembly, the Security Council, the United Nations High Commissioner for Refugees (UNHCR), the International Organization for Migration (IOM), and the UNFCCC—in particular, paragraph 14(f) of the Cancún Adaptation Framework[49]—in conceptualizing, regulating, and operationalizing protection and assistance strategies in relation to climate-related movement. The chapter concludes that there is a need to strengthen legal, policy, and institutional frameworks underpinned by basic human rights principles. If interventions are not sufficiently well coordinated and/or pre-planned, ad hoc assistance may become the default response of the international community. Interventions must be proactive, not just remedial, and

[44] See Steven R Fischer, *A History of the Pacific Islands* (Palgrave Macmillan, 2002) Ch 1, on ancient Pacific mobility.

[45] Interview with President Anote Tong (Tarawa, Kiribati, 12 May 2009).

[46] ibid; interview with David Lambourne, Solicitor-General of Kiribati (Tarawa, Kiribati, 8 May 2009). See further Ch 7.

[47] For example, during the League of Nations period, refugee instruments selected specific national groups of refugees as subjects of international protection, even though it was recognized that there were many other groups of people in need of such protection: see Guy S Goodwin-Gill and Jane McAdam, *The Refugee in International Law* (3rd edn, Oxford University Press, 2007) 7.

[48] Sultana Kamal, Executive Director of Ain O Salish Kendra (ASK) (Dhaka, Bangladesh, 21 June 2010); see also Peter Penz, 'International Ethical Responsibilities to "Climate Change Refugees"' in McAdam (n 20).

[49] See 'Draft Decision -/CP.16: Outcome of the Work of the Ad Hoc Working Group on Long-Term Cooperative Action under the Convention' (2010) <http://unfccc.int/files/meetings/cop_16/application/pdf/cop16_lca.pdf#page=3> accessed 15 March 2011.

comprehensive and multi-sectoral approaches are essential. A one-size-fits-all approach to climate-related movement will be inadequate.

Chapter 9 draws together the themes of Chapters 7 and 8 to consider the overarching normative principles of international cooperation, humanity, and dignity that should underpin State and institutional responses to cross-border displacement in the context of climate change and disasters. Placing the needs and rights of affected individuals at the centre of analysis mandates that legal and policy responses are human-rights-focused.

There is an increasing coalescence of institutional support for the creation of a set of guiding principles, similar in nature to the Guiding Principles on Internal Displacement, to provide a guiding framework in cross-border scenarios.[50] This is coupled with a noticeable shift in emphasis at the international level away from 'climate change' displacement, to a focus on 'disasters', of which climate change-related disasters are a sub-category. This sits more comfortably with the conceptual problems of isolating 'climate change' as a cause of movement, discussed throughout the book. However, although the development of such an instrument could helpfully address one aspect of climate-related movement—displacement by rapid-onset disasters—it is important that this does not overshadow responses to slow-onset movement. While the needs of those displaced by a sudden-onset disaster may be more pressing in the short-term, the longer-term consequences may be graver for those facing slower, but more permanent, threats to their homes. Addressing the impacts of movement from slow-onset climate change impacts is where the greater challenge lies.

When a person contemplates moving away from the impacts of climate change, the first question he or she will ask is 'where can I go?'. But the second question will be 'how will I be treated when I get there?'. It is imperative that human rights principles are brought to bear not only on issues relating to entry and admission, but also on the treatment and status of people in the host State. Respect for human dignity demands nothing less.

[50] Bellagio Deliberations (n 36) para 13.

1

Conceptualizing Climate Change-Related Movement

> I am very passionate because I do meet people who make me very angry when they make, whether wittingly or unwittingly, frenzied statements. But I've learned that you don't scream; you talk very rationally and people will listen to you. I screamed the first couple of years. I shouted at the UN; no-one listened. I think you've got to be a bit more rational, because people will join you.[1]

I. Introduction

There is now a considerable amount of literature about how the phenomenon of climate change-related movement should be 'conceptualized'—that is, how it should be understood as an area of academic inquiry and a subject of legal and policy development. The purpose of this chapter is not to rehearse the debates on this issue, but rather to synthesize contemporary approaches[2] and explain why the 'lens' through which the phenomenon is viewed can dramatically change the way it is perceived and regulated. Conceptualization is therefore key to presenting the 'issue' to be tackled, and devising appropriate policy responses to address it. As Chapter 8 demonstrates, it is also central to locating appropriate institutional responses.

According to the United Nations High Commissioner for Refugees (UNHCR), it is becoming increasingly difficult to categorize displaced people because of the combined impacts of conflict, the environment, and economic pressures.[3] The degree to which it can be said that an 'increasing number of people [are] displaced as a direct result of climate change'[4] is therefore controversial. That is not to say that the phenomenon of climate change-related displacement is not real, but rather that it is much more

[1] Interview with President Anote Tong (Tarawa, Kiribati, 12 May 2009).

[2] See Stephen Castles, 'Afterword: What Now? Climate-Induced Displacement after Copenhagen' in Jane McAdam (ed), *Climate Change and Displacement: Multidisciplinary Perspectives* (Hart Publishing, 2010).

[3] See remarks made by the UN High Commissioner for Refugees, António Guterres, in an interview with *The Guardian*: Julian Borger, 'Conflicts Fuelled by Climate Change Causing New Refugee Crisis, Warns UN', *The Guardian* (London, 17 June 2008) <http://www.guardian.co.uk/environment/2008/jun/17/climatechange.food> accessed 2 December 2009.

[4] Angela Williams, 'Turning the Tide: Recognizing Climate Change Refugees in International Law' (2008) 30 Law & Policy 502, 504.

complex than is often assumed. The reasons for this are explored both empirically and theoretically throughout this book. They relate primarily to the role of climate change as a proximate 'cause' of displacement, and the degree to which individuals and households consider movement to be a rational response to adversity at a given point in time. Migration decisions—even those pertaining to 'forced' migration—are multifaceted and debunk common assumptions about linear causes and effects.[5]

First, the growing body of empirical research shows that in most cases, climate-related movement is likely to be predominantly internal and/or gradual.[6] This accords with migration patterns generally, which indicate that while there are around 200 million international migrants, there are approximately 740 million internal migrants.[7] Of course, there *will* be some cross-border movement on account of climate change impacts, but not in the magnitude often predicted,[8] nor necessarily in the nature of refugee-like 'flight'.[9] As one study explains: 'It is important to provide a corrective to mainstream policy thinking that migration is by definition international and/or long term or permanent, even if internal migration is usually less politically and ethnically charged and requires more national than international policy responses.'[10] This is very important when it comes to devising appropriate legal and policy responses, which must be attuned to the reality of movement.

Secondly, it is inherently fraught to speak of 'climate change' as the 'cause' of human movement, even though its impacts may exacerbate existing socio-economic or environmental vulnerabilities. Rather, climate change will have an 'incremental impact', '*add[ing]* to existing problems' and 'compound[ing] existing threats'.[11] As one government official in the so-called 'sinking island State' of Kiribati observed, climate change overlays pre-existing pressures—overcrowding, unemployment,

[5] Graeme J Hugo, 'Environmental Concerns and International Migration' (1996) 30 International Migration Review 105. See Ch 6 for examples of the kinds of issues people consider.

[6] For a detailed study of 23 countries, see EACH-FOR, 'Synthesis Report' (14 May 2009) 72 <http://www.each-for.eu/documents/EACH-FOR_Synthesis_Report_090515.pdf> accessed 2 November 2010; see also Foresight, *Migration and Global Environmental Change: Future Challenges and Opportunities: Final Project Report* (The Government Office for Science, London 2011); Frank Laczko and Christine Aghazarm (eds), *Migration, Environment and Climate Change: Assessing the Evidence* (IOM, 2009); Richard Black, Dominic Kniveton and Kerstin Schmidt-Verkerk, 'Migration and Climate Change: Towards an Integrated Assessment of Sensitivity' (2011) 43 Environment and Planning A 431; Clionadh Raleigh, Lisa Jordan, and Idean Salehyan, 'Assessing the Impact of Climate Change on Migration and Conflict', Working Paper for Social Dimensions of Climate Change (World Bank, 2008).

[7] United Nations Development Programme (UNDP), *Overcoming Barriers: Human Mobility and Development—Human Development Report 2009* (UNDP, 2009) 1–2.

[8] For alarmist figures, see, eg, Christian Aid, *Human Tide: The Real Migration Crisis* (Christian Aid, 2007). See also Friends of the Earth Australia, *A Citizen's Guide to Climate Refugees* (rev edn, Friends of the Earth Australia, 2007); Robin McKie, 'Climate Change Will Cost a Billion People Their Homes, Says Report' *The Observer* (London, 28 November 2010) 11; International Federation of the Red Cross (IFRC) and Red Crescent Societies, *World Disasters Report: Focus on Recovery* (IFRC and Red Crescent Societies, 2001); Norman Myers, 'Environmental Refugees in a Globally Warmed World' (1993) 43 BioScience 752; Norman Myers, 'Environmental Refugees: An Emergent Security Issue' (13th Economic Forum, Prague, 23–27 May 2005).

[9] See, eg, Laczko and Aghazarm (n 6); EACH-FOR (n 6).

[10] Black, Kniveton, and Schmidt-Verkerk (n 6) 446.

[11] Interview with Saber Chowdhury MP, Member of the All Parliamentary Committee on Climate Change, Bangladesh (Dhaka, Bangladesh, 21 June 2010).

environmental, and development concerns—which means that it may provide a 'tipping point' that would not have been reached in its absence.[12] The nature of people movement will therefore vary greatly depending upon a range of context-specific variables which may interact in different ways at different times.[13] As Kibreab notes, the effects of climate change on human movement are 'spatially and socially differentiated'.[14]

In order to appropriately theorize these issues, this chapter draws on fieldwork undertaken in three countries which have become synonymous with the notion of climate change-related displacement—Kiribati, Tuvalu, and Bangladesh. These empirical studies challenge certain understandings of the phenomenon, and inform the way in which climate change-related movement is conceptualized in this book. They necessarily shape the legal analysis in Chapters 2, 3, and 4, and the solutions proffered in Chapters 5 and 6.

II. The Nature of Displacement

The way a phenomenon is conceptualized is central to the way its regulation is approached. Responses to human trafficking, for instance, will differ depending on whether the issue is viewed through a criminal justice or a human rights lens. Similarly, how (and to what extent) international law and institutions respond to climate-related human movement will depend in part on: (a) whether such movement is perceived as voluntary or involuntary; (b) the nature of the trigger (a rapid-onset disaster versus a slow-onset process); (c) whether international borders are crossed; (d) the extent to which there are political incentives to characterize movement as linked to climate change or not; and (e) whether movement is driven or aggravated by human factors, such as discrimination.[15]

[12] Interview with I-Kiribati Solicitor-General David Lambourne (Tarawa, Kiribati, 8 May 2009). However, in Tuvalu, there is a concern that if climate drivers are overshadowed by other factors such as general poverty, which have traditionally not given rise to a protection response by third States, efforts to achieve funding for adaptation and migration options for the future will be stymied. This was the impression given in the author's interview with Enele Sopoaga, Secretary for Foreign Affairs, Tuvalu (Funafuti, Tuvalu, 25 May 2009).

[13] Black, Kniveton and Schmidt-Verkerk (n 6) 435; Foresight (n 6).

[14] Gaim Kibreab, 'Climate Change and Human Migration: A Tenuous Relationship?' (2009) 20 Fordham Environmental Law Review 357, 377.

[15] See, eg, Irene Khan, *The Unheard Truth: Poverty and Human Rights* (WW Norton and Co, 2009). How an issue is characterized can cut both ways, of course. While some Pacific leaders have highlighted the existential threat that climate change poses to their States (see further below), countries like the United States, Australia, and the Member States of the European Union have highlighted the security threats that climate change pose to themselves, including the threat of climate migration: see, eg, Peter Schwartz and Doug Randall, 'An Abrupt Climate Change Scenario and its Implications for United States National Security' (October 2003); CNA Corporation, 'National Security and the Threat of Climate Change' (2007); Prime Minister Kevin Rudd, 'The First National Security Statement to the Australian Parliament' (4 December 2008) <http://pmrudd.archive.dpmc.gov.au/node/5424> accessed 18 October 2011; Defence White Paper, *Defending Australia in the Asia Pacific Century: Force 2030* (Commonwealth of Australia, 2009); Council of the European Union, 'Climate Change and International Security' (Report from the Commission and the Secretary-General/High Representative to the European Council) Doc No 7249/08 (3 March 2008).

In 2008, the former United Nations (UN) Secretary-General's Representative on the Human Rights of Internally Displaced Persons, Walter Kälin, developed a framework setting out the diversity of scenarios that could be encompassed within the very wide concept of environmental displacement.[16] This was subsequently adopted by the UN's Inter-Agency Standing Committee (IASC) Working Group on Migration/Displacement and Climate Change, comprised of the UN's main humanitarian agencies and representatives from other relevant organizations.[17] Not all scholars accept this way of categorizing climate change-related movement, and a variety of other frameworks have also been proposed.[18] Any typology is necessarily artificial because its rigid compartmentalization does not account for the fluidity between categories. Bearing that in mind, Kälin's typology nonetheless provides a useful illustration of the different types of movement.

1. *The increase of hydro-meteorological disasters*, such as flooding, hurricanes, typhoons, cyclones, and mudslides, leading predominantly to internal displacement. Often, such displacement will be across relatively short distances.[19]
2. Government-initiated *planned evacuation* of areas at high risk of disasters. This is likely to lead to permanent internal displacement.
3. *Environmental degradation and slow-onset disasters*, such as reduced water availability, desertification, recurrent flooding, and increased salinity in coastal zones. Kälin explains: 'Such deterioration may not necessarily cause displacement, but it may prompt people to consider "voluntary" migration as a way to adapt to the

[16] Walter Kälin, 'The Climate Change–Displacement Nexus' (16 July 2008) <http://www.brookings.edu/speeches/2008/0716_climate_change_kalin.aspx?p=1> accessed 2 December 2008. This was subsequently refined and published as Walter Kälin, 'Conceptualising Climate-Induced Displacement' in McAdam (n 2) 81.

[17] Inter-Agency Standing Committee (IASC) Working Group on Migration/Displacement and Climate Change, 'Displacement and Climate Change: Towards Defining Categories of Affected Persons' (20 September 2008) Working Paper First Draft. See also the six-pronged typology in International Organization for Migration (IOM) Policy Brief, 'Migration, Climate Change and the Environment' (May 2009). See further Ch 8.

[18] See François Gemenne, 'Environmental Changes and Migration Flows: Normative Frameworks and Policy Responses' (PhD thesis, Institut d'Etudes Politiques de Paris and University of Liège, 2009) 116–17, 124–5, 140–5, 161–4. For example, Gemenne queries the utility of categorizing environmental changes in this way. He argues that movement patterns 'depend more on policy responses than on the type of change involved' (163), and that it is impossible to encapsulate the wide variety of changes in clear-cut categories. Instead, he proposes viewing changes as part of different continua, including the geographical extent of the change (local to global); the degree of human responsibility for the change; and the pace of the change. It is important to note, though, that his approach refers to *environmental* change, which is broader than the *climate* change typology outlined above. See also François Gemenne, 'How They Became the Human Face of Climate Change: Research and Policy Interactions in the Birth of the "Environmental Migration" Concept' in Etienne Piguet, Antoine Pécoud, and Paul de Guchteneire (eds), *Migration and Climate Change* (Cambridge University Press and UNESCO Publishing, 2011). Renaud and others have developed a 'decision framework for determining environmental migrant sub-categories' which identifies the ways in which the environment—through rapid-onset events, on the one hand, and slow-onset hazards and loss of ecosystem services, on the other—may impact on movement decisions. Usefully, they also analyse the extent to which rehabilitation of the affected area may influence return: Fabrice G Renaud and others, 'A Decision Framework for Environmentally Induced Migration' (2011) 49 International Migration 5, 16. See also Jon Barnett and Michael Webber, *Accommodating Migration to Promote Adaptation to Climate Change* (Commission on Climate Change and Development, 2009) 17ff.

[19] See Ch 6.

changing environment and be a reason why people move to regions with better living conditions and income opportunities. However, if areas become uninhabitable over time because of further deterioration, finally leading to complete desertification, permanent flooding of coastal zones or similar situations, population movements will amount to forced displacement and become permanent.'[20] Humanitarian agencies note that '[t]he scope and scale of displacement due to slow-onset disasters, such as drought, remains a guess at best.'[21]

4. *Small island countries at risk of disappearing* because of rising seas. At the point at which a territory is no longer habitable (eg because of the inability to grow crops or obtain fresh water), permanent relocation to other countries would be necessary even if the country were not yet under water. Kälin notes that current international law provides no protected status for such people, and even if they were to be treated as 'stateless', 'current legal regimes are hardly sufficient to address their very specific needs'.[22] For example, although small island countries (such as Kiribati and Tuvalu) emit less than 1 per cent of global greenhouse gases, their small physical size, exposure to natural disasters and climate extremes, very open economies, and low adaptive capacity make them particularly susceptible, and less resilient, to climate change.[23] This is discussed further in Chapter 5.
5. *Risk of conflict* over essential resources. Even though the humanitarian community is used to dealing with conflict, and people displaced by conflict may be eligible for protection as refugees or assistance as internally displaced persons (IDPs), resource-based conflicts 'may be particularly challenging' at the operational level. In particular, where resource scarcity cannot be resolved, 'it will be extremely difficult to reach peace agreements providing for an equitable solution. The likely outcome is both conflict and the displacement of a protracted nature.'[24] Conflict is likely to be social conflict, rather than armed conflict.[25]

Each type of scenario described above involves different kinds of pressures and impacts, which will affect the time, speed, and size of movement. Thus, at various

[20] Walter Kälin, 'Conceptualising Climate-Induced Displacement' (n 16) 85. See also Jane McAdam and Ben Saul, 'Displacement with Dignity: International Law and Policy Responses to Climate Change Migration and Security in Bangladesh' (2010) 53 German Yearbook of International Law 233 on secondary movement.

[21] IASC, 'Quick Guide to Climate Change Adaptation' (2010) 1, referring to OCHA and IDMC, 'Monitoring Disaster Displacement in the Context of Climate Change: Findings of a Study by the United Nations Office for the Coordination of Humanitarian Affairs and the International Displacement Monitoring Centre' (September 2009).

[22] Kälin, 'The Climate Change–Displacement Nexus' (n 16).

[23] Nobuo Mimura and others, 'Small Islands' in Martin L Parry and others (eds), *Climate Change 2007: Impacts, Adaptation and Vulnerability: Contribution of Working Group II to the Fourth Assessment Report of the Intergovernmental Panel on Climate Change* (Cambridge University Press, 2007) 692–3. The report additionally lists the impacts of globalization, pressures on infrastructure, a scarcity of fresh water and, in the Pacific, internal and external political and economic processes, including the imposition of western adaptation models which are not readily transposable to the island context. These features have resulted in some small island countries being recognized by the UN as Least Developed Countries (LDC) or small-island developing States (SIDS).

[24] ibid.

[25] See McAdam and Saul (n 20); see further Ch 8.

points in time, the role of climate change in individual or household decisions to move may be stronger or weaker, and will interact with other reasons for moving (including structural, institutional, or network influences).[26] Barnett and Webber posit a similar typology but based on the spatial (internal/international) and temporal (temporary/permanent) aspects of movement 'stimulated' by climate change. They identify seven categories: internal/international labour migration (movement from declining conditions at home); internal/international displacement by rapid-onset disasters (typically movement across short distances and for a temporary period); internal/international permanent migration (as the consequence of stressors exacerbated by incremental changes and slow-onset processes, such as drought); and relocating communities (to reduce exposure to climate risks). Of these, they regard labour migration as offering 'the best potential for harnessing the power of migration to promote adaptation to climate change'.[27]

Since such a wide range of scenarios can be caught under the 'climate displacement' umbrella, no single legal or policy response is appropriate or able to address them all. However, existing legal frameworks (conceived broadly as encompassing protection jurisprudence and State practice, not just formal legal instruments) seem better equipped to respond to disaster-related movement, and less able to accommodate pre-emptive movement on account of slower-onset processes.

A. Does climate change 'cause' movement?

Although scientists now attest that '[m]ost of the observed increase in globally averaged temperatures since the mid-twentieth century is *very likely* due to the observed increase in anthropogenic greenhouse gas concentrations', which has '*very likely*... contributed to a rise in mean sea level',[28] it does not necessarily follow that climate change can be described as the sole cause of human movement, or that those who move would attribute their personal circumstances to it.[29] Studies reveal that decisions to move or to stay are influenced by the overall socio-economic situation of those concerned.[30] For example, studies have shown that droughts in parts of Africa resulted in decreases in international and long-distance migration,

[26] Black, Kniveton, and Schmidt-Verkerk (n 6) 435.

[27] Barnett and Webber (n 18) 18. For a tabular summary of what policy responses would look like for each, see 42–4.

[28] See respectively IPCC, 'Summary for Policymakers' 10 and Gabriele C Hegerl and others, 'Understanding and Attributing Climate Change' in Susan Solomon and others (eds), *Climate Change 2007: The Physical Science Basis: Contribution of Working Group I to the Fourth Assessment Report of the Intergovernmental Panel on Climate Change* (Cambridge University Press, 2007) 729.

[29] See Colette Mortreux and Jon Barnett, 'Climate Change, Migration and Adaptation in Funafuti, Tuvalu' (2009) 19 Global Environmental Change 105. The present author's own fieldwork in Kiribati and Tuvalu in 2009 reinforces this.

[30] Ruth Haug, 'Forced Migration, Processes of Return and Livelihood Construction among Pastoralists in Northern Sudan' (2002) 26 Disasters 70; Elisabeth Meze-Hausken, 'Migration Caused by Climate Change: How Vulnerable Are People in Dryland Areas?' (2000) 5 Mitigation and Adaptation Strategies for Global Change 379; Rita Afsar, 'Internal Migration and the Development Nexus: The Case of Bangladesh'(Refugee and Migratory Movements Research Unit, Dhaka and DFID, 2003) <http://www.migrationdrc.org/publications/working_papers/WP-C2.pdf> accessed 1 September 2010; Dominic Kniveton and others, 'Climate Change and Migration: Improving

with food scarcity and increased food prices forcing people to spend money on basic needs rather than moving. By contrast, short-distance migration increased as women and children sought work to supplement household incomes through remittances.[31] Consistent with findings in traditional refugee literature, the poorest or most vulnerable may not have any choice but to stay put, because they may not have the economic ability, social networks, health, or skills to move.

As was noted in the Chairperson's Summary of the high-level 2011 Nansen Conference on Climate Change and Displacement in the 21st Century, 'it is not necessarily the temperature increase itself that poses the largest challenge in terms of human mobility, but the *associated changes in, and combined effects of*, precipitation patterns (drought and flooding), storms, and sea level rise; loss of biodiversity, and ecosystem services; *and* resulting health risk, food and livelihood insecurity'.[32] Political, structural, and idiosyncratic factors, such as ethnicity, family size, productivity in potential destination zones, the agrarian state of rural areas, schooling opportunities, social networks (in destinations), and the availability of food and water in other areas all impact on decision-making about 'the degree to which migration constitutes a possible strategy'.[33]

In other words, the combined changes to the climate and—importantly—their interaction with pre-existing stressors will determine if and when people move. An alternative way of understanding this is by reference to people's adaptive capacity and resilience. Communities that have the resources to put adaptation measures in place, including disaster risk reduction and management policies and sustainable development practices, may have less need to move permanently if disaster strikes. Their 'displacement' may take the form of 'evacuation' (which is itself a disaster risk reduction strategy).

Similarly, the extent to which relief and rehabilitation is available to those displaced by a sudden-onset disaster, for example, will affect whether and how quickly they can return home and rebuild. As Warner notes, policy interventions—and the timing of them—will play a major role in shaping outcomes and will determine whether migration is a form of adaptation, or a sign of a failure to adapt.[34]

Methodologies to Estimate Flows' (2008) IOM Migration Research Series No 33, 35; Mortreux and Barnett (n 29); Laczko and Aghazarm (n 6); McAdam (n 2); EACH-FOR (n 6).

[31] Sabine Henry, Bruno Schoumaker, and Cris Beauchemin, 'The Impact of Rainfall on the First Out-Migration: A Multi-Level Event-History Analysis in Burkina Faso' (2004) 25 Population and Environment 423. See also Michelle Leighton, 'Desertification and Migration' in Pierre Marc Johnson, Karel Mayrand and Marc Pacquin (eds), *Governing Global Desertification* (Ashgate, 2006); Fabrice Renaud and others, 'Control, Adapt or Flee: How to Face Environmental Migration?' (2007) InterSecTions No 5/2007, 19; Afsar (n 30) 2.

[32] Chairperson's Summary, 'Nansen Conference on Climate Change and Displacement in the 21st Century' (Oslo, 6–7 June 2011) para 4 (emphasis added).

[33] James Morrissey, 'Rural–Urban Migration in Ethiopia' (2008) 31 Forced Migration Review 28, 29.

[34] Koko Warner, 'Assessing Institutional and Governance Needs Related to Environmental Change and Human Migration' (Study Team on Climate-Induced Migration, German Marshall Fund of the United States, June 2010) 8. See also Renaud and others (n 18) 19. For examples of national policies, see Martha Thompson and Izaskun Gaviria, 'Weathering the Storm: Lessons in Risk Reduction from Cuba' (Oxfam America, 2004) 20; Refugees International, 'Confronting

People are less likely to move permanently (or internationally) where information and resources are available to help remedy their plight (such as drought or flood relief packages, housing assistance, employment opportunities, social security, and so on).[35]

International movement may be considered undesirable because it severs a person's connections with their land, family, culture, and community, whether in small Pacific island countries or rural Bangladesh. People's skills, education, and form of livelihood may also not be well recognized or valued elsewhere. In effect, therefore, individuals 'try to calculate the *relative advantages* of moving against the relative advantages of remaining behind'.[36]

Adaptation aims to reduce the risk of harmful climate change impacts on communities. While this is often conceived of as requiring new strategies and protective measures to be put in place, it also requires a rethink of *existing* planning and development practices. As one Bangladeshi official explained, poor development practices, combined with over-farming of land, overcrowding, deforestation, and settlements in environmentally-fragile/disaster-susceptible areas (generated largely by extreme poverty) mean that people are placed in conditions where they are inherently more vulnerable to disasters and slower-onset processes linked to a changing climate than if those underlying problems were addressed.[37] The implication is that governments must address climate change adaptation holistically, not merely in terms of erecting physical structures like sea-walls or financing the development of climate-resistant crops.[38]

Some researchers suggest that it is arbitrary to identify 'climate change' as a driver of forced migration, while omitting other causes such as poverty, general conflict, or lack of opportunity (especially since they may impact on the lives of even more people).[39] They instead prefer the term 'survival migration',[40] which shifts attention away from particular causes of movement and towards a more comprehensive understanding of the multiple stressors that may compel people to leave their homes. Of course, there is danger that this notion is all-encompassing

Climate Displacement: Learning from Pakistan's Floods' (November 2010) <http://www.refugeesinternational.org/sites/default/files/ConfrontingClimateDisplacement.pdf> accessed 7 July 2011.

[35] See Barnett and Webber, 'Migration as Adaptation: Opportunities and Limits' in McAdam (n 2) 37. There may be generational differences, however. The EACH-FOR study found that longer-term, permanent movement is starting to become more common than temporary or seasonal relocation, especially among younger people: EACH-FOR, 'Synthesis Report' (n 6) 5.

[36] Morrissey (n 33) 29 (emphasis added).

[37] Interview with Abu M Kamal Uddin, Comprehensive Disaster Management Programme (CDMP) (Dhaka, Bangladesh, 16 June 2010). See further, Ch 6.

[38] See, eg, *Budayeva v Russia* App nos 15339/02, 21166/02, 20058/02, 11673/02, and 15343/02 (ECtHR, 20 March 2008).

[39] See Alexander Betts and Esra Kaytaz, 'National and International Responses to the Zimbabwean Exodus: Implications for the Refugee Protection Regime' (2009) UNHCR New Issues in Refugee Research, Research Paper No 175; Alexander Betts, 'Towards a "Soft Law" Framework for the Protection of Vulnerable Irregular Migrants' (2010) 22 International Journal of Refugee Law 209; Kibreab (n 14). Kälin also suggests that 'it is conceptually sounder to look at sudden-onset disasters as a cause of displacement, and not to limit the focus to those triggered by global warming': Walter Kälin, 'Conceptualising Climate-Induced Displacement' (n 16) 85.

[40] See Betts and Kaytaz (n 39).

(merely another term for 'forced migration') and thus conceptually indistinct and unworkable in practice. However, it is useful to bear in mind its rationale when thinking about international policy responses, and which institutions should be responsible for addressing climate-related movement.[41] It raises questions about whether it is appropriate to differentiate between displaced people who deserve 'protection' on account of climate change, and those who are victims of 'mere' economic or environmental hardship. For example, in urban slums in Bangladesh, it is difficult to distinguish between those who move on account of general poverty from those who are affected by climate change.[42] In particular, one might query the appropriateness of singling out a category of 'climate-displaced people' in an international treaty (as some have advocated),[43] especially since much of the 'responsibility' for movement resulting from these other drivers may be similarly attributable to international structures (economic and political) that perpetuate an ever-growing divide between rich and poor countries,[44] rather than to acts or omissions of the government in the country of origin. Additionally, it is worth noting that the Chairperson's Summary of the Nansen Conference on Climate Change and Displacement in 2011 reflected the view that '[f]rom a protection perspective, there is no compelling reason to distinguish between displacement due to climate-related and other disasters'.[45] In other words, should displacement be addressed in terms of what drives it, or rather in terms of the needs of those who move?

Complex causality arguments are also supported by the science of climate change. As climate scientists Huber and Gulledge explain, asking whether climate change has 'caused' a particular event is a nonsensical and 'fundamentally unanswerable question',[46] since 'no particular short-term event can be conclusively attributed to climate change'.[47] Rather, 'climate' represents 'the average of many weather events over a span of years',[48] and varying averages over time define climate *change*.[49] Statistical trends point to an increased frequency and/or severity

[41] See Ch 8.

[42] Interview with SM Munjurul Hannan Khan, Deputy Secretary, Ministry of Environment and Forests and National Focal Point for the UNFCCC and IPCC, Bangladesh (Dhaka, Bangladesh, 15 June 2010). Discussions with slum dwellers showed that some had moved on account of environmental degradation, but this was a very small sample and no firm conclusions can be drawn from it: interviews in Shonamia *bosti* (slum) (Dhaka, Bangladesh, 18 June 2010). A Bangladeshi MP noted that: 'If poverty continues to be a major issue, and climate change intensifies poverty, then indirectly it's causing it, but not by itself': Interview with Saber Chowdhury MP (n 11).

[43] See Ch 7.

[44] Stephen Castles, 'Environmental Change and Forced Migration: Making Sense of the Debate' (2002) UNHCR New Issues in Refugee Research, Working Paper No 70, 11.

[45] Chairperson's Summary (n 32) para 5.

[46] Daniel G Huber and Jay Gulledge, 'Extreme Weather and Climate Change: Understanding the Link and Managing the Risk' (Center for Climate and Energy Solutions, December 2011) 2 <http://www.c2es.org/docUploads/white-paper-extreme-weather-climate-change-understanding-link-managing-risk.pdf> accessed 3 December 2011. See also Richard JT Klein and Robert J Nicholls, 'Assessment of Coastal Vulnerability to Climate Change' (1999) 28 Ambio 182, 183.

[47] Huber and Gulledge (n 46) 4. Accordingly, the distinction in forced migration scholarship between so-called 'climate events' and 'climate processes' that pervades some of the social sciences literature is scientifically flawed: eg Oli Brown, 'Migration and Climate Change' (2008) IOM Migration Research Series No 31, 17ff.

[48] Huber and Gulledge (n 46) 2.

[49] ibid. '[W]hat used to be a 500-year event may become a 100-year or 10-year event, so that most people will experience such events within their lifetimes': 6.

of extreme weather events, which is consistent with global warming. While individual events cannot be predicted, the likelihood of their occurrence can.[50] Huber and Gulledge liken this to 'playing with loaded dice, where the dice are weighted to roll high numbers more frequently'.[51] Accordingly, they argue that a probability-based risk management framework is the most appropriate one for analysing the link between climate change and extreme weather events, since it enables policy-makers to better understand how risk is changing so that prevention and adaptation strategies can be prioritized.[52]

It is therefore conceptually problematic and empirically flawed to suggest that climate change alone *causes* migration. It is best described as a threat multiplier which exacerbates existing socio-economic or environmental vulnerabilities.

B. How many people will move?

The question 'how many people are likely to be displaced by climate change?' has been extensively debated and remains a matter of controversy.[53] Because there is no agreed definition of a 'climate migrant' or 'climate refugee',[54] there are both conceptual and methodological problems about 'who' is counted. Disagreement is focused less on precise numerical estimates, and more on whether—and how—the phenomenon can be quantified.

To begin with, existing bureaucratic categories do not 'count' climate-related movement. There is no such thing as a 'climate change' migration or protection visa in the immigration law of any country. Although there is scope in the domestic law of countries like Finland and Sweden to take environmental degradation into

[50] Modelling simulations of Russia's 2010 heatwave 'found no evidence that climate change has so far increased the risk of such an event but did find that continued warming is very likely to produce frequent heat waves of a similar magnitude later this century. Hence, regardless of the cause of that particular heat wave, the risk of similar events in the future can be expected to rise with continued warming of the global climate': ibid, 10, citing Randall Dole and others, 'Was There a Basis for Anticipating the 2010 Russian Heat Wave?' (2011) 38 Geophysical Research Letters L06702.

[51] Huber and Gulledge (n 46) 7. 'Risk is the accumulation of all future possibilities weighted by their probabilities of occurrence': 6.

[52] ibid, 6ff. 'Risk management eschews futile attempts to forecast individual chaotic events and focuses on establishing long-term risk certainty; that is, an understanding of what types of risks are increasing and what can be done to minimize future damages. An understanding of the meaning of risk and how it relates to changes in the climate system is crucial to assessing vulnerability and planning for a future characterized by rising risk': 2.

[53] See, eg, Myers 'Environmental Refugees in a Globally Warmed World' (n 8); Myers 'Environmental Refugees: An Emergent Security Issue' (n 8); cf, Richard Black, 'Environmental Refugees: Myth or Reality?' (2001) UNHCR New Issues in Refugee Research, Working Paper No 34; Castles (n 44). See also Oli Brown, 'The Numbers Game' (2008) 31 Forced Migration Review 8.

[54] For a definition of 'environmental migrants', see IOM, 'Discussion Note: Migration and the Environment', 94th Session, Doc No MC/INF/288 (1 November 2007) para 6. For other definitions, see examples listed in Renaud and others (n 31) 13. Gemenne criticizes the IOM definition for failing to acknowledge the multicausality of movement: Gemenne (2009) (n 18) 150. See also the definition provided by EACH-FOR: Olivia Dun, François Gemenne and Robert Stojanov, 'Environmentally Displaced Persons: Working Definitions for the EACH-FOR Project' (2007) <http://www.each-for.eu/documents/Environmentally_Displaced_Persons_-_Working_Definitions.pdf> accessed 1 July 2011. Kniveton and others (n 30) 31 define people displaced by climate change as a sub-set of the IOM category. For an article revisiting Renaud and others' conception, see Renaud and others (n 18).

account in assessing asylum claims,[55] visas have never been issued on this basis. Even if they were, it is unlikely that they would be counted separately from the overall refugee/complementary protection/humanitarian intake. Secondly, people who are eligible to migrate under regular labour, education, or family migration programmes will tend to access those existing visa categories where possible. For example, highly skilled migration from Bangladesh is likely to increase as internal rural–urban movement places acute pressure on the infrastructure of cities like Dhaka, 'pushing' those who can (the relatively wealthy and well-educated) to move abroad.[56] Availing themselves of existing migration visas, it is unlikely that their movement will ever be cast as climate change-related, even though it might be indirectly driven by it. Finally, most climate-related movement is predicted to be internal, and few countries keep count of who moves where and why within their borders since it is ordinarily the right of citizens and residents to move freely within a country.[57] This means that climate change-related movement may remain a largely invisible phenomenon in bureaucratic and legal terms.

It is also very difficult to factor qualitative elements into numerical estimates. Current predictions 'are fraught with numerous methodological problems and caveats'[58] because of difficulties in identifying: (a) the extent to which 'climate change', as opposed to other factors, drives migration; (b) the pace at which such change may force people to move; (c) the extent to which movement is already an adaptation strategy employed by the community (eg cyclical movement in flood-prone areas) and can continue; (d) the level of adaptation and recovery assistance available within the country; (e) the degree to which movement will occur within a country or across international borders; (f) the degree to which movement will be permanent or temporary; (g) whether there are pre-existing migration options for the particular community; and (h) the unpredictable degree to which carbon emissions will be reduced.

Emphasizing the complex determinants involved in mobility decisions and the importance of factoring in resilience and adaptive capacity, is a hallmark of what is described in the literature as the minimalist, or sceptical, approach.[59] It is a view that is characteristic of those who work in the areas of refugee protection and/or migration, whether academically or institutionally.[60] Proponents see a danger in creating policy that does not recognize the complexity of human movement, and they emphasize the need to consider empirical evidence of past and contemporary migration patterns in assessing likely future movement, as well as the role of adap-

[55] See Ch 4. [56] Confirmed in interview with Saber Chowdhury MP (n 11).

[57] Universal Declaration of Human Rights (adopted 10 December 1948) UNGA res 217A (III) ('UDHR') Art 13; International Covenant on Civil and Political Rights (adopted 16 December 1966, entered into force 23 March 1976) 999 UNTS 171 ('ICCPR') Art 12.

[58] Frank Biermann and Ingrid Boas, 'Preparing for a Warmer World: Towards a Global Governance System to Protect Climate Refugees' (2007) Global Governance Working Paper No 33, 9 <http://www.sarpn.org.za/documents/d0002952/Climate_refugees_global_governance_Nov2007.pdf> accessed 4 July 2011.

[59] Adherents are not sceptics of climate change, but rather of climate change as *the* cause of movement. For an excellent overview of the various approaches, see Gemenne (2011) (n 18).

[60] For early proponents of this approach, see Black (n 53) and Castles (n 44).

tation in preventing some movement.[61] The arguments in this book reflect this approach.

By contrast, the maximalist,[62] or alarmist,[63] approach has gained considerable public attention. It is typified by organizations and scholars from an environmental studies background. Its proponents see climate change-related movement as part of a bigger discourse that highlights the dangers of climate change generally.[64] Emphasizing that very large numbers of people will be displaced by climate change (and sometimes linking this to security concerns)[65] is a way of demonstrating just how destructive climate change will be.

Emblematic of the alarmist approach is the work of social scientist, Norman Myers. In 1993, Myers wrote a paper suggesting that 150 million people could be displaced by climate change by the middle of this century.[66] His methodology was fairly crude: he identified parts of the world thought to be vulnerable to sea-level rise and an increase in extreme weather events, and calculated the number of people likely to be affected by them based on anticipated population growth in those regions over the coming decades.[67] In 2005, he revised his estimate, suggesting that it could be up to 200 million, and in 2007, in an interview with Christian Aid, he suggested that the figure could be closer to 250

[61] See, eg, Graeme Hugo, 'Climate Change-Induced Mobility and the Existing Migration Regime in Asia and the Pacific' in McAdam (n 2) 9. Black, Kniveton, and Schmidt-Verkerk (n 6) analyse the main drivers of *existing* migration in particular contexts, and then examine their sensitivity to climate change. Barnett and Webber (n 18) 3 state: 'while climate change is likely to result in some increase above baseline rates of mobility in the next 40 years, many of the most dire risks arising from this increased mobility are amenable to management through aid, development, and migration policies'.

[62] See, eg, Astri Suhrke, 'Environmental Degradation and Population Flows' (1994) 47 Journal of International Affairs 473, 474. For an overview of the proponents, see James Morrissey, 'Environmental Change and Forced Migration: A State of the Art Review' (2009) Refugee Studies Centre, University of Oxford, Background Paper <http://www.rsc.ox.ac.uk/events/environmental-change-and-migration/EnvChangeandFmReviewWS.pdf> accessed 4 August 2011.

[63] See analyses in Gemenne (2009) (n 18).

[64] This is the genesis of the term 'environmental refugee' in official discourse. El-Hinnawi used the term in a 1985 report for the United Nations Environment Program to highlight the potentially devastating impacts of unchecked development and pollution on human society: Essam El-Hinnawi, *Environmental Refugees* (United Nations Environment Programme, 1985).

[65] See, eg, Thomas F Homer-Dixon, 'On the Threshold: Environmental Changes as Causes of Acute Conflict' (1991) 16 International Security 76.

[66] Myers, 'Environmental Refugees in a Globally Warmed World' (n 8) 758. See also Norman Myers, 'Environmentally-Induced Displacements: The State of the Art' in IOM/UNHCR, *Environmentally-Induced Population Displacements and Environmental Impacts Resulting from Mass Migration* (IOM/UNHCR International Symposium, Geneva, 1996); Norman Myers, 'Environmental Refugees' (1997) 19 Population and Environment 167; Norman Myers, 'Environmental Refugees: A Growing Phenomenon of the 21st Century' (2002) 357 Philosophical Transactions of the Royal Society: Biological Sciences 609; Myers, 'Environmental Refugees: An Emergent Security Issue' (n 8).

[67] Dominic Kniveton is very critical of the crude counting method by which Myers reached this estimate: Dominic Kniveton, 'Climate Change and Migration: Developing Methodologies', Environment, Forced Migration and Social Vulnerability Conference (Bonn, 9 October 2008). See also Kniveton and others (n 30) 29; Jeff Crisp, '"Who Has Counted the Refugees?" UNHCR and the Politics of Numbers' (1999) UNHCR New Issues in Refugee Research, Working Paper No 12.

million.[68] In its report and media briefings, Christian Aid misleadingly implied that one billion people would be displaced by climate change by 2050, including within that figure many other forms of displacement, again based on crude methodologies.[69]

Problematically, such figures continue to be cited in media and policy debates despite their lack of empirical justification.[70] For example, in the run-up to the Cancún climate negotiations in December 2010, *The Observer* ran an article entitled 'Climate Change Will Cost a Billion People Their Homes, Says Report'. It opened with the alarming revelation that: 'A special report, to be released at the start of climate negotiations in Cancún, Mexico, will reveal that up to a billion people face losing their homes in the next 90 years because of failures to agree curbs on carbon emissions.'[71] The 'report' referred to was in fact a series of papers resulting from the Tyndall Centre for Climate Change Research's conference 'Four Degrees and Beyond'. Whereas most analysis of climate change impacts is based on a two degree Celcius temperature rise, that conference considered the ramifications of a rise of at least four degrees.

The paper on climate-related movement was written by François Gemenne, whose work is empirically based and well reasoned. He hypothesized that in the event of global warming of four degrees Celcius or more, patterns of movement could be significantly different since temperatures could threaten people's very survival.[72] When one examines his original paper, it is clear, however, that at no point did he suggest that a billion people would be displaced. Rather, in a section entitled 'water stress', he refers to remarks by the Intergovernmental Panel on Climate Change (IPCC) that 'freshwater availability in Central, South, East and Southeast Asia, particularly in large river basins, is projected to decrease due to climate change which, along with population growth and increasing demand arising from higher standards of living, could adversely affect more than a billion people by the 2050s'.[73] This is *not* a statistic about population movement and certainly does not justify the newspaper's headline.[74] Misreporting, accompanied by a portrayal of the phenomenon which fails to appreciate its nuances and complexity, means that public debate is simplistic and often ill-informed.

[68] Christian Aid (n 8).

[69] ibid, 5–6.

[70] For example, an article published in early 2011, referring to statements by the American Association for the Advancement of Science, stated that: 'Climate change is forcing a lot of people—50 million by 2020, to be exact—to skip town': Jenara Nerenberg, 'Science Panel Says World Will See 50 Million Climate Change Refugees by 2020' (22 February 2011) <http://www.fastcompany.com/1730475/climate-change-forces-50-millionpeople-to-migrate-libya-tunisia-egypt> accessed 23 February 2011.

[71] McKie (n 8).

[72] François Gemenne, 'Climate-Induced Population Displacements in a 4°C+ World' (2011) 369 Philosophical Transactions of the Royal Society A 182. See also Barnett and Webber (n 18) 4.

[73] IPCC, 'Summary for Policymakers' in Parry and others (n 24) 12–13, cited in Gemenne (n 70) 185.

[74] For a response to *The Observer* piece, see Jane McAdam, 'Number of Climate Refugees Overstated', *Sydney Morning Herald* (Sydney, 10 December 2010) <http://www.smh.com.au/opinion/politics/number-of-climate-refugees-overstated-20101209-18rab.html> accessed 10 December 2010.

Myers' numerical estimates have been described as relying 'on Malthusian logic infused with theories of environmental change'[75] and criticized as too simplistic in their method of calculation,[76] since they do not account for any adaptation or human resilience. As Barnett and Webber observe: 'All estimates of the increased number of migrants that are likely to be stimulated by climate change are based on very broad scale assessments of exposure to risk, rather than on systematic evidence about the sensitivity of migration patterns to environmental changes', and 'they cannot account for the ways in which adaptation may offset climate impacts.'[77]

As noted above, part of the problem in compiling accurate statistics is how to account adequately for unknown variables, such as precisely when the effects of climate change are likely to be felt most acutely, and the level of investment, planning, and resources that will be committed to try to counter them.[78] But perhaps the most difficult variable to account for is human adaptive capacity or 'resilience'.[79] It is well documented in the refugee literature that it cannot always be anticipated when people will move in response to external triggers such as war or persecution: some flee instantly, some move later, others never move. The line between movement that is 'voluntary' and 'forced' is also very blurred, and many choices will involve a delicate mix of both elements in different proportions.

In the absence of a more rigorous dataset, however, Myers' figures have become the yardstick adopted in much of the literature, often without question.[80] For instance, Sir Nicholas Stern, in his authoritative scientific review of climate change in 2007, described Myers' estimates of 200 million as based on 'conservative assumptions'.[81] Some global policymakers have also adopted language that suggests that climate change may lead to large-scale forced migration. For example, in mid-2009, Kofi Annan issued a report which described 'millions of people' being 'uprooted or permanently on the move as a result' of climate change, with '[m]any more millions' to follow.[82] These tend to overshadow the more sobering remarks of other organizations, including the IPCC, that numerical estimates are 'at best, guesswork'.[83]

[75] Morrissey (n 62) 4. [76] Kniveton and others (n 30).

[77] Barnett and Webber (n 18) 14. For a summary of various figures on populations of concern by region to 2050, see 13.

[78] Nicholas Stern, *The Economics of Climate Change: The Stern Review* (Cambridge University Press, 2007) 112.

[79] See Jessica G Fritze and others, 'Hope, Despair and Transformation: Climate Change and the Promotion of Mental Health and Wellbeing' (2008) 2 International Journal of Mental Health Systems 13. Note, too, that historically people moved throughout the Pacific in order to adapt to changing climatic patterns: Steven R Fischer, *A History of the Pacific Islands* (Palgrave Macmillan, 2002) xvi, 37–8, 44. However, at that time, immigration controls did not hamper movement in the same way that they do today. [80] cf, Black (n 53); Castles (n 2).

[81] Stern (n 78) 77.

[82] *The Anatomy of a Silent Crisis: Human Impact Report: Climate Change* (Global Humanitarian Forum, 2009) ii <http://www.eird.org/publicaciones/humanimpactreport.pdf> accessed 7 December 2009.

[83] Parry and others (n 23) 365 ('Box 7.2. Environmental Migration'). See also Cecilia Tacoli, 'Not Only Climate Change: Mobility, Vulnerability and Socio-Economic Transformations in Environmentally Fragile Areas of Bolivia, Senegal and Tanzania', Human Settlements Working Paper Series (International Institute for Environment and Development, 2011).

Trumpeting such figures in the absence of defensible methodologies can have a deleterious effect. When the empirical evidence does not support them, the very motivation behind the maximalist approach—engendering action—is undercut, because people perceive that there is no clear evidence to support the climate displacement phenomenon. This was seen in 2010, when *New Scientist* reported that Kiribati and Tuvalu were growing, not disappearing.[84] Some media commentators suggested that this undermined Pacific island claims for assistance with resettling their people.[85] Similarly, the Carteret Islanders in Papua New Guinea have argued for a number of years now that their islands are at risk of disappearing as a result of rising sea levels caused by climate change, yet other scientific theories suggest that the islands are subsiding as a result of natural processes.[86]

In early 2011, the UN was lambasted in the media for comments made in 2005 that there would be 50 million 'climate refugees' by 2010.[87] Reports cited a press release issued by the UN University's Institute for Environment and Human Security on UN Day for Disaster Reduction in 2005, which opened with the statement: 'Amid predictions that by 2010 the world will need to cope with as many as 50 million people escaping the effects of creeping environmental deterioration, UN University experts say the international community urgently needs to define, recognize and extend support to this new category of "refugee".'[88] The UN Environment Programme posted a map on its website showing the at-risk places where these people were living. Though these figures were subsequently retreated from, the damage was done. Climate sceptic blogs cited it as evidence that climate change was not real and that the UN could not be trusted.

These three examples highlight the dangers of portraying climate change as the sole driver of displacement. The first example overlooks the fact that underlying

[84] Wendy Zukerman, 'Shape-Shifting Islands Defy Sea-Level Rise', *New Scientist* (London, 2 June 2010) 10 referring to research by Paul Kench and Arthur Webb.

[85] Rowan Callick, 'Coral Islands Left High and Dry', *The Australian* (Australia, 11 June 2010) <http://www.theaustralian.com.au/news/features/coral-islands-left-high-and-dry/story-e6frg6z6-1225878132101> accessed 12 June 2010.

[86] John Campbell, 'Climate-Induced Community Relocation in the Pacific: The Meaning and Importance of Land' in McAdam (n 2) 68. See also Mortreux and Barnett (n 29); Jon Barnett and John Campbell, *Climate Change and Small Island States: Power, Knowledge and the South Pacific* (Earthscan, 2010) 173–4.

[87] Gavin Atkins, 'What Happened to the Climate Refugees?' (11 April 2011) <http://asiancorrespondent.com/52189/what-happened-to-the-climate-refugees/> accessed 28 June 2011; Axel Bojanowski, 'UN Embarrassed by Forecast on Climate Refugees', *Spiegel Online* (18 April 2011) <http://www.spiegel.de/international/world/0,1518,757713,00.html> accessed 27 April 2011; 'Climate Refugees, Not Found', *Wall Street Journal* (New York, 21 April 2011) <http://online.wsj.com/article/SB10001424052748704658704576274470237832478.html?mod=googlenews_wsj> accessed 10 June 2011; Amos Aikman, 'World Still Waiting for "50 Million Climate Refugees by 2010"', *The Australian* (Australia, 21 April 2011) <http://www.theaustralian.com.au/national-affairs/world-still-waiting-for-50-million-climaterefugees-by-2010/story-fn59niix-1226042490227> accessed 10 June 2011.

[88] United Nations University, 'As Ranks of "Environmental Refugees" Swell Worldwide, Calls Grow for Better Definition, Recognition, Support' (12 October 2005) <http://www.ehs.unu.edu/file/get/3916> accessed 28 June 2011. The UNU's Institute for Environment and Human Security has since contributed an enormous amount towards the considered, evidence-based analysis of climate-related movement. See also Bojanowski (n 87) for the background story.

socio-economic conditions mean that some migration from Pacific atolls such as Kiribati and Tuvalu is inevitable—as it has been historically. Thus, whether the islands are 'growing' or 'disappearing' will have little impact on the need for migration options. Similarly, the third example highlights the danger of adopting figures derived from problematic methodologies, especially when there is no reliable mechanism to 'count' those who move for climate-related reasons.

Acknowledging the *multi*causal nature of movement means that studies like these do not discredit discussions about projected movements, and do not set back research (and policy development) on the issue. This stance is not inconsistent with recognizing the vulnerability of such countries to climate change.[89]

III. The Importance of Context

The impacts of climate change on mobility must be understood within a broader social and historical context. As the case studies in this book of Kiribati, Tuvalu, and Bangladesh illustrate,[90] examining climate change-related movement without an appreciation of historical demographic and migration patterns will lead to policy formation in a vacuum. As Silverman has observed in the context of Pacific relocations, even if a particular movement 'seem[s] unique in the history of a single group, we might find them to be recurrent as we enlarge the scale of analysis to a colonial system or a regional mobility system'.[91] Thus, while the driver of 'climate change' may be new (or at least the labelling of it), it resonates with more familiar patterns of movement and adaptation to change. Recognizing the longstanding role that migration has played as an adaptation mechanism in certain regions helps to inform and contextualize current policy debates, and may also help to diffuse some of the more sensational approaches that are at times invoked.[92]

[89] See also Barnett and Campbell (n 86) 173.

[90] See Chs 5–7.

[91] Martin G Silverman, 'Introduction: Locating Relocation in Oceania' in Michael D Lieber (ed), *Exiles and Migrants in Oceania* (University of Hawaii, 1977) 7.

[92] For example, there is a very long history of labour migration from Pacific islands to neighbouring countries: see, eg, John Connell, 'Paradise Left? Pacific Island Voyagers in the Modern World' in James T Fawcett and Benjamin V Cariño (eds), *Pacific Bridges: The New Immigration from Asia and the Pacific Islands* (Center for Migration Studies, 1987); Lieber (n 91); Fischer (n 79). 'Tuvaluans have long had the habit of journeying far from home. The limited resources of the group have encouraged many to go away in search of employment. Others have gone away to preach the Gospel [encouraged by the London Missionary Society, whose policy was to use indigenous teachers to spread Christianity in the Pacific], or to study, and some have been carried away accidentally while sailing between islands. Still others have gone to seek adventure.... The sea was a highway, and not just a barrier, to far off lands well before the development of air transport linked Tuvalu closely to the outside world': Simati Faaniu, 'Travellers and Workers' in Simati Faaniu and others, *Tuvalu: A History* (Institute of Pacific Studies and Extension Services, USP and the Ministry of Social Services, 1983) 121–2. At any one time, several hundred Tuvaluans are being educated overseas, which is disproportionately more than any other country in the region. More recently, circular migration from Kiribati and Tuvalu has developed as men work as merchant seamen on overseas shipping lines, generating significant remittances: John Connell, 'Losing Ground? Tuvalu, the Greenhouse Effect and the Garbage Can' (2003) 44 Asia Pacific Viewpoint 89, 94.

A. Migration in the Pacific

Long before climate change became an issue of concern, the governments of Kiribati and Tuvalu were actively lobbying countries in the region for migration opportunities.[93] Demographic pressures on these already environmentally fragile atolls, which had experienced rapid urbanization and internal migration within a 20 year period,[94] meant that permanent migration abroad was a live public policy issue. As early as 1984, a report by Australia's foreign aid agency, AusAID, observed that enhancing migration opportunities for Tuvaluans would be the most useful form of aid.[95] A decade later, the then Prime Minister of Tuvalu stated in relation to students overseas: 'We want them to come back, but certainly we cannot have everybody, even if they are graduates. There will come a time when we can only take back a portion of our population. The rest—we will have to assist them in obtaining employment overseas and we need to prepare people for when that time comes.'[96]

During the 1990s, Tuvalu consistently sought migration opportunities to Australia as a means of relieving pressure from overcrowding and to assist with the return of migrants working in the phosphate mining industry in Nauru.[97] Although the Prime Minister first highlighted Tuvalu's vulnerability to climate change in 1990 at the second World Climate Conference, stating that the meeting 'could make the difference between Tuvalu's imminent demise and its continued existence',[98] climate change was not linked to migration. This was still the case in 1997.[99]

Yet, by the early 2000s, lobbying for migration assistance was recast as a sudden, urgent request by the world's first 'climate refugees',[100] and the long history of migration from Tuvalu was ignored. The background of overcrowding, resource constraints, and other socio-economic pressures was entirely overlooked in the new 'climate displacement' narrative. This was not merely misrepresentation by the media, but a shift in emphasis by Tuvalu itself.[101] At the international climate conference in November 2000 (COP6), the Prime Minister of Tuvalu stated: 'This

[93] Note the Pacific Islands Forum, when Niue offered to resettle a small number of Tuvaluan families. Since the people of Niue are New Zealand citizens, these families were ultimately able to acquire New Zealand citizenship: interview with Enele Sopoaga, Secretary for Foreign Affairs, Tuvalu (n 12).

[94] Connell (n 92) 93.

[95] *Report of the Committee to Review the Australian Overseas Aid Program* (Commonwealth of Australia, 1984), cited in Connell (n 92) 94.

[96] Connell (n 92) 94, citing remarks in *Pacific Report* (4 April 1994) 4.

[97] ibid, 95.

[98] 'Address by the Rt Hon Bikenibeu Paeniu, Prime Minister of Tuvalu', cited in Jill Jäger and HL Ferguson (eds), *Climate Change: Science, Impacts and Policy Proceedings of the Second World Climate Conference* (Cambridge University Press, 1991).

[99] Connell (n 92) 97, citing the National Communication Report submitted to the UNFCCC meeting in Kyoto in December 1997, in *Pacific News Bulletin* (December 1997) 7.

[100] See Connell (n 92) 97–9. See also Carol Farbotko, 'Tuvalu and Climate Change: Constructions of Environmental Displacement in the *Sydney Morning Herald*' (2005) 87 B Geografiska Annaler 279.

[101] This was evidenced in my fieldwork, when certain interviewees asked me not to focus on underlying social issues but only on climate change as a driving factor of movement. They were concerned that emphasizing the multicausality of movement could undermine the political traction gained by focusing on climate change.

concern is so serious for our people, that the Cabinet, in which I am a member, has been exploring the possibility of buying land in a near-by country, in case we become refugees to the impacts of climate change.'[102] In July 2001, the Tuvaluan government approached Australia requesting special immigration assistance to enable some of its citizens to relocate on account of climate change impacts.[103]

Connell describes this as an example of the 'garbage can' effect: where once isolated phenomena become systematically inter-related.[104] As Connell notes, Tuvalu's diverse environmental problems were essentially 'laid at the door of climatic change, at a time when the country [wa]s experiencing problems of return migration from Nauru and of achieving economic development, and aspirations (fuelled by overseas experience) continue[d] to rise'. In effect, new problems were 'grafted onto old ones and given a single cause',[105] as 'climate change' became the master narrative.

On the one hand, this narrative has helped to generate international attention for the plight of small island States. It has attracted significant media space and has been a call to action for a number of non-governmental organizations (NGOs) and institutions.[106] It has not, however, resulted in significant commitments by the international community on mitigation or adaptation assistance for the governments of affected islands.[107] Indeed, Connell goes so far as to suggest that the discourse has 'diverted attention from the real need both to transform those policies in metropolitan states that continue to contribute to global warming and to develop appropriate environmental management policies within atoll states'.[108]

On the other hand, the narrative of vulnerability has generated a number of problems. First, it resonates with the alarmist approach to climate change and displacement. In addition to the concerns expressed above about that approach, the particular focus on small islands taps into a longer discourse about islands as 'sites

[102] Cited in Barnett and Campbell (n 86) 170.

[103] Senate Foreign Affairs, Defence and Trade Committee, *A Pacific Engaged: Australia's Relations with Papua New Guinea and the Island States of the South-West Pacific* (Commonwealth of Australia, 2003) para 6.78. See also David Wroe, 'Australia Refuses to Throw Lifeline to Drowning Tuvalu', *Sydney Morning Herald*, (19 July 2001) <http://clickthrough.net/bboard/q-and-a-fetch-msg.tcl?msg_id=005oMF> accessed 20 May 2011, which quotes Tuvalu's Acting Secretary for Foreign Affairs and Trade, Mr Pusinelli Lafai, as saying: 'the statement was hardly out of their [the delegates'] mouths before the Australian delegation shut it up...Australia is absolutely against opening up any dialogue.' The Department of Immigration and Multicultural and Indigenous Affairs stated: 'While Australia is committed to fulfilling its humanitarian obligations as a member of the international community, it should give priority to those who require assistance urgently. The threat to Tuvalu's population from a possible rise in sea-level would appear to be a long-term one which should be monitored closely but is not one that requires urgent action at this time': Submission of the Department of Immigration and Multicultural and Indigenous Affairs (Submission No 44) to Senate Foreign Affairs, Defence and Trade References Committee Inquiry into Australia's Relationship with Papua New Guinea and Other Pacific Island Countries (July 2002) para 3.51. The report of the Committee suggests that 'long-term' was envisaged as 30 to 40 years: Senate Committee Report, para 6.79.

[104] Connell (n 92) 94, referring to Michael D Cohen, James G March and Johan P Olsen, 'A Garbage Can Model of Organizational Choice' (1972) 17 Administrative Science Quarterly 1.

[105] Connell (n 92) 98.

[106] For a list of poignant examples, see table in Barnett and Campbell (n 86) 168.

[107] ibid, 155.

[108] Connell (n 92) 105.

of backwardness, insularity, constraint, fragility and weakness'.[109] In blaming the global system, Tuvaluans are rendered disempowered victims of it. Dominant narratives do not leave room for people 'to define the problem in their own terms; to apply their own systems of knowledge; to implement the solutions that are appropriate to their needs and values and which accommodate uncertainty; and to make knowledge claims of equal value to those of science'.[110]

Secondly, and linked to the point above, the increasing 'scientization' of environmental knowledge serves to delegitimize the perceptions and perspectives of local laypeople, and may undermine local capacity to intervene.[111] Despite awareness for at least the past 20 years that small Pacific islands are particularly vulnerable to the impacts of climate change, there have been very few local empirical studies of vulnerability and adaptation. This is notwithstanding a wealth of local knowledge to draw upon.[112] 'Experts' have squeezed out opportunities for local approaches, and their reliance on models, which 'homogenizes people, see social life as the sum of rational individual actions, assume "culture" is separable from other aspects of society, and assume that nature and society are independent "facts"',[113] is underpinned by certain assumptions of the world which may not resonate in a Pacific context. This approach tends to 'produce knowledge of the Pacific Islands that is alien to, and alienates, people in the region'.[114]

Thirdly, there is inadequate research and policy development in areas that might enable people to continue living the lives that they value.[115] In other words, the dominant discourse shapes donor priorities, which in turn determines where funds are allocated.[116] There is anecdotal evidence of projects being recast as 'climate change-related' simply to tap into funding earmarked for that purpose—whether in an institutional research context or in securing funds for projects on the ground. This is not always disingenuous: if funding is sorely needed for medical treatment in a small Pacific island, and it is likely to be forthcoming if its need can be linked to climate impacts (a policy objective set by the funding agency), then one can understand why it might be represented in that way. In other contexts, though, Barnett and Campbell suggest that 'the vulnerability of the islands is a symbol used by researchers who need problems to investigate, journalists who need problems to sell,

[109] Barnett and Campbell (n 86) 2. This historically served to legitimize colonial interventions in the region.

[110] ibid.

[111] Connell (n 92) 103.

[112] Barnett and Campbell (n 86) 80.

[113] ibid, 81, referring to James D Proctor, 'The Meaning of Culture in Global Environmental Change: Retheorizing Culture in Human Dimensions Research' (1998) 8 Global Environmental Change 227; Simon Shackley and Clair Gough, 'The Use of Integrated Assessment: An Institutional Analysis Perspective' (2002) Tyndall Centre for Climate Change Working Paper 14 <http://www.tyndall.ac.uk/sites/default/files/wp14.pdf> accessed 4 July 2011.

[114] Barnett and Campbell (n 86) 81, referring to Jon Barnett, 'Climate Change Science and Policy, As If People Mattered', in Karen O'Brien, Asunción L St Clair, and Berit Kristoffersen (eds), *Climate Change, Ethics and Human Security* (Cambridge University Press, 2010).

[115] Barnett and Campbell (n 86) 1.

[116] On the importance of delivering aid in a manner that *enhances* small island States' capacity to adapt to climate change impacts, see Jon Barnett, 'The Effect of Aid on Capacity to Adapt to Climate Change: Insights from Niue' (2008) 60 Political Science 31.

and NGOs who need problems to solve'.[117] While they may be well intentioned, the effect is frequently to deprive affected communities of agency (hence the idea of people 'unable to act until their islands sink and they are forced to flee and become refugees'[118]) and—simultaneously—capacity to make their own decisions. The result of this approach at the level of international policy is that adaptation responses and 'solutions' are determined independently of understandings of local context.[119]

Finally, a fatalistic outlook may mean that inadequate infrastructure, resource management, and adaptation policies are put in place. If a country's very existence is questioned, there is a risk that this could be self-fulfilling, in the sense that it may generate the view that adaptation is pointless. This, in turn, could impact negatively on foreign aid, on which Tuvalu and Kiribati are heavily reliant, and thereby further reduce adaptive capacity.[120] This is of concern to government officials in Tuvalu, for example,[121] although in Kiribati this does not seem to be perceived as a pressing issue at present.[122] There is also a risk that if the need to move is internalized by locals, unsustainable development practices may ensue and 'the impacts of climate change [may] materialise more through the idea of climate change than through material changes in ecosystems driven by climatic processes'.[123] If migration is divorced from the broader sphere of development, then the effects of climate change might be hastened by depleting communities of their population and culture.[124]

Providing aid in a manner that is appropriate, that does not itself undermine local development, and that can encourage the sustainability of livelihoods once the initial emergency phase has past are crucial (yet often overlooked) factors in donor government aid programmes.[125] There is evidence that '[f]or every dollar invested in disaster risk reduction and preparedness, roughly seven dollars are saved in disaster response'.[126] Thus, just as Warner sees governance structures as having a direct impact on the ways in which people move[127]—the timing of that movement, where they move to, and whether they move back—how funding is expended similarly affects those decisions. Donor States often direct humanitarian aid towards strategic concerns rather than the absolute needs of the displaced,[128] which means that there has been insufficient funding for emergencies in parts of

117 Barnett and Campbell (n 84), 177. 118 ibid. 119 ibid, 177–8.

120 ibid, 7, 42, 170.

121 Interview with Enele Sopoaga, Secretary for Foreign Affairs, Tuvalu (n 12).

122 Interviews with various government officials in Tarawa, Kiribati, May 2009.

123 Mortreux and Barnett (n 29) 106 referring to Jon Barnett and W Neil Adger, 'Climate Dangers and Atoll Countries' (2003) 61 Climatic Change 321, 329–30.

124 Mortreux and Barnett (n 29).

125 Charles Ehrhart and others, *Humanitarian Implications of Climate Change: Mapping Emerging Trends and Risk Hotspots* (2nd edn, OCHA and CARE, 2009) 3–4. 126 ibid, 3.

127 See Warner (n 34), discussed in Ch 8.

128 Elizabeth Ferris, 'Natural Disaster- and Conflict-Induced Displacement: Similarities, Differences and Inter-Connections', prepared for the Annual Meeting of the Society for Applied Anthropology, Panel: 'Expanding the Borders of Disaster-Caused Displacement Research' (27 March 2008) 5. One study concluded that in 2,337 cases of US disaster assistance between 1964 and 1995, foreign policy and domestic factors were 'the overriding determinant': A Cooper Drury, Richard S Olson, and Douglas A Van Belle, 'The Politics of Humanitarian Aid: US Foreign Disaster Assistance 1964–1995' (2005) 67 Journal of Politics 454, 461. See also UNHCR, *The State of the World's Refugees: Human Displacement in the New Millennium* (Oxford University Press, 2006) 100.

the world that are of little significance to donor States' interests. The creation of the UN Central Emergency Response Fund in 2006 was intended to be a response to this problem.[129] However, affected States are facing significant difficulties in accessing international funds.[130]

Thus, the common representation of climate change in small island developing States 'is a discursive formation that *limits* understanding and action to address the interests of people living in islands'.[131] Recent changes in approach by the government of Tuvalu would suggest that these negative impacts are now being realized. The emphasis is now on adaptation to enable people to remain at home.[132]

In 2008 and 2009, Tuvaluan Prime Minister Apisai Ielemia made a series of statements to the following effect:

> While Tuvalu faces an uncertain future because of climate change, it is our view that Tuvaluans will remain in Tuvalu. We will fight to keep our country, our culture and our way of living. We are not considering any migration scheme. We believe if the right actions are taken to address climate change, Tuvalu will survive.[133]

In 2010, the new Prime Minister, Maatia Toafa, also argued against relocation: 'We should be able to say to the country identity is very important, you have your people, the same land and your culture.'[134] This shift in approach does not downplay the effects of climate change, but rather reasserts the population's right to remain at home and focus efforts on *in situ* adaptation strategies. Migration and relocation are reluctantly acknowledged as options of last resort, but few officials want to characterize them as elements of government strategy or policy.[135] This is interesting, given that migration experts regard migration as a form of adaptation. As discussed in Chapters 6 and 7, migration can facilitate livelihoods elsewhere for those who move, and provide an economic benefit to those who remain through remittances (at both the household and national

[129] See, eg, 'Strengthening of the Coordination of Emergency Humanitarian Assistance of the United Nations' UNGA res 60/124 (8 March 2006). See generally, Barnett and Webber (n 18) 33–4.

[130] Sean Dorney, 'Kiribati Struggling to Access Climate Change Funding', *ABC Asia Pacific News* (Australia, 4 August 2011) <http://abcasiapacificnews.com/stories/201108/3285417.htm> accessed 4 August 2011.

[131] Barnett and Campbell (n 86) 1 (emphasis added).

[132] Some officials described migration as being an option at the back of the government's mind, with adaption at the forefront: Kelesoma Saloa, Prime Minister's Private Secretary (Funafuti, Tuvalu, 25 May 2009); Enele Sopoaga, Secretary for Foreign Affairs, Tuvalu (n 12).

[133] 'Prime Minister Apisai Ielemia Says Climate Change Threatens Tuvalu's Survival' (European Parliament Press Release, 10 December 2009) <http://www.europarl.europa.eu/news/expert/infopress_page/028–66101-341-12-50–903-20091207IPR66100-07-12–2009-2009-false/default_en.htm> accessed 13 December 2009, quoting the Tuvaluan Prime Minister's comments to the Development Committee on 10 December 2009. It was reported that he made similar remarks at the December 2008 UNFCCC conference in Poznan: Nic Maclellan, 'Rudd Leaves Island Nations for Dead', *New Matilda* (14 December 2009) <http://newmatilda.com/2009/12/14/rudd-leaves-pacific-island-nations-dead> accessed 2 July 2011. See similar comments made by the government of Nauru when it was proposed that its population relocate to Australia in the 1960s, discussed in Ch 5.

[134] 'Tuvalu's Prime Minister Optimistic His Government Will Be Stable', *Radio Australia News* (6 October 2010) <http://www.radioaustralianews.net.au/stories/201010/3030712.htm?desktop> accessed 20 May 2011. See further, Ch 5.

[135] This was evident in fieldwork conducted in Tuvalu in 2009.

levels).[136] Additionally, as an adaptation strategy, migration may alleviate pressure on resources at home, and thereby enable a smaller population to remain for longer.

IV. The 'Invisibility' of Climate Change-Related Movement

Finally, it is important to draw attention to the deliberate (but perhaps largely unselfconscious) parameters of the climate displacement literature. Research focusing on the international legal and protection gaps relating to climate change and displacement typically overlooks (a) the plight of people who do not move; and (b) the climate change impacts on States which already have secure migration options. This is because the 'problem' this scholarship seeks to address is the absence of formal pathways for lawful migration or protection from displacement, which creates a geographical bias towards countries which lack a migration safety net. Citizens of the Marshall Islands and the Federated States of Micronesia have the option of migrating to the United States, pursuant to the Compact of Free Association.[137] The people of Tokelau, Niue, and the Cook Islands are New Zealand citizens and can, accordingly, move there should they need to do so. Although movement away from their homes involves very similar issues around exposure to risk, vulnerability, and loss of land,[138] their predicament does not feature as prominently in climate change displacement debates as that of Tuvalu, Kiribati, and the Maldives, since they already have a 'solution'.[139] Yet, the option to move does not resolve underlying and fundamental questions relating to identity, culture, and self-determination, and the loss of 'home', in particular. Interventions by the Marshallese, for example, reveal the same concerns and anxieties as those expressed by the people of Tuvalu and Kiribati.[140] It is therefore important not to view policy options that allow for migration or relocation as complete 'solutions'. They may move people out of harm's way, but they bring with them a host of other issues which need to be addressed through a human rights framework.

136 See Alex Julca, 'Multidimensional Re-creation of Vulnerabilities and Potential for Resilience in International Migration' (2011) 49 International Migration 30, 34, 41.

137 Compact of Free Association Act of 1985, PL 99–239, 99th Congress (14 January 1986) <http://www.doi.gov/oia/compact/RMICompact.pdf>. There are no guarantees of permanent residence or citizenship, however.

138 See especially Ch 5.

139 But see Ruth Laugesen, 'New Zealand's Forgotten Isles', *Listener* (11 July 2009) 32, who argues that 'by choosing to stay under New Zealand's wing, Tokelau has also chosen international invisibility.'

140 The Marshall Islands has argued that '[w]ithout concerted action and the support of the international community, the Marshall Islands might prove to be among the first nations to experience wholesale threats to its very statehood': *Draft Report of the Working Group on the Universal Periodic Review: Republic of the Marshall Islands*, UN Doc A/HRC/WG.6/9/L.10 (15 November 2010) para 21. See also interventions at 'Threatened Island Nations: Legal Implications of Rising Seas and a Changing Climate' (Conference at Columbia University, New York, 23–25 May 2011) <http://www.law.columbia.edu/centers/climatechange/resources/threatened-island-nations> accessed 1 August 2011.

Additionally, while a need to move may be a sign of vulnerability,[141] the inability to move may signal even greater exposure to harm. As has been noted in migration studies generally, movement may be impossible 'because of systemic, structural and individual reasons', including 'domestic and familial obligations and responsibilities, disability and illness, age, education and skills, and an absence or lack of access to networks and relationships'.[142] This is pertinent to normative frameworks that seek to articulate the needs of the displaced, since those who remain behind may have similar (or even greater) needs requiring redress. Barnett and Webber suggest that the numbers of people who cannot migrate may be far in excess of those who do, 'and so may pose a far larger humanitarian problem, even though this problem will be more spatially and temporally diffuse than events where displaced people are concentrated in specific locations'.[143]

Finally, as noted above, those who move for reasons of climate change impacts are not accounted for in existing legal or bureaucratic categories. This means that even where climate change plays a role in migration decisions, it will not be understood as 'climate' migration/displacement. This also means that the data on such movement is scattered.

This brings us back to the question of conceptualization. The 2010 floods in Pakistan—the worst disaster ever faced by the UN—were described predominantly as a 'natural' disaster,[144] and not in the context of 'climate change' or 'environmental' displacement.[145] If responses to *sudden*-onset events such as floods or cyclones remain couched in the language of 'humanitarian disaster', then it is even more improbable that longer-term movements from *slow*-onset impacts will gain any attention at all as climate change-driven movements. As Gemenne notes, '[e]nvironmental migration as a social phenomenon is generally apprehended through its definition, which bears high responsibility for the development of normative framework and policy responses'.[146] The construction of the problem substantially shapes legal and policy interventions, and for the reasons outlined above, climate-related movement may in many respects remain an invisible phenomenon in legal and bureaucratic terms.

[141] It may also be a sign of resilience and adaptation, depending on the circumstances.

[142] Uma Kothari, 'Migration and Chronic Poverty' (2002) Institute for Development Policy and Management, University of Manchester, Working Paper No 16, 14 <http://cprc.abrc.co.uk/pubfiles/16Kothari.pdf> accessed 4 July 2011. See also Arjan de Haan, 'Livelihoods and Poverty: The Role of Migration—A Critical Review of the Migration Literature' (1999) 35 Journal of Development Studies 1.

[143] Barnett and Webber (n 18) 12–13.

[144] See, eg, 'Exodus of Pakistanis Excels as Flood Defences Fail', *Sydney Morning Herald* (Sydney, 28–29 August 2010) 18.

[145] See, eg, 'Pakistan Facing "Serious" Threat of Epidemic Disease', *BBC News* (24 August 2010) <http://www.bbc.co.uk/news/world-south-asia-11068259?print=true> accessed on 24 August 2010; 'Pakistanis Should Live Away From Flood Areas: UN Agency', *Right Vision News* (22 August 2010).

[146] Gemenne (2009) (n 18) 148. See also Warner (n 34) 4, discussed further in Ch 7.

V. Conclusion: A Human Rights Approach

Some may argue that not focusing squarely on climate change in the context of human movement misses an opportunity to leverage funding and assistance, given its special ethical and political sway. By contrast to general poverty or disadvantage, responsibility for climate change can be more easily attributed to the actions of a wide range of States (morally, if not legally).[147] Furthermore, the countries likely to be most severely affected by its impacts are those which have contributed among the least to global carbon emissions.[148] The sense of injustice this invokes makes it a powerful focal point for activism and social change. For example, the Director of the leading human rights NGO in Bangladesh argues that those displaced by climate change should be given a special focus because climate change is 'not natural',[149] and they are victims of other governments' failure to reduce greenhouse gas emissions. Similarly, interviews in Tuvalu revealed that some government officials worry that highlighting the complex and multifaceted dimensions of movement will detrimentally shift the focus away from climate change, and that the magnitude of that problem warrants maintaining attention solely on its impacts.[150]

And yet, as Khan has observed, 'although floods and cyclones play havoc with the lives and livelihoods of poor people, their insecurity of livelihood is not simply a consequence of nature. To a greater extent it is a result of bad laws and policies, unscrupulous employers and corrupt officials.'[151] Thus, the interplay between the global and the national, the 'natural' and the 'man-made', and cause and effect is complex.

In my view, the most effective responses will consider climate change-related movement within a broader human rights matrix. Arguably, an approach which views climate change as one of a multitude of possible drivers of movement—and which advocates for solutions to those wider problems—opens up more opportunities for solutions, institutional knowledge, and capacity. Indeed, one of the problems with conceptualizing climate change-related movement as a discrete phenomenon is that a plethora of material pertaining to other 'non-refugee' forced movement may be too hastily overlooked. In some instances, the parallels with flight from armed conflict or generalized violence may be great; in others, lessons may be able to be learnt from refugee resettlement practices or development-forced migration and resettlement. Although focusing on the complexity of climate change-related movement requires a nuanced understanding, it represents a more sober, and ultimately justifiable, approach.

147 See Ch 4.

148 Barnett and Campbell (n 86) 10, referring to data compiled by the World Resources Institute Climate Analysis Indicators Tool <http://cait.wri.org> accessed 10 August 2011.

149 Interview with Sultana Kamal, Founder of Ain o Salish Kendra (ASK) (Dhaka, Bangladesh, 21 June 2010).

150 'Far too much effort is being diverted from the real issue of climate change, far too much ... my fear is that the issue is too urgent for all sorts of diversion': interview with Enele Sopoaga, Minister for Foreign Affairs, Tuvalu (n 12).

151 Khan (n 15) 74.

2

The Relevance of International Refugee Law

I. Introduction

The term 'climate change refugee' is sometimes used to describe people who will be forced to leave their homes as a result of climate change impacts. While this label may be useful from a political perspective, in that it highlights some of the most extreme ways in which climate change will affect human society, it is both legally and conceptually flawed.

This chapter examines the extent to which international refugee law may apply to people displaced by the impacts of climate change. Though there are some circumstances in which it will be applicable, it is, by and large, an inappropriate normative framework for responding to the needs of those forced to move on account of environmental or climate change impacts. Nevertheless, refugee law has some useful standards to offer any new protection-oriented instrument, such as the assessment of potential, future harm.

A. Background

The notion of an 'ecological refugee' first appeared in 1948,[1] but its more recent and first 'official' derivation was a United Nations Environment Programme (UNEP) report in 1985 by El-Hinnawi.[2] El-Hinnawi was not seeking to mount any kind

[1] William Vogt, *Road to Survival* (William Sloane Associates, 1948) in François Gemenne, 'How They Became the Human Face of Climate Change: Research and Policy Interactions in the Birth of the "Environmental Migration" Concept' in Etienne Piguet, Antoine Pécoud, and Paul de Guchteneire (eds), *Migration and Climate Change* (Cambridge University Press and UNESCO Publishing, 2011) 227.

[2] Essam El-Hinnawi, *Environmental Refugees* (United Nations Environment Programme, 1985). See also the work of Lester Brown of the Worldwatch Institute in the 1970s, and a 1984 briefing document by the International Institute for Environment and Development, *Environmental Refugees* (Earthscan, 1984), referred to in Gemenne (n 1) 228. For discussion on the history and development of the concept, see Richard Black, 'Environmental Refugees: Myth or Reality?' (2001) UNHCR New Issues in Refugee Research, Working Paper No 34; Norman Myers and Jennifer Kent, *Environmental Exodus: An Emergent Crisis in the Global Arena* (Climate Institute, 1995); Astri Suhrke and Annamaria Visentin, 'The Environmental Refugee: A New Approach' [1991] Ecodecision 73; James Morrissey, 'Environmental Change and Forced Migration: A State of the Art Review' (2009) Refugee Studies Centre, University of Oxford, Background Paper <http://www.rsc.ox.ac.uk/events/environmental-change-and-migration/EnvChangeandFmReviewWS.pdf> accessed 4 August 2011.

of legal—or even ethical[3]—argument about the extension of refugee law to people displaced for environmental reasons.[4] Rather, he used the term to highlight the potentially devastating impacts of unchecked development and pollution,[5] in much the same way as environmental lobby groups today use the language of 'climate refugees' to draw attention to the most deleterious aspects of greenhouse gas emissions.[6]

On the whole, forced migration scholars reject the apparent simplicity of this representation.[7] While the concept of 'environmental' or 'climate change' refugees may provide a useful advocacy tool to generate attention and mobilize civil society around the dangers of global warming, it can also contribute to misunderstandings about the likely patterns, timescale, and nature of climate change-related movement. That is not to say that this approach is disingenuous, but rather that it is important to be alert to the particular objectives it seeks to promote.

Even as a merely descriptive term, the 'climate change refugee' label is at best preemptive, and at worst offensive to those to whom it is ascribed.[8] In the small Pacific island States of Kiribati and Tuvalu, the refugee label is resoundingly rejected at both the official and personal levels.[9] This is because it is seen as invoking a sense of helplessness and a lack of dignity which contradicts the very strong sense of Pacific

[3] Contrast Andrew Shacknove, 'Who is a Refugee?' (1985) 95 Ethics 274 with James C Hathaway, 'Is Refugee Status Really Elitist? An Answer to the Ethical Challenge' in Jean-Yves Carlier and Dirk Vanheule (eds), *Europe and Refugees: A Challenge?* (Kluwer Law International, 1997); Matthew E Price, *Rethinking Asylum: History, Purpose, and Limits* (Cambridge University Press, 2009).

[4] El-Hinnawi (n 2) 4. Kibreab argued that the term 'environmental refugee' was 'invented at least in part to depoliticise the causes of displacement, so enabling states to derogate their obligation to provide asylum'. He argued that the term 'environmental' detaches the individual's circumstances from any underlying socio-political causes that might indeed fall within the ambit of the 1951 Refugee Convention, and thus may be a way of receiving States seeking to avoid their obligations towards people whose plight can, at least on one view, be characterized in this way. However, in my view this overstates the genesis of the term. See Gaim Kibreab, 'Environmental Causes and Impact of Refugee Movements: A Critique of the Current Debate' (1997) 21 Disasters 20, 21.

[5] El-Hinnawi (n 2).

[6] See, eg, Friends of the Earth Australia, *A Citizen's Guide to Climate Refugees* (rev edn, Friends of the Earth Australia, 2007). For an analysis of the 'refugee' terminology, see below.

[7] See, eg, UNHCR Expert Roundtable on Climate Change and Displacement, 'Summary of Deliberations on Climate Change and Displacement' (Bellagio, 22–25 February 2011) ('Bellagio Deliberations'), 'Main Messages', 1: 'The terms of "climate refugee" and "environmental refugee" should be avoided as they are inaccurate and misleading.' See also 'Chairperson's Summary', Nansen Conference on Climate Change and Displacement in the 21st Century (Oslo, 6–7 June 2011) para 21.

[8] See Jane McAdam and Maryanne Loughry, 'We Aren't Refugees', *Inside Story* (30 June 2009) <http://inside.org.au/we-arent-refugees/> accessed 1 July 2009; Anote Tong, President of Kiribati, 'Address' (Australian National University, Canberra, 19 June 2008) <http://news.anu.edu.au/?p=437> accessed 20 August 2008; François Gemenne, 'Environmental Changes and Migration Flows: Normative Frameworks and Policy Responses' (PhD thesis, Institut d'Etudes Politiques de Paris and University of Liège, 2009), who also discusses the rejection of the term by people who fled from Hurricane Katrina in the United States.

[9] See remarks by Pelenise Alofa Pilitati, Chair of the Church Education Association, Kiribati in 'Climate Refugees', *Australia Talks* (ABC Radio National, Australia, 3 August 2009) <http://www.abc.net.au/rn/australiatalks/stories/2009/2641241.htm> accessed 14 December 2009. And yet the language persists: 'Climate Refugees in Australia "Inevitable"', *ABC News* (11 December 2009) <http://www.abc.net.au/news/stories/2009/12/11/2769403.htm?site=news> accessed 14 December 2009.

pride. As the President of Kiribati explained, 'when you talk about refugees—climate refugees—you're putting the stigma on the victims, not the offenders'.[10]

> We don't want to lose our dignity. We're sacrificing much by being displaced, in any case. So we don't want to lose that, whatever dignity is left. So the last thing we want to be called is 'refugee'. We're going to be given as a matter of right something that we deserve, because they've taken away what we have.[11]

Rather than regarding 'refugees' as people with resilience, who have actively fled situations of violence or conflict, they are perceived as passive victims, waiting helplessly in camps and relying on handouts, with no prospects for the future.[12] Some men explained that being described as a 'refugee' would signal a failure on their behalf to provide for and protect their family. The people of Tuvalu and Kiribati do not want to be seen in this way. When they speak of their own possible movement to countries like Australia or New Zealand, they describe the importance of being regarded as active, valued members of a community who can positively contribute to it.[13] Similarly, in Bangladesh, despite an initial embracing of the 'refugee' terminology, it is now viewed by advocacy groups as inappropriate.[14]

Interestingly, these concerns highlight some of the central failures of the international protection regime, most notably the fact that an absence of political will to implement the principle of burden-sharing is currently leaving millions of refugees in protracted situations with no hope of durable solutions.[15] However, their discomfort also stems from the fact that refugees flee from their own government, whereas the people of Kiribati and Tuvalu have no desire to escape from their countries. They say it is the actions of other States that will ultimately force their movement, not the actions of their own leaders. Indeed, if anything, the persecutor in such cases might be better described as the 'international community', and industrialized States in particular—the very States to which movement might be

[10] Interview with President Anote Tong, President of Kiribati (Tarawa, Kiribati, 12 May 2009).

[11] ibid.

[12] The Tuvaluan and I-Kiribati languages do not have a word for 'refugee', because the concept is foreign to their communal cultures: see interviews with Etita Morikao (Funafuti, Tuvalu, 25 May 2009) and Isala T Isala (Funafuti, Tuvalu, 27 May 2009). See also the comment by the President of the Maldives: 'We do not want to leave the Maldives, but we also do not want to be climate refugees living in tents for decades', cited in Randeep Ramesh, 'Paradise Almost Lost: Maldives Seek to Buy a New Homeland', *The Guardian* (London, 10 November 2008) <http://www.guardian.co.uk/environment/2008/nov/10/maldives-climate-change> accessed 4 August 2011. See also the comments of Kiribati's Foreign Secretary, Tessie Lambourne: 'We are proud people. We would like to relocate on merit and with dignity' cited in Laurie Goering, 'Kiribati Officials Plan for "Practical and Rational" Exodus from Atolls', *Reuters AlertNet* (9 December 2009) <http://www.alertnet.org/db/an_art/60714/2009/11/9-181804-1.htm> accessed 13 December 2009. In the specific context of climate change, the President of Kiribati also invoked the language of responsibility: 'When you talk about refugees—climate refugees—you're putting the stigma on the victims, not the offenders': interview with President Anote Tong (n 10).

[13] This is not unique: many refugees describe similar feelings.

[14] See Ch 7.

[15] Ch 7 picks up on this observation as a reason why it is unlikely that there is sufficient political will to conclude a new treaty dealing specifically with climate change-related displacement.

sought, whose failure to cut greenhouse gas emissions has led to the predicament now being faced.[16]

II. The Relevance of the 1951 Refugee Convention

The term 'refugee' is a legal term of art. The legal definition of a 'refugee' and the rights and entitlements it entails are set out in the 1951 Refugee Convention relating to the Status of Refugees, read in conjunction with its 1967 Protocol. A 'refugee' is defined as someone who:

> owing to well-founded fear of being persecuted for reasons of race, religion, nationality, membership of a particular social group or political opinion, is outside the country of his nationality and is unable, or owing to such fear, is unwilling to avail himself of the protection of that country; or who, not having a nationality and being outside the country of his former habitual residence as a result of such events, is unable or, owing to such fear, is unwilling to return to it.[17]

At the outset, it is important to point out that while international law defines a 'refugee' in a particular way, this does not mean that people outside this definition are unworthy of protection, or necessarily denied it.[18] Definitions serve an instrumental purpose. They are bureaucratic labels that delimit rights and obligations, and may seek to bolster some kind of ethical claim to protection or assistance as well.[19] Indeed, the creation of a definition inevitably leads to a testing of its boundaries, and establishes the parameters for re-evaluating and re-defining what it should be. In some ways it is stultifying, for it entrenches a particular historical, instrumental, or political view as a legal threshold, which becomes the benchmark for further development. On the other hand, it provides a starting point to which States are willing to agree, and from which subsequent solutions and developments may stem. Compellingly, and importantly, legal definitions bind States in a way that descriptive labels cannot. The key point here is that the law does not answer or resolve the fundamental problems of definitional debates—it simply provides a set of criteria from which certain rights and obligations may flow.

There are a number of obstacles that make it very difficult to argue that people displaced by the impacts of climate change are refugees within the meaning of the Refugee Convention.

[16] See Intergovernmental Panel on Climate Change (IPCC), *Climate Change: The IPCC Scientific Assessment: Final Report of Working Group I* (Cambridge University Press, 1990) 8 (fn omitted); IPCC, *Climate Change 2007: Synthesis Report: Summary for Policymakers* (Cambridge University Press, 2007) 5, 6, 12, 13. See section II below.

[17] Convention relating to the Status of Refugees (adopted 28 July 1951, entered into force 22 April 1954) 189 UNTS 137, Art 1A(2), read in conjunction with Protocol relating to the Status of Refugees (adopted 31 January 1967, entered into force 4 October 1967) 606 UNTS 267 (together 'Refugee Convention').

[18] See Ch 3.

[19] As Gemenne notes, '[e]nvironmental migration as a social phenomenon is generally apprehended through its definition, which bears high responsibility for the development of normative framework and policy responses': Gemenne (n 8) 148.

First, the refugee definition only applies to people who have already crossed an international border. As discussed in Chapter 1, much of the anticipated movement in response to climate change will be internal, and thus will not meet this preliminary requirement. This criterion also means that the Refugee Convention does not facilitate direct resettlement from the country of persecution. Thus, any proposal to extend the Refugee Convention regime for people displaced by climate change impacts, whether through opening up that instrument to renegotiation or drafting a protocol, would have this same limitation.[20]

Secondly, there are difficulties in characterizing 'climate change' as 'persecution'. While some have sought to argue the case,[21] I find their arguments unconvincing. 'Persecution' entails violations of human rights that are particularly serious, either because of their inherent nature or because of their repetition (for example, an accumulation of breaches which, individually, would not be so severe but which together constitute a serious violation).[22] It remains very much a question of degree and proportion. Whether something amounts to 'persecution' is assessed according to the nature of the right at risk, the nature and severity of its restriction or impairment, and the likelihood of the restriction or impairment eventuating in the individual case.[23]

Although adverse climate impacts such as rising sea levels, salination, and increases in the frequency and severity of extreme weather events (eg storms, cyclones, and floods) are harmful, and in some cases fatal, they do not meet the threshold of 'persecution' as this is currently understood in international and domestic law.[24] This is so even if climate change-related harms are understood as breaches of socio-economic rights. The idea that persecution can be economic in nature 'is not a modern construct or radical notion; rather there is evidence that from the earliest days of its operation some types of socio-economic claims were considered to fall within the purview of the Refugee Convention definition'.[25] For example, in small island developing States such as Kiribati and Tuvalu, there are

[20] See further, Ch 7. Interestingly, the Director of the Intergovernmental Committee on Refugees established by the 1938 Evian Conference was charged with undertaking 'negotiations to improve the present conditions of exodus (of refugees from Germany and Austria) and to replace them by orderly emigration'. Orderly departure was also proposed (and eventually adopted), as an alternative to the departure of refugees from Vietnam by boat. See Ch 9 on comprehensive plans of action. See generally Guy S Goodwin-Gill and Jane McAdam, *The Refugee in International Law* (3rd edn, Oxford University Press, 2007) 2.

[21] See, eg, Jessica B Cooper, 'Environmental Refugees: Meeting the Requirements of the Refugee Definition' (1998) 6 New York University Environmental Law Journal 480; Christopher M Kozoll, 'Poisoning the Well: Persecution, the Environment, and Refugee Status' (2004) 15 Colorado Journal of International Environmental Law and Policy 271.

[22] See Council Directive (EC) 2004/83 on Minimum Standards for the Qualification and Status of Third Country Nationals or Stateless Persons as Refugees or as Persons Who Otherwise Need International Protection and the Content of the Protection Granted [2004] OJ L304/12 ('Qualification Directive') Art 9. It may include a threat to life or liberty, significant physical harassment or ill-treatment, or significant economic hardship or denial of access to basic services or denial of capacity to earn a livelihood, where such hardship or denial threatens the applicant's capacity to subsist: Migration Act 1958 (Cth) (Australia), s 91R(2).

[23] See Goodwin-Gill and McAdam (n 20) 92.

[24] On which, see ibid, 90–134.

[25] Michelle Foster, *International Refugee Law and Socio-Economic Rights: Refuge from Deprivation* (Cambridge University Press, 2007) 88 (fn omitted).

real concerns about the long-term viability of economic, social, and cultural rights given that they may be disrupted as climate change slowly erodes the territories' habitability. Both Kiribati and Tuvalu experience significant overcrowding that is only set to worsen as population naturally increases. Unemployment is high in both countries. Climate change threatens to reduce habitable land in a number of ways, including through coastal erosion and increased salination. This will impact upon agricultural capacity and, in turn, is likely to lead to greater urbanization (through movement from outer islands) and increased pressure on the labour market. There are also negative health consequences as people become increasingly reliant on imported processed foods.[26]

In refugee law, for deprivation to move beyond the 'mere' non-realization of a right to a violation of a right in a manner that amounts to persecution, a discriminatory element is required.[27] In other words, it needs to be shown that the persecutor is engaging in such acts because of an attribute—real or perceived—of the person being persecuted, rather than simply being a random attack. That attribute must be linked to at least one of the five Convention grounds. Thus, it is not poverty or lack of economic opportunity alone that renders a person in need of international protection, but rather the requirements that: (a) the level of deprivation amounts to 'persecution'; (b) that persecution is on account of one of the five Convention grounds; and (c) the person's government is unable or unwilling to shield that person from such persecution.

There is nothing implicit in the Refugee Convention that would preclude recognition of environmental harms amounting to persecution provided that the requisite elements of Article 1A(2) could be established. However, it exceeds the presently recognized bounds of the Convention to argue that climate change per se would meet the requisite threshold. In many cases, environmental harms may be bound up in other practices that are persecutory, thereby avoiding the need to base a claim solely on such grounds.[28]

Accordingly, a refugee claim based generally on the broad impact of climate change will not succeed.[29] Rather, there must be a differential impact as against the rest of society at large (in other words, because the group is marginalized). For example, one must be able to demonstrate that one is poor because government policy, inaction, or discrimination treats one group in society differently from others. Indeed, as the New Zealand Refugee Status Appeals Authority (RSAA) affirmed in relation to very poor refugee claimants from Tuvalu, they could not be refugees if they had not been treated differently from anyone else:

> This is not a case where the appellants can be said to be differentially at risk of harm amounting to persecution due to any one of these five grounds. All Tuvalu citizens face the same environmental problems and economic difficulties living in Tuvalu. Rather, the appellants are unfortunate victims, like all other Tuvaluan citizens, of the forces of nature leading to

26 See Ch 1.

27 See Foster (n 25) 104–5 and cases discussed there.

28 See, eg, Bellagio Deliberations (n 7) para 8; Nansen Chairperson's Summary (n 7) para 20; discussion in Ch 3.

29 For parallel jurisprudence on socio-economic deprivation, see Foster (n 25) 310 and relevant cases cited there.

the erosion of coastland and the family property being partially submerged at high tide. As for the shortage of drinkable water and lack of hygienic sewerage systems, medicines and appropriate access to medical facilities, these are also deficiencies in the social services of Tuvalu that apply indiscriminately to all citizens of Tuvalu and cannot be said to be forms of harm directed at the appellants for reason of their civil or political status.[30]

Claims based generally on 'climate change' do not meet this persecution mould. Part of the problem in the climate change context is identifying a 'persecutor'.[31] For example, the governments of Kiribati and Tuvalu are not responsible for climate change as a whole, nor are they developing policies which increase its negative impacts on particular sectors of the population. Indeed, the Tuvaluan and I-Kiribati governments remain willing to protect their citizens, although the extent of their ability to do so over time is unclear. As noted above, one might argue that the 'persecutor' in such a case is the 'international community', and industrialized countries in particular, whose failure to reduce greenhouse gas emissions has resulted in the predicament now confronting them.[32] These are the very countries to which movement might be sought if the land becomes unsustainable. This delinking of the actor of persecution from the territory from which flight occurs is a complete reversal of the traditional refugee paradigm: whereas Convention refugees flee their own government (or private actors that the government is unable or unwilling to protect them from),[33] a person fleeing the effects of climate change is not escaping his or her government, but rather is seeking refuge from—yet within—countries that have contributed to climate change.[34]

The Australian Refugee Review Tribunal (RRT) has rejected the argument that this amounts to 'persecution' for the purposes of the Refugee Convention:

In this case, the Tribunal does not believe that the element of an attitude or motivation can be identified, such that the conduct feared can be properly considered persecution for reasons of a Convention characteristic as required.... There is simply no basis for concluding that countries which can be said to have been historically high emitters of carbon dioxide

[30] *Refugee Appeal No 72189/2000*, RSAA (17 August 2000) para 13; see also *Refugee Appeal No 72179/2000* (31 August 2000) cited in Foster (n 25) 310. See also the quote from *Subramaniam v Immigration and Naturalization Service*, 724 F Supp 799, 801 (9th Cir, 1989) in Foster (n 25) 288.

[31] This is also said to be a limitation in economic claims: see Foster (n 25) 9. Harding says that refugees generally attract greater international sympathy than economic migrants because there is an identifiable persecutor, as opposed to a general degree of economic difficulty that prevails in some parts of the world: see Jeremy Harding, *The Uninvited: Refugees at the Rich Man's Gate* (Profile Books, 2000) 122 cited in Foster (n 25) 9.

[32] See IPCC, *Climate Change: The IPCC Scientific Assessment* (n 16) 8 (fn omitted); IPCC, *Climate Change 2007: Synthesis Report* (n 16) 5, 6, 12, 13.

[33] The language of Art 1A(2) of the Refugee Convention is that the refugee is 'unable or, owing to such fear, is unwilling to avail himself of the protection of that country'. The drafters stated that '"unable" refers primarily to stateless refugees but includes also refugees possessing a nationality who are refused passports or other protection by their own government': see United Nations Economic and Social Council, *Report of the Ad Hoc Committee on Statelessness and Related Persons (Lake Success, New York, 16 January–16 February 1950)*, UN Doc E/1618 (17 February 1950) 39. A number of domestic courts have also stated that it extends to situations in which the government is either non-existent, ineffective, or colluding with the persecutors: see, eg, *Zalzali v Minister of Employment and Immigration* [1991] FCJ No 341.

[34] See analysis of State responsibility and the remoteness of harm: Ch 3.

or other greenhouse gases, have any element of motivation to have any impact on residents of low lying countries such as Kiribati, either for their race, religion, nationality, membership of any particular social group or political opinion.[35]

Finally, even if the impacts of climate change could be characterized as 'persecution', the Refugee Convention requires such persecution to be *for reasons of* an individual's race, religion, nationality, political opinion, or membership of a particular social group. Persecution alone is not enough. The difficulty in the present context is that the impacts of climate change are largely indiscriminate, rather than tied to particular characteristics such as a person's background or beliefs. Although climate change affects some countries more adversely than others by virtue of their geography and resources, the reason it does so is not premised on the nationality or race of their inhabitants. An argument that people affected by its impacts could constitute a 'particular social group' would be difficult to establish, because the law requires that the group must be connected by a fundamental, immutable characteristic other than the risk of persecution itself.[36]

As McHugh J explained in *Applicant A*, although a shared fear may help to define a group, it is the particular attribute ascribed to them, rather than the persecutory acts themselves, that serves to 'create' them as a particular social group:

> the actions of the persecutors may serve to identify or even cause the creation of a particular social group in society. Left-handed men are not a particular social group. But, if they were persecuted because they were left-handed, they would no doubt quickly become recognizable in their society as a particular social group. Their persecution for being left-handed would create a public perception that they were a particular social group. But it would be the attribute of being left-handed and not the persecutory acts that would identify them as a particular social group.[37]

Superior courts around the world have explained that the Refugee Convention does not cover people in search of better living conditions or victims of natural disasters, even though 'both of these cases might seem deserving of international sanctuary'.[38] It has expressly recognized that this is so, 'even when the home state is *unable* to provide assistance'.[39] This is because the Refugee Convention has a more limited scope. As the High Court of Australia has observed, the requirement of 'persecution' limits the Convention's 'humanitarian scope and does not afford universal protection to asylum seekers. No matter how devastating may be epidemic, natural disaster or famine, a person fleeing them is not a refugee within the terms of the Convention.'[40] People fleeing 'natural disasters and bad economic conditions' fall outside the Convention.[41] The House of Lords in the United Kingdom has also observed that the Convention does not provide protection in all cases:

35 *0907346* [2009] RRTA 1168 (10 December 2009) para 51.

36 Goodwin-Gill and McAdam (n 20) 79–80; *Applicant A v Minister for Immigration and Ethnic Affairs* (1997) 190 CLR 225, 341 (Dawson J).

37 *Applicant A* (n 36) 264 (McHugh J). Nor does the size of a group necessarily preclude the possibility of refugee status being recognized.

38 *Canada (Attorney General) v Ward* [1993] 2 SCR 689, 732.

39 ibid (emphasis added).

40 *Applicant A* (n 36) 248 (Dawson J).

41 *Minister for Immigration v Haji Ibrahim* (2000) 204 CLR 1, 48–9 (Gummow J).

The applicant may have a well-founded fear of threats to his life due to famine or civil war or of isolated acts of violence or ill-treatment for a Convention reason which may be perpetrated against him. But the risk, however severe, and the fear, however well founded, do not entitle him to the status of a refugee. The Convention has a more limited objective, the limits of which are identified by the list of Convention reasons and by the principle of surrogacy.[42]

Since 2000, there have been a small number of cases in Australia and New Zealand where people from Tuvalu, Kiribati, and Tonga have sought to argue they should receive refugee protection from climate change impacts.[43] Earlier cases considered whether refugee protection could be granted to applicants from Bangladesh on the basis of natural disasters,[44] or on account of drought and destitution in Fiji.[45] All claims failed on the basis that the harm feared did not amount to persecution, and there was no differential impact on the applicant.

Nonetheless, there remain limited exceptions where exposure to climate impacts or environmental degradation might amount to persecution for a Convention reason. Examples could include cases where:

- 'victims of natural disasters flee because their government has consciously withheld or obstructed assistance in order to punish or marginalize them on one of the five [Convention] grounds';[46]
- government policies target particular groups reliant on agriculture for survival, in circumstances where climate change is already hampering their subsistence;
- a government induces famine by destroying crops or poisoning water, or contributes to environmental destruction by polluting the land and/or water;[47]

[42] *Horvath v Secretary of State for the Home Department* [2001] 1 AC 489, 499–500 (Lord Hope).

[43] NZ cases: *Refugee Appeal No 72719/2001*, RSAA (17 September 2001) (Tuvalu); *Refugee Appeal No 72313/2000*, RSAA (19 October 2000) (Tuvalu); *Refugee Appeal No 72314/2000*, RSAA (19 October 2000) (Tuvalu); *Refugee Appeal No 72315/2000*, RSAA (19 October 2000) (Tuvalu); *Refugee Appeal No 72316/2000*, RSAA (19 October 2000) (Tuvalu); *Refugee Appeal Nos 72179–72181/2000*, RSAA (31 August 2000) (Tuvalu); *Refugee Appeal Nos 72189–72195/2000*, RSAA (17 August 2000) (Tuvalu); *Refugee Appeal No 72185/2000*, RSAA (10 August 2000) (Tuvalu); *Refugee Appeal No 72186/2000*, RSAA (10 August 2000) (Tuvalu). Australian cases: *1004726* [2010] RRTA 845 (30 September 2010) (Tonga); *0907346* [2009] RRTA 1168 (10 December 2009) (Kiribati); *N00/34089* [2000] RRTA 1052 (17 November 2000) (Tuvalu); *N95/09386* [1996] RRTA 3191 (7 November 1996) (Tuvalu); *N96/10806* [1996] RRTA 3195 (7 November 1996) (Tuvalu); *N99/30231* [2000] RRTA 17 (10 January 2000) (Tuvalu); *V94/02840* [1995] RRTA 2383 (23 October 1995) (Tuvalu).

[44] *Mohammed Motahir Ali v Minister of Immigration, Local Government and Ethnic Affairs* [1994] FCA 887 (Bangladesh and natural disasters in the refugee context); *N93/00894* [1996] RRTA 3244 (14 November 1996) (Bangladesh, generalized socio-economic circumstances and natural disasters). See also consideration of natural disasters and poor economic conditions in *1004725* [2010] RRTA 845 (30 September 2010) (Tonga).

[45] *Refugee Appeal No 70965/98*, RSAA (27 August 1998) (Fiji); *Refugee Appeal No 70959/98*, RSAA (27 August 1998) (Fiji).

[46] UNHCR, 'Forced Displacement in the Context of Climate Change: Challenges for States under International Law', Submission to the 6th Session of the Ad Hoc Working Group on Long-Term Cooperative Action under the Convention (20 May 2009) 9–10. This was endorsed in the Bellagio Deliberations (n 7) para 8.

[47] *SERAC and CESR v Nigeria, Decision regarding Communication No 155/96*, ACHPR/COMM/A044/1, '15th Annual Activity Report of African Commission on Human and Peoples' Rights' (7

- a government refuses to accept aid from other States when it is in need, such as in the aftermath of a disaster;[48]
- a government does not establish appropriate measures for prevention of disasters.[49]

In each case, of course, a persecutory element for Convention reasons would need to be shown. However, in most cases, people displaced by climate change are unlikely to receive protection as refugees.

III. The Relevance of Regional Refugee Instruments: OAU Convention and Cartagena Declaration

The regional Organization of African Unity (OAU) Convention in Africa and the Cartagena Declaration in Latin America contain broader refugee definitions than the 1951 Refugee Convention.[50] The OAU Convention includes as refugees *inter alia* people who are displaced on account of 'events seriously disturbing the public order', and it has been queried whether this could encompass environmental catastrophes such as famine and drought.[51] Edwards argues that such an interpretation is theoretically possible, but notes that even though people fleeing such catastrophes are 'frequently given refuge on the territory of neighbouring States (eg Congolese fleeing the eruption of Mount Nyiragongo in January 2002 sought refuge in Rwanda), receiving States rarely declare that they are acting pursuant to their OAU Convention obligations.'[52] This is significant because the explanation

May 2002) Annex V, para 44. Cooper argues that desertification in the African Sahel and the nuclear explosion at Chernobyl constitute persecution: see Cooper (n 21) 480.

[48] See, eg, International Law Commission (ILC), 'Report of the International Law Commission on the Work of its Sixty-Second Session (3 May–4 June and 5 July–6 August 2010)', UN Doc A/65/10, 327, draft Art 5.

[49] See, eg, *Budayeva and Others v Russian Federation* App nos 15339/02, 21166/02, 20058/02, 11673/02, and 15343/02 (ECtHR, 20 March 2008).

[50] Organization of African Unity Convention Governing the Specific Aspects of Refugee Problems in Africa (adopted 10 September 1969, entered into force 20 June 1974) 1001 UNTS 45 ('OAU Convention') (definition reflected in the national law of at least South Africa, Tanzania, Kenya, Malawi, Mozambique, and Namibia); Cartagena Declaration on Refugees (adopted by the Colloquium on the International Protection of Refugees in Central America, Mexico, and Panama, 22 November 1984) in 'Annual Report of the Inter-American Commission on Human Rights' (1984–85) OAS Doc OEA/Ser.L/V/II.66/doc.10, rev 1, 190–3 ('Cartagena Declaration') (definition reflected in the national law of at least Argentina, Bolivia, Colombia, El Salvador, Guatemala, Nicaragua, and Panama).

[51] Alice Edwards, 'Refugee Status Determination in Africa' (2006) 14 African Journal of International and Comparative Law 204, 225–7; Walter Kälin, 'Conceptualising Climate-Induced Displacement' in Jane McAdam (ed), *Climate Change and Displacement: Multidisciplinary Perspectives* (Hart Publishing, 2010) 88–9; cf James C Hathaway, *The Law of Refugee Status* (Butterworths, 1991) 16–21, who argues that the definition 'does not...suggest that victims of natural disasters or economic misfortune should become the responsibility of the international community': 17. See also Michelle Schwartz, 'International Legal Protection for Victims of Environmental Abuse' (1993) 18 Yale Journal of International Law 355, 380.

[52] Edwards (n 51) 227.

a State gives for acting in a particular way is relevant to ascertaining whether it supports or rejects a liberal interpretation of the treaty. Kälin similarly sees the potential for sudden-onset disasters to be characterized in this way, but sees it as 'rather unlikely that the States concerned would readily accept such an expansion of the concept beyond its conventional meaning of public disturbances resulting in violence'.[53] Thus, Edwards suggests that, at most, the general practice of hosting people displaced by environmental events 'may be seen as contributing to the development of a right of temporary protection on humanitarian grounds under customary international law, rather than under treaty'.[54]

However, if refuge were sought on account of riots in the aftermath of a disaster, triggered by the government's failure to provide assistance, Kälin suggests that the OAU Convention would apply. By analogy, he argues that the same analysis applies to Article III(3) of the Cartagena Declaration with respect to 'refugees' who 'have fled their country because their lives, safety or freedom have been threatened by generalized violence, foreign aggression, internal conflicts, massive violation of human rights or other circumstances which have seriously disturbed public order'.[55] Kälin's example highlights the difficulties of attributing movement to 'climate change'—at what point does it become too indirect to be considered a driver of movement, and should this matter in terms of the protection and assistance granted?

By contrast to the 1951 Refugee Convention, which assesses the risk of potential future harm, both regional instruments seem to require evidence of an *actual* threat: protection is premised on having already been compelled to leave because of it.[56] Thus, their utility as tools for providing pre-emptive protection is limited.

IV. The Relevance of Refugee Law Concepts

Although the Refugee Convention's direct applicability may be limited, it embodies a number of conceptual constructs that could be usefully transposed to the climate change-related displacement context.

First, refugee law is forward-looking in its assessment of harm. While past persecution can provide an important indication of future risk of persecution (going to the 'well-foundedness' of fear), it is not a requirement for a refugee claim to be substantiated. In other words, the 'refugee' concept recognizes anticipatory flight. However, refugee jurisprudence implicitly places limits on how pre-emptive flight may be. While the assessment of risk of persecution is prospective—and the risk

[53] Kälin (n 51) 88.

[54] Edwards (n 51) 227. UNHCR similarly made clear that its assistance activities for people displaced by the Boxing Day tsunami did not fall within its formal protection mandate, but rather constituted 'time-limited humanitarian assistance' requested especially by the UN Secretary-General: UNHCR, 'Note on International Protection', UN Doc A/AC.96/1008 (4 July 2005) para 36 in Edwards (n 51) 227.

[55] Kälin (n 51) 88–9.

[56] OAU Convention, Art 1(2): 'was compelled to leave'; Cartagena Declaration, Art III(3): 'who have fled their country because . . . '.

of harm can be less than a 50 per cent chance[57]—the fear must be plausible and reasonable in all the circumstances.[58] This includes consideration of 'the relation between the nature of the persecution feared and the degree of likelihood of its happening'.[59] This turns in part on the imminence of harm if the person is returned.[60] It is this element of timing that poses particular difficulties for pre-emptive movement away from the slow-onset impacts of climate change.[61]

Secondly, persecution does not need to be the sole reason for a refugee's flight. Provided that the requisite elements of the refugee definition can be established, any other reasons for movement are beyond the scope of the decision-maker's inquiry. In the present context, provided *a* link to climate change could be established, other reasons for flight would not be irrelevant to the assessment. However, there remain conceptual difficulties with framing protection needs in this way, given (a) the way that climate change functions as a threat multiplier and is interlinked with other underlying stressors (did the event cause movement, or people's underlying vulnerabilities? Should this matter?); and (b) the scientific impossibility of isolating climate change as a cause of an individual extreme weather event.[62] Although climate change increases the risk of particular kinds of weather events occurring more often and with greater intensity, such that virtually any climate-related natural disaster could satisfy the 'climate change' threshold, the question remains whether it is legally and morally defensible to focus only on climate-related, as opposed to geophysical, disasters. This is the rationale that underpins moves at the international level towards a broader, disaster-oriented protection instrument, rather than one that is focused solely on climate change.[63]

V. Conclusion

International refugee law is a cumbersome framework for addressing flight from climate change-related impacts. It was devised for a different context and will in most cases be an inappropriate vehicle for responding to environmental displacement.[64] Despite the 'temptation to start with definitions that would be derivative

[57] *Chan v Minister for Immigration and Ethnic Affairs* (1989) 169 CLR 379, 389 (Dawson J), citing *INS v Cardozo-Fonseca* 480 US 421 (1987) 431.

[58] See ibid.

[59] Goodwin-Gill and McAdam (n 20) 64.

[60] See UNHCR, 'Handbook on Procedures and Criteria for Determining Refugee Status under the 1951 Convention and the 1967 Protocol relating to the Status of Refugees', UN Doc HCR/IP/4/Eng/Rev.1 (2nd edn 1992) para 42: 'In general, the applicant's fear should be considered well-founded if he can establish, to a reasonable degree, that his continued stay in his country of origin has become intolerable to him for the reasons stated in the definition, or would for the same reasons be intolerable if he returned there.'

[61] See further, Ch 3 on timing.

[62] See Ch 1.

[63] See, eg, Nansen Principles (appended to Chairperson's Summary) (n 7); António Guterres, United Nations High Commissioner for Refugees, 'Statement' (Nansen Conference on Climate Change and Displacement in the 21st Century, Oslo, 6 June 2011) 4; Ch 9.

[64] For example, *Refugee Appeal No 76374*, RSAA (28 October 2009) (Myanmar) found the applicant to be a Convention refugee on account of the highly politicized nature of disaster-relief work in which she was involved in the aftermath of Cyclone Nargis. While her work responded to an environmental disaster, environmental degradation was not the basis of her claim per se.

of existing concepts',[65] it does not adequately address the time dimension of pre-emptive and staggered movement, nor the maintenance of culture and statehood in cases where whole communities may have to relocate.[66] In other contexts, it may be inappropriate because movement is only internal, and there the Guiding Principles on Internal Displacement will be instructive.[67] However, certain aspects of refugee law—its standard of proof ('well-founded fear'), the durable solutions it envisages, its protective rights-based framework and the status it creates, and its institutional oversight by the United Nations High Commissioner for Refugees—may be helpful in crafting responses to climate change-related movement. These elements are discussed in the following chapters.

[65] Walter Kälin, 'Keynote Address: Climate Change, Migration Patterns and the Law' (International Association of Refugee Law Judges 8th World Conference, Cape Town, 28 January 2009) 1.

[66] See Ch 5.

[67] Guiding Principles on Internal Displacement, UN Doc E/CN.4/1998/53/Add.2 (11 February 1998).

3

Climate Change-Related Movement and International Human Rights Law: The Role of Complementary Protection

I. Introduction

It is a trite observation that climate change will impact upon people's enjoyment of their human rights. Coastal erosion, flooding, drought, and sea-level rise, together with more frequent and intense severe weather events, such as storms and cyclones, will affect agriculture, infrastructure, services, and the continued habitability of certain parts of the world. This, in turn, may threaten rights such as the right to life, health, housing, culture, means of subsistence, and, in extreme cases, self-determination.[1] The most drastic impacts of climate change are likely to be felt in the poorest parts of the world where human rights protection is often weak.[2] Starting from a place of disadvantage hampers responsive capacity: poor levels of education, technical capacity, resource availability, and institutional support make lobbying for assistance and adaptation difficult.

There are three main reasons why international human rights law is of importance to the present analysis. First, it sets out minimum standards of treatment that States must afford to individuals within their territory or jurisdiction, and provides a means of assessing which rights are compromised by climate change and which national authorities have primary responsibility for responding to those rights at risk. Secondly, if those rights are at risk, human rights law may provide a legal basis on which protection may be sought (and granted) in another State (known as 'complementary protection').[3] Thirdly, if relocation occurs, human rights law requires minimum standards of treatment to be observed in the host State, and is thus relevant to the legal status afforded to those displaced.

[1] See generally, Human Rights Council, 'Report of the Office of the United Nations High Commissioner for Human Rights on the Relationship between Climate Change and Human Rights', UN Doc A/HRC/10/61 (15 January 2009) Annex.

[2] Stephen Humphreys, 'Introduction: Human Rights and Climate Change' in Stephen Humphreys (ed), *Human Rights and Climate Change* (Cambridge University Press, 2010) 1.

[3] Jane McAdam, *Complementary Protection in International Refugee Law* (Oxford University Press, 2007).

Human rights law has expanded States' protection obligations beyond the 'refugee' category, to include (at least) people at risk of arbitrary deprivation of life, torture, or cruel, inhuman or degrading treatment or punishment. This is known in international law as 'complementary protection', because it describes human rights-based protection that is complementary to that provided by the 1951 Refugee Convention.[4] The European Union (EU),[5] Canada,[6] the United States (US),[7] New Zealand,[8] Hong Kong,[9] Mexico,[10] and Australia[11] all have systems of complementary protection in place to implement these international law obligations.

Although, in theory, any human rights violation may give rise to a *non-refoulement* obligation,[12] in most cases 'it will be virtually impossible for an applicant to establish that control on immigration was disproportionate to any breach' of a human right.[13] This is because unlike the absolute prohibition on returning someone to inhuman or degrading treatment, for example, most other human rights provisions permit a balancing test between the interests of the individual and the State, thus placing protection from *refoulement* out of reach in all but the most exceptional cases.[14]

[4] Convention relating to the Status of Refugees (adopted 28 July 1951, entered into force 22 April 1954) 189 UNTS 137, Art 1A(2), read in conjunction with Protocol relating to the Status of Refugees (adopted 31 January 1967, entered into force 4 October 1967) 606 UNTS 267.

[5] Council Directive (EC) 2004/83 on Minimum Standards for the Qualification and Status of Third Country Nationals or Stateless Persons as Refugees or as Persons Who Otherwise Need International Protection and the Content of the Protection Granted [2004] OJ L304/12 ('Qualification Directive') Arts 2(e), 15.

[6] Immigration and Refugee Protection Act, SC 2001, c 27, s 97.

[7] Immigration and Nationality Act of 1952, 8 CFR §§ 208.16, 208.17. This is CAT-based protection only: Convention against Torture and Other Cruel, Inhuman or Degrading Treatment or Punishment (adopted 10 December 1984, entered into force 26 June 1987) 1465 UNTS 85 ('CAT').

[8] Immigration Act 2009, No 51 (NZ) ss 130, 131.

[9] CAT-based protection only. Refugee status determination is conducted by UNHCR. See further, Kelley Loper, 'Human Rights, *Non-Refoulement* and the Protection of Refugees in Hong Kong' (2010) 22 International Journal of Refugee Law 404.

[10] Law on Refugees and Complementary Protection (December 2010): Decreto por el que se expide la Ley sobre Refugiados y Protección Complementaria y se reforman, adicionan y derogan diversas disposiciones de la Ley General de Población <http://www.dof.gob.mx/nota_detalle.php?codigo=5175823&fecha=27/01/2011> accessed 28 January 2011. Thank you to Jeremy Dicker for this reference. Mexico is the first country in Latin America to grant complementary protection: 'UNHCR Welcomes Breakthrough: Mexico Legislation on Protection' (UNHCR, 10 December 2010) <http://www.unhcr.org/4d025a8a6.html> accessed 21 January 2011.

[11] In September 2011, legislation on complementary protection was passed by the Australian Parliament to amend the Migration Act 1958 (Cth): Migration Amendment (Complementary Protection) Act 2011 (Cth).

[12] *R v Special Adjudicator, ex parte Ullah* [2004] UKHL 26, paras 24–5 (Lord Bingham), 49–50 (Lord Steyn), 67 (Lord Carswell).

[13] *Kacaj v Secretary of State for the Home Department* [2002] EWCA Civ 314, para 26.

[14] A State's general lack of resources cannot be used to justify inhuman or degrading treatment, however. For example in *Kalashnikov v Russia* (2002) 36 EHRR 587, the European Court of Human Rights rejected Russia's argument that squalid prison conditions did not violate Art 3 because they were a result of Russia's economic difficulties, and were experienced generally by detainees in Russia. See also Katharina Röhl, 'Fleeing Violence and Poverty: *Non-Refoulement* Obligations under the European Convention of Human Rights' (2005) UNHCR New Issues in Refugee Research, Working Paper No 111. But see UK Discretionary Leave, discussed in section III.D below.

It is therefore common for a violation of a socio-economic right—for example, violation of the right to an adequate standard of living—to be re-characterized as a form of inhuman treatment, which *is* a right giving rise to international protection.[15] However, courts have carefully circumscribed the meaning of 'inhuman or degrading treatment' so that it cannot be used as a remedy for general poverty, unemployment, or lack of resources or medical care except in the most exceptional circumstances.[16]

Although existing jurisprudence does not *preclude* climate change impacts from being recognized as a source of inhuman treatment, it would need to be substantially developed before such harms would fall clearly within the scope of this concept.[17] It is also important to note that in a removal case, an internal flight alternative may be considered a reasonable option.[18] Finally, a judicialized approach to expanding protection has inherent constraints, the limits of which need to be appreciated. While judicial interpretations may lead to broadened interpretations of the law, they are reliant on good test cases, cogent legal arguments by counsel for expansion, and acceptance by Parliament (which may be able to legislate to overturn the implications of the decision for the future).[19]

[15] The South African constitutional court has rejected the use of civil and political rights as a fallback for social and economic rights in the domestic context: *Soobramoney v Minister of Health KwaZulu Natal* 1998 (1) SA 765. There, it relied on the constitutional right to health instead of the constitutional right to life. Indian courts have conflated the constitutional right to life (a 'first generation' right, Art 21) with social and economic ('second generation') rights. The Indian position, in part, has come about because the constitutional protection given to a first generation right is higher than that given to a second generation right. Socio-economic rights under the Indian constitution are protected as directive principles and State policy, not as enforceable rights as such. However, Indian courts have greatly expanded the constitutional right to life to make socio-economic rights justiciable.

[16] *D v United Kingdom* (1997) 24 EHRR 423; *N v United Kingdom* [2008] ECHR 453; *HLR v France* (1997) 26 EHRR 29, para 42. See also the views of Committee against Torture, as in *AD v The Netherlands* Communication No 96/1997 (12 November 1999) UN Doc CAT/C/23/D/96/1997 (24 January 2000) para 7.2. See discussion in Guy S Goodwin-Gill and Jane McAdam, *The Refugee in International Law* (3rd edn, Oxford University Press, 2007) 350–1. The developing jurisprudence on re-characterizing socio-economic violations as civil and political ones has been described as an 'extension upon an extension': *AJ (Liberia) v Secretary of State for the Home Department* [2006] EWCA Civ 1736, para 12 referring to *N v Secretary of State for the Home Department* [2003] EWCA Civ 1369, paras 37, 46 in Michelle Foster, '*Non-Refoulement* on the Basis of Socio-Economic Deprivation: The Scope of Complementary Protection in International Human Rights Law' [2009] New Zealand Law Review 257, 266.

[17] *Ullah* (n 12); Human Rights Committee, 'General Comment 15: The Position of Aliens under the Covenant' (11 April 1986) para 5. See also Human Rights Committee, 'General Comment 18: Non-Discrimination' (10 November 1989).

[18] This principle is contained in Qualification Directive, Art 8 and applies both to refugees and to beneficiaries of subsidiary protection. See also *Januzi v Secretary of State for the Home Department* [2006] UKHL 5; *Rasaratnam v Canada (Minister of Employment and Immigration)* [1992] 1 FC 706 (CA) (Canada); *Ranganathan v Minister of Citizenship and Immigration* [2001] 2 FC 164 (Canada); *SZATV v Minister for Immigration and Citizenship* (2007) 233 CLR 18 (Australia); *SZFDV v Minister for Immigration and Citizenship* (2007) 233 CLR 51 (Australia); Refugee Appeal No 71684/99 (29 October 1999) (New Zealand); 8 CFR §208.13(b)(3)—reasonableness of internal relocation (asylum) (US); 8 CFR §208.16(b)(3) —reasonableness of internal relocation (withholding of removal) (US). See further section III.C.(3) below.

[19] This depends on the status of the court (eg national legislatures cannot enact laws that would conflict with rulings of the European Court of Human Rights) and the basis on which the claim is brought (eg constitutionally-protected rights cannot be simply overturned by Parliament).

This chapter examines the extent to which existing international and regional standards on complementary protection offer protection options for those forcibly displaced across international borders as a result of climate change-related impacts. It assesses the degree to which a progressive interpretation of the law is required to expand protection opportunities for people displaced by climate change, and the extent to which it might already accommodate them. It examines: (a) whether, when, and to what extent certain socio-economic forms of harm may be regarded as triggering the principle of *non-refoulement*; (b) whether they can do so independently, or whether they need to be re-characterized as violations of civil and political rights already recognized as mandating this (such as a violation of the right to life); and (c) whether they may form part of the progressive development of the principle of *non-refoulement*, as foreshadowed by international treaty monitoring bodies, the European Court of Human Rights, and the House of Lords (now Supreme Court). The final part of the chapter examines whether States may be held responsible for climate change under international environmental law, and if so, whether this provides another cause of action for people displaced on account of climate change impacts.

The key human rights to consider in the complementary protection context are: (a) the right to life (sometimes expressed in the removal context as the right not to be subjected to arbitrary deprivation of life); and (b) the right not to be subjected to torture or cruel, inhuman or degrading treatment or punishment. While these are not necessarily the only rights which entail a *non-refoulement* obligation,[20] they are the two which are clearly recognized in international law as giving rise to such an obligation and which have been incorporated into a number of domestic complementary protection regimes.

Although the following sections consider 'international' and 'regional' standards separately, much of the analysis about the nature and scope of the rights applies equally to both contexts. The jurisprudence of the European Court of Human Rights is discussed in particular detail since it is the most developed in this area and provides the most extensive reasoning about the scope and content of human rights-based *non-refoulement*.

II. Right to Life

A. International (ICCPR, Article 6)

The right to life is protected in Article 3 of the Universal Declaration of Human Rights (UDHR),[21] Article 6 of the International Covenant on Civil and Political Rights (ICCPR), Article 6 of the Convention on the Rights of the Child (CRC),[22] and in all regional human rights treaties.[23] It has been described by the United

[20] See section IV. below.

[21] Universal Declaration of Human Rights (adopted 10 December 1948) UNGA res 217A (III) ('UDHR').

[22] Convention on the Rights of the Child (adopted 20 November 1989, entered into force 2 September 1990) 1577 UNTS 3 ('CRC').

[23] European Convention on Human Rights (formally the Convention for the Protection of Human Rights and Fundamental Freedoms) (drafted 4 November 1950, entered into force 3

Nations (UN) Human Rights Committee as the 'supreme right' which is 'basic to all human rights'.[24] It is non-derogable and is recognized as entailing a *non-refoulement* obligation.[25]

The right to life is closely connected to other human rights. The right to an adequate standard of living (including adequate food, clothing, housing, and the continuous improvement of living conditions)[26] and the right not to be deprived of a means of subsistence[27] have been argued to be necessary components of the right to life, which are compromised where global warming leads to the destruction of people's ability to hunt, fish, gather, or undertake subsistence farming.[28] The UN Commission on Human Rights has observed that the right 'encompasses existence in human dignity with the minimum necessities of life'.[29]

Similarly, the CRC links the right to life to States' duty 'to ensure to the maximum extent possible the survival and development of the child'.[30] The Committee on the Rights of the Child has explained the need to view and implement the right to life holistically, 'through the enforcement of all the other provisions of the Convention, including rights to health, adequate nutrition, social security, an adequate standard of living, [and] a healthy and safe environment'.[31] This is also reflected in the Inter-Agency Standing Committee (IASC) Operational Guidelines on the Protection of Persons in Situations of Natural Disasters.[32]

September 1953) ETS No 5 ('ECHR') Art 2; American Convention on Human Rights (adopted 22 November 1969, entered into force 18 July 1978) 1144 UNTS 123, Art 4; African Charter on Human and Peoples' Rights (adopted 17 June 1981, entered into force 21 October 1986) 21 ILM 58 (1982) Art 4; Arab Charter on Human Rights (adopted 22 May 2004, entered into force 15 March 2008) Art 5.

[24] Human Rights Committee, 'General Comment No 6: The Right to Life (Art 6)' (30 April 1982) para 1; Human Rights Committee, 'General Comment No 14: Nuclear Weapons and the Right to Life (Art 6)' (9 November 1984) para 1.

[25] *Ahani v Canada* Communication No 1051/2002 (29 March 2004) UN Doc CCPR/C/80/D/1051/2002 (15 June 2004); Human Rights Committee, 'General Comment No 31: The Nature of the General Legal Obligation Imposed on States Parties to the Covenant', UN Doc CCPR/C/21/Rev.1/Add.13 (26 May 2004) para 12. See also Committee on the Rights of the Child, 'General Comment No 6 (2005): Treatment of Unaccompanied and Separated Children Outside Their Country of Origin', UN Doc CRC/GC/2005/6 (1 September 2005) para 27.

[26] International Covenant on Economic, Social and Cultural Rights (adopted 16 December 1966, entered into force 3 January 1976) 993 UNTS 3 ('ICESCR') Art 11.

[27] International Covenant on Civil and Political Rights (adopted 16 December 1966, entered into force 23 March 1976) 999 UNTS 171 ('ICCPR') Art 1(2); ICESCR, Art 1(2). See, eg, F Menghistu, 'The Satisfaction of Survival Requirements' in BG Ramcharan (ed), *The Right to Life in International Law* (Martinus Nijhoff, 1985) 68.

[28] See, eg, 'Petition to the Inter American Commission on Human Rights Seeking Relief from Violations Resulting from Global Warming Caused by Acts and Omissions of the United States' (7 December 2005) <http://www.earthjustice.org/library/legal_docs/petition-to-the-inter-american-commission-on-human-rights-on-behalf-of-the-inuit-circumpolar-conference.pdf> accessed 10 December 2009.

[29] UN Commission on Human Rights, 'Human Rights and Extreme Poverty', Human Rights res 2005/16 (14 April 2005) para 1(b).

[30] CRC, Art 6(2).

[31] Committee on the Rights of the Child, 'General Comment No 7 (2005): Implementing Child Rights in Early Childhood', UN Doc CRC/C/GC/7/Rev.1 (20 September 2006).

[32] IASC Operational Guidelines on the Protection of Persons in Situations of Natural Disasters (Brookings–Bern Project on Internal Displacement 2011) <http://www2.ohchr.org/english/issues/idp/docs/OperationalGuidelines.pdf> accessed 14 July 2011.

Importantly, the right to life includes an obligation to take positive measures to protect it, which may be relevant in considering whether a country of origin is in fact taking steps to improve such things as healthcare and nutrition. A useful analogy for the climate change context may be provided by the UN Human Rights Committee's remarks on the threat to life posed by nuclear weapons. Nuclear weapons may not only cause death directly, but also indirectly by contaminating the environment with radiation. Similarly, the impacts of climate change experienced (for example) through salt-water intrusion into fresh water supplies could, on this reasoning, be interpreted as a threat to the right to life. In each case, however, the severity and extent of the harm would determine whether the right to life had been violated.[33]

An analysis of the views expressed by the UN Human Rights Committee in relation to individual complaints suggest that the following criteria apply to Article 6 cases:

- the risk to life must be actual or imminent;[34]
- the applicant must be personally affected by the harm;[35]
- environmental contamination with proven long-term health effects may be a sufficient threat, however there must be sufficient evidence that harmful quantities of contaminants have reached, or will reach, the human environment;
- a hypothetical risk is insufficient to constitute a violation of the right to life; and
- cases challenging public policy will, in the absence of an actual or imminent threat, be considered inadmissible.[36]

[33] Asia Pacific Forum of National Human Rights Institutions (APF), *Human Rights and the Environment: Reference Paper* (2007) 57–8, 62–3, 66 cited in APF, *Human Rights and the Environment: Final Report and Recommendations* (2007). In October 2011, for example, Tuvalu and Tokelau each declared a state of emergency because of severe water shortages. Although Australia and New Zealand responded quickly by sending emergency desalination plants, rehydration packs, technical support, and water supplies, in certain conditions (where, for example, a sustainable, ongoing supply of water could not be guaranteed), the right to life would be threatened. See, eg, Kirsty Needham, 'Australia Helps as Tuvalu Runs Dry', *Sydney Morning Herald* (Sydney, 7 October 2011) 3; 'Tokelau, Tuvalu Declare State of Emergency', *ABC News* (5 October 2011); Jen Lacey, 'Need Water Now: Australia Sends Desal Plant to Tuvalu', *ABC News* (10 October 2011) <http://www.abc.net.au/local/stories/2011/10/10/3335802.htm?site=sydney> accessed 10 October 2011; 'Water Situation Getting More Desperate on Tuvalu', *Pacific Beat* (ABC Radio National, Australia, 10 October 2011) <http://www.radioaustralia.net.au/pacbeat/stories/201110/s3335901.htm> accessed 10 October 2011.

[34] See section VI.A. below on timing.

[35] See the difficulties this has caused in the context of armed conflict pursuant to Art 15(c) of the EU Qualification Directive: Hélène Lambert and Theo Farrell, 'The Changing Character of Armed Conflict and the Implications for Refugee Protection Jurisprudence' (2010) 22 International Journal of Refugee Law 237; Jean-François Durieux, 'Salah Sheekh is a Refugee: New Insights into Primary and Subsidiary Forms of Protection' (2008) Refugee Studies Centre Working Paper Series No 49 <http://www.rsc.ox.ac.uk/publications/working-papers/RSCworkingpaper49.pdf> accessed 10 August 2011; Jane McAdam, 'Individual Risk, Armed Conflict and the Standard of Proof in Complementary Protection Claims: The European Union and Canada Compared' in James C Simeon (ed), *Critical Issues in International Refugee Law: Strategies for Interpretative Harmony* (Cambridge University Press, 2010).

[36] APF, 'Final Report' (n 33) 19–20.

B. Regional (ECHR, Article 2)

Although the European Court of Human Rights has confirmed that Article 2 of the ECHR may be relied upon to prevent removal,[37] no removal case has succeeded solely on this ground.[38] Article 2 is generally raised in conjunction with Article 3, and if a violation of the latter is found, then the analysis of Article 2 typically falls away.[39]

(1) Right to life and the environment[40]

The negative impacts on human rights that flow from natural disasters are not usually the result of intentional government policies, but rather inadequate planning, preparation, and response mechanisms. As the UN Secretary-General has observed, 'the risks and potential for disasters associated with natural hazards are largely shaped by prevailing levels of vulnerability and measures taken to prevent, mitigate and prepare for disasters'.[41] Accordingly, it is useful to consider whether States have legal obligations to ensure that effective disaster risk management policies are put in place, and the extent to which the destruction of the environment might be understood as a threat to the right to life.

Despite formally recognizing the links between human rights and the environment since at least 1972, when States adopted the non-binding Stockholm Declaration at the UN Conference on the Human Environment,[42] the precise legal relationship between the two remains in the developmental stages. Although the International Court of Justice (ICJ) has acknowledged that the realization of human rights is largely dependent on whether the physical environment is capable of sustaining people as rights-bearers,[43] and some rights are dependent on a

[37] For example in *Z and T v United Kingdom* App no 27034/05 (ECtHR, 28 February 2006) 6: the Art 3 analysis from *Soering v United Kingdom* (1989) 11 EHRR 439 'applies equally to the risk of violations of [Art] 2'. See also *Ullah* (n 12) para 40 (Lord Steyn): 'If article 3 may be engaged it is difficult to follow why, as a matter of logic, article 2 could be peremptorily excluded. There may well be cases where article 3 is not applicable but article 2 may be'.

[38] A breach of Art 2 has only been found in one removal case, and on that occasion it was in conjunction with Art 3: *Bader and Kanbor v Sweden* App no 13284/04 (ECtHR, 8 November 2005).

[39] For example *Tatete v Switzerland* App no 41874/98 (ECtHR, 6 July 2000) (settlement reached); *D v United Kingdom* (n 16) (decided on Art 3); *Mamatkulov v Turkey* (2005) 41 EHRR 494 (Grand Chamber). Thus, in *D v United Kingdom*, for example, the court did not regard the Art 2 claim as unfounded in principle, but thought it unnecessary to review the Art 2 claim separately from Art 3. In *Dehwari v The Netherlands* App no 37014/97 (European Commission of Human Rights) 29 October 1998) para 61, the European Commission of Human Rights stated that Art 2 would preclude removal 'in circumstances in which the expelling State knowingly puts the person concerned at such high risk of losing his life as for the outcome to be a near-certainty'. This overstates the threshold required by a 'real risk' of harm. In any event, the Commission held that the expulsion 'would amount to inhuman treatment within the meaning of Article 3 of the Convention'.

[40] See further, section VII.

[41] UNGA 'Report of the Secretary General, International Cooperation on Humanitarian Assistance in the Field of Natural Disasters, from Relief to Development', UN Doc A/60/227 (12 August 2005) para 1, referred to in IASC Operational Guidelines (n 32).

[42] Declaration of the United Nations Conference on the Human Environment (adopted 16 June 1972) UN Doc A/CONF.48/14/Rev.1 (1972) ('Stockholm Declaration').

[43] *Case Concerning the Gabčikovo-Nagymaros Project* (Separate Opinion of Judge Weeramantry) [1997] ICJ Rep 7, 91; see also Dinah Shelton, 'Environmental Rights' and Philip Alston, 'Peoples' Rights:

particular quality of environment, there is not yet an individual or collective 'right to a healthy environment' in international law.[44]

In a Separate Opinion in the *Gabčikovo-Nagymaros Project case* in the ICJ, Judge Weeramantry noted that the protection of the environment is 'a vital part of contemporary human rights doctrine, for it is [an indispensable requirement] ... for numerous human rights such as the right to health and the right to life itself'.[45] While this statement does not recognize an independent human right to a safe environment, it provides a persuasive, logical underpinning to the realization of human rights more generally, premising their fulfilment on the threshold question of whether the physical environment is capable of sustaining humans as rights-bearers. The European Court has similarly acknowledged that the right to a healthy environment is linked to the right to life, and that environmental damage can affect the rights to life, property, home, and private life.[46] In particular, the obligation to protect the right to life may also include protection from environmental harm.[47] In *Öneryildiz v Turkey*, the court stated that Article 2 of the ECHR entails 'above all a primary duty ... to put in place a legislative and administrative framework designed to provide effective deterrence against threats to the right to life'.[48] So far, this issue has not arisen in a removal case, however.

In *Budayeva v Russia*, the European Court of Human Rights held that the State's duty to protect life extends to protection from natural disasters where the risk is known.[49] Competent authorities must:

- enact and implement laws dealing with all relevant aspects of disaster risk mitigation and set up the necessary mechanisms and procedures;
- take the necessary administrative measures, including supervising potentially dangerous situations;

Their Rise and Fall' in Philip Alston (ed), *Peoples' Rights* (Oxford University Press, 2001); Meeting of Experts on Human Rights and the Environment, Final Text (16 January 2002) (see also background papers at <http://www2.ohchr.org/english/issues/environment/environ/conclusions.htm>). Ramcharan suggests that the right to life implies a right to environment and a concomitant obligation on States 'to take effective measures to prevent and to safeguard against the occurrence of environmental hazards which threaten the lives of human beings': BG Ramcharan, 'The Concept and Dimensions of the Right to Life' in Ramcharan (n 27) 13.

[44] APF, *Final Report* (n 33), 26. See also Shelton (n 43); Alston (n 43); Meeting of Experts on Human Rights and the Environment and background papers (n 43); Patricia Birnie, Alan Boyle, and Catherine Redgwell, *International Law and the Environment* (3rd edn, Oxford University Press, 2009) 271–302; Susan Glazebrook, 'Human Rights and the Environment' (2009) 40 Victoria University of Wellington Law Review 293.

[45] *Case Concerning the Gabčikovo-Nagymaros Project* (n 43) 91 (Separate Opinion of Judge Weeramantry). See section VII.

[46] See cases cited in Loukis G Loucaides, 'Environmental Protection through the Jurisprudence of the European Convention on Human Rights' in Loukis G Loucaides, *The European Convention on Human Rights: Collected Essays* (Martinus Nijhoff, 2007), beginning with *Arrondelle v United Kingdom* (1980) 19 DR 186 (noise pollution cases offensive smells); *Sergio Euben Lopez Burgos v Uruguay* Communication No R.12/52 (29 July 1981) UN Doc Supp No 40 (A/36/40), 176 (1981); *Guerra and Others v Italy* (1998) 26 EHRR 357; *Fadeyeva v Russia* App no 55723/00 (ECtHR, 9 June 2005).

[47] See *Öneryildiz v Turkey* (2005) 41 EHRR 20, paras 71–2.

[48] ibid, para 89.

[49] *Budayeva v Russia* App nos 15339/02, 21166/02, 20058/02, 11673/02, and 15343/02 (ECtHR, 20 March 2008).

- inform the population about possible dangers and risks;
- evacuate potentially affected populations;
- conduct criminal investigations and prosecute those responsible for having neglected their duties in case of deaths caused by a disaster; and
- compensate surviving relatives of victims killed as a consequence of neglecting these duties.[50]

This case concerned a complaint against a Contracting State for not establishing sufficient early warning or defence systems against foreseeable disasters. There is nothing in principle which would prevent its reasoning from extending to removal cases where there is a real risk that the applicant would be affected by natural disasters in a State that failed to mitigate against them. However, the court has emphasized that the burden placed on the State must be reasonable, which means that consideration must be given, 'in particular, to the operational choices which [it] must make in terms of priorities and resources'.[51] This necessarily takes into consideration the economic capacity of the State in question.[52] It would also appear that the burden is less onerous in cases of natural, as opposed to human-induced, harms, presumably on the basis that the former are less easy to predict and control. The factors to be taken into account include the 'origin of the threat and the extent to which one or the other risk is susceptible to mitigation'.[53] The burden would be more stringent if it were a 'recurring calamity affecting a distinct area developed for human habitation or use'.[54] Questions about the degree of interference, necessity, and balance also characterize the case law.[55]

In a complementary protection claim, the focus is the potential 'harm' to the applicant if he or she is removed. Thus, the relevant question is the extent to which the receiving State is able and willing to mitigate against that harm, whatever its cause. Since the European Court of Human Rights has been inclined to allow the State a higher degree of latitude where the cause of harm is 'natural', such as a landslide, it may actually be more beneficial to an applicant to acknowledge the multicausality of climate change-related impacts, rather than trying to pinpoint 'climate change' as *the* cause of harm. In other words, the combination of environmental, social, economic, and political factors, which draw on human-made as

[50] Walter Kälin and Claudine Haenni Dale, 'Disaster Risk Mitigation: Why Human Rights Matter' (2008) 31 Forced Migration Review 38, 39.

[51] *Budayeva v Russia* (n 49) para 135: 'this results from the wide margin of appreciation States enjoy'.

[52] If a State's resources and capacity were pushed to the limit, such that it could not protect sufficiently against climate change-related harms, could it argue that the international community has a duty to cooperate and provide it with assistance? For example, if States cannot provide shelter on their own, then does the duty require other States through funding *or other means* to ensure those rights are met? See Ch 9.

[53] *Budayeva v Russia* (n 49) para 137.

[54] ibid.

[55] See Alan Boyle, 'Human Rights and the Environment: A Reassessment' (UNEP paper revised, 2010) 20 <http://www.unep.org/environmentalgovernance/LinkClick.aspx?fileticket=GccCLN-brmg%3D&tabid=2046&language=en-US> accessed 27 June 2011, referring in particular to *Fredin v Sweden* (1991) EHCR Series A/192; *Pulp Mills Case (Argentina v Uruguay)* (Provisional Measures) [2006] ICJ Rep 13, 133; *case Concerning the Gabčikovo-Nagymaros Project* (n 43) 77–8.

well as natural vulnerabilities, may better substantiate an Article 2 or 3 claim than one based solely on the impacts of 'climate change'.

In other regions, the Inter-American Commission on Human Rights has recognized that realization of the right to life is necessarily linked to and dependent on the physical environment.[56] It has found that forcibly displacing indigenous people from their land could breach the right to life if it causes indignity.[57] Similarly, the African Commission on Human and Peoples' Rights has found a breach of the rights to health and life as a result of repeated eviction and displacement from lands in Mauritania, which were confiscated by the government.[58]

Additionally, regional human rights treaties in Africa and Latin America specifically recognize the right to a safe environment.[59] Article 24 of the 1981 African Charter on Human and Peoples' Rights provides that all peoples 'shall have the right to a general satisfactory environment favourable to their development'. This has been interpreted by the African Commission on Human and Peoples' Rights as obliging States to 'take reasonable and other measures to prevent pollution and ecological degradation, to promote conservation, and to secure an ecologically sustainable development and use of natural resources'.[60] In particular, the Commission has held that:

> an environment degraded by pollution and defaced by the destruction of all beauty and variety is as contrary to satisfactory living conditions and development as the breakdown of the fundamental ecological equilibria is harmful to physical and moral health.[61]

That this provision is expressed as a right attaching to 'peoples', rather than as an individual human right, means that the right is focused on safeguarding specific groups sharing a common cultural heritage (such as indigenous peoples), supporting international human rights principles that seek to safeguard the cultural and linguistic integrity of such groups.

56 'Report on the Situation of Human Rights in Ecuador' (1997) OEA/Ser.L/V/II.96, Doc 10, Rev 1, ch 8; *Yanomami case* (Case 7615 of 5 March 1985) in Inter-American Commission on Human Rights, 'Annual Report (1984–85) OEA/Ser.L/V/II.66, Doc 10, Rev 1.

57 *Yakye Axa v Paraguay* (Inter-American Court of Human Rights, Judgment of 17 June 2005). In *The Massacres of Ituango v Colombia* (Inter-American Court of Human Rights, Judgment of 1 July 2006), the court found that forced eviction, displacement, and house destruction by paramilitaries violated the right to property.

58 *Malawi African Association v Mauritania* (2000) AHRLR 149 (African Commission on Human and Peoples' Rights) pursuant to the African Charter on Human Rights, Arts 16 and 4 (respectively). See also *Centre for Minority Rights Development (Kenya) and Minority Rights Group International on behalf of Endorois Welfare Council v Kenya*, African Commission on Human and Peoples' Rights (4 February 2010) 276/2003, para 216.

59 Additional Protocol to the American Convention on Human Rights in the Area of Economic, Social and Cultural Rights (adopted 17 November 1988, entered into force 16 November 1999) OAS Treaty Series 69 ('Protocol of San Salvador') Art 11; 1981 African Charter on Human and Peoples' Rights (adopted 27 June 1981, entered into force 21 October 1986) OAU Doc CAB/LEG/67/3 rev.5, 21 ILM 58 (1982) Art 24.

60 *The Social and Economic Rights Action Centre and the Centre for Economic and Social Rights v Nigeria* Communication No 155/96 (27 October 2001) para 54 <http://www1.umn.edu/humanrts/africa/comcases/155-96.html> accessed 10 August 2011. This is known as the *Ogoniland case*.

61 ibid, para 51. Boyle (n 55) 5 notes that its exceptional basis in Art 24 has to be remembered, since no other treaty contains anything comparable.

The 1988 Additional Protocol to the American Convention on Human Rights in the Area of Economic, Social and Cultural Rights provides in Article 11 that: 'Everyone shall have the right to live in a healthy environment and to have access to basic public services. The States Parties shall promote the protection, preservation, and improvement of the environment.' However, in the American context the right is not subject to individual petition to the Inter-American Commission on Human Rights, thus hampering the ability of individuals to seek a remedy for a violation.[62]

C. Domestic frameworks

It would appear that protection of the right to life is an under-utilized provision in the removal context and that it is ripe for progressive development. This may begin to occur in domestic contexts, since non-return to arbitrary deprivation of life is one of the complementary protection grounds in Canadian, New Zealand, and Australian legislation. German law also includes in its complementary protection provisions protection against return to a concrete and considerable danger to life, person, or liberty (although the requirement of a 'concrete' danger may be hard to meet in cases of pre-emptive movement in response to slow-onset climate impacts).[63]

On its face, Canadian law does not appear to offer much scope to protect people at risk of climate change-related impacts. Its complementary protection provisions on risk to life and risk of cruel and unusual treatment preclude protection being granted if the harm feared is generalized or based on a country's inability to provide adequate health or medical care.[64] The Legal Services division of the Immigration and Refugee Board has noted that a 'claim based on natural catastrophes such as drought, famine, earthquakes, etc. will not satisfy the definition as the risk is generalized.'[65] Nevertheless, a nine year old child successfully argued that his return to Haiti would put his life at risk since his biological family was unknown, and he was at risk of becoming homeless and prey to prostitution if returned.[66]

[62] Protocol of San Salvador, Art 19(6).

[63] Residence Act of 30 July 2004 (Federal Law Gazette I, p 1950), last amended by the Act on Implementation of Residence- and Asylum-Related Directives of the European Union of 19 August 2007 (Federal Law Gazette I, p 1970) (Germany) s 60(7).

[64] Immigration and Refugee Protection Act, s 97(b). It would, however, be different if the country refused to provide such care to particular classes of people: Legal Services, Immigration and Refugee Board of Canada (IRB), 'Consolidated Grounds in the *Immigration and Refugee Protection Act*: Persons in Need of Protection: *Risk to Life or Risk of Cruel and Unusual Treatment or Punishment*' (15 May 2002) 11.

[65] IRB (n 64) 9. See also *Sinnappu v Canada (Minister of Citizenship and Immigration)* [1997] 2 FC 791 (TD) para 38 where an expert on the earlier regime Post-Claim Determination Class Risk Review (PDRCC), which had similar bases for protection, stated that the requirement that the risk be one that is not faced generally by other individuals would apply only in extreme situations such as a generalized disaster of some sort that would involve all the inhabitants of a given country.

[66] *Re MQF* [2004] RPDD No 87. See also *FB (Lone Women—PSG—Internal Relocation—AA (Uganda) Considered) Sierra Leone* [2008] UKAIT 00090, para 81, in which the AIT ultimately found that the applicant was not at risk of prostitution, 'the most degrading form of destitution and one that is most alien to universal principles of human dignity'. Although that was a case about internal relocation, one might seek to extend the argument to the non-removal context in relation to Arts 2 or 3 of the ECHR.

Similarly, in the United Kingdom (UK), the Asylum and Immigration Tribunal (AIT) has acknowledged in the context of an internal flight alternative that 'if survival comes at a cost of destitution, beggary, crime or prostitution, then that is a price too high'.[67] This is also supported by AIT case law on the protective scope of Article 3 of the ECHR in relation to poor socio-economic conditions, discussed below.

In the non-removal context, the following domestic constitutional law cases have developed the socio-economic elements of the right to life. They are included here to illustrate the potential scope of that right. In India, the constitutional protection of the right to life has been held to include the right to a clean environment;[68] the right to food and freedom from malnutrition;[69] the protection of human dignity;[70] the right to education;[71] and the right to health.[72] In a case concerning evictions without notice from homes that were later demolished, the Supreme Court of Bangladesh has also drawn on the constitutional right to life to imply a 'right to livelihood'.[73] The Supreme Court of Pakistan has found that the constitutional right to life implies the right to a healthy environment.[74] The High Court of Botswana has held that the termination of water, food, and health services, and forced evictions from traditional lands, violate the constitutional right to life.[75]

III. Cruel, Inhuman or Degrading Treatment

A. International (ICCPR, Article 7)

Article 7 of the ICCPR prohibits torture and cruel, inhuman or degrading treatment or punishment. The standard approach of the UN Human Rights Committee is to regard these forms of ill-treatment as falling on a sliding scale, or hierarchy, with torture the most severe manifestation. The distinction between torture and inhuman treatment is thus one of degree. The UN Human Rights Committee considers it undesirable 'to draw up a list of prohibited acts or to establish sharp distinctions between the different kinds of punishment or treatment; the distinctions

[67] *FB* (n 66) para 39. The facts did not preclude her relocation in this case.

[68] *Rural Litigation and Entitlement Kendra, Dehradun v State of Uttar Pradesh* AIR 1985 SC 652 (Supreme Court of India); *Subhash Kumar v State of Bihar* AIR 1991 SC 426 (Supreme Court of India); *Attakoya Thangal v Union of India* (1990) KLT 580; *Koolwal v State of Rajasthan* AIR 1998 Raj 2.

[69] *People's Union for Civil Liberties v Union of India* WP (Civil) No 196/2001.

[70] *Francis Coralie Mullin v The Administrator Union Territory of Delhi* 1981 (1) SCC 608, 746: 'the right to life includes the right to live with human dignity and all that goes along with it', which includes the right to a 'healthy environment with minimal disturbance of [the] ecological balance'.

[71] *Unni Krishnan JP v State of Andhra Pradesh* (1993) 1 SCC 645.

[72] *Paschim Banga Khet Mazdoor Samity and Others v State of West Bengal* AIR 1996 SC 2426 (Supreme Court of India).

[73] *ASK v Government of Bangladesh* Writ No 3034 of 1999. As in India, Bangladesh's Constitution provides a directive principle, rather than a justiciable right, to shelter (Art 15).

[74] *Shehla Zia v WAPDA* PLD 1994 SC 693; *General Secretary, West Pakistan Salt Miners Labor Union (CBA) Khewra, Jhelum v The Director, Industries, and Mineral Development ('Salt Miners case')*

[75] *Sesana v Attorney-General* (2006) AHRLR 183 (High Court of Botswana).

depend on the nature, purpose and severity of the treatment applied'.[76] For that reason, it commonly fails to determine precisely which aspect of Article 7 of the ICCPR has been violated, and there is accordingly very little jurisprudence from that body about the nature of each type of harm.

Article 7 contains a *non-refoulement* obligation,[77] although a violation of this provision from a proposed removal has been substantiated on the facts only once.[78] By contrast, Article 3 of the ECHR—which protects against torture and inhuman or degrading treatment or punishment—is a frequently utilized provision which has significantly developed the human rights-based *non-refoulement* jurisprudence in the European Court of Human Rights. It is for this reason that decisions from that jurisdiction form the bulk of the discussion.

B. Regional (ECHR, Article 3)

Since the case of *Soering v United Kingdom*, Article 3 of the ECHR has been recognized as precluding removal to a place where an applicant would face a real risk of being subjected to torture, or inhuman or degrading treatment or punishment.[79] Article 3 is absolute, and the European Court of Human Rights has consistently affirmed that it cannot be balanced against the public interest or any other matter, irrespective of the applicant's criminal or personal conduct.[80]

Inhuman treatment must attain 'a minimum level of severity' and involve 'actual bodily injury or intense physical or mental suffering'.[81] Importantly, for the present context, it does not need to be deliberate.[82] Degrading treatment 'humiliates or debases an individual, showing a lack of respect for, or diminishing, his or her human dignity, or arouses feelings of fear, anguish or inferiority capable of breaking an individual's moral and physical resistance'.[83] A lack of intent to humiliate will not conclusively rule out a violation of Article 3.[84]

[76] Human Rights Committee, 'General Comment No 20: Replaces General Comment 7 concerning Prohibition of Torture and Cruel Treatment or Punishment (Art 7)' (10 March 1992) para 4.

[77] Human Rights Committee, 'General Comment No 31' (n 25) para 12; *Kindler v Canada* Communication No 470/1991 (30 July 1993) UN Doc CCPR/C/48/D/470/1991 (11 November 1993) para 6.2; *ARJ v Australia* Communication No 692/1996 (28 July 1997) UN Doc CCPR/C/60/D/692/1996 (11 August 1997) para 6.8; *Ahani v Canada* (n 25). See also Committee on the Rights of the Child, 'General Comment No 6' (n 25) para 27.

[78] *Ng v Canada* Communication No 469/1991 (5 November 1993) UN Doc CCPR/C/49/D/469/1991 (7 January 1994) in Sarah Joseph, Jenny Schultz, and Melissa Castan, *The International Covenant on Civil and Political Rights: Cases, Materials, and Commentary* (2nd edn, Oxford University Press, 2004) para 9.80. This is in contrast to the multiple findings by the Committee against Torture with respect to Art 3 of CAT.

[79] *Soering v United Kingdom* (n 37).

[80] *Chahal v United Kingdom* (1996) 23 EHRR 413; *Saadi v United Kingdom* (2007) 44 EHRR 50. See the concerns of the dissenting judges in *N v United Kingdom* (n 16) with respect to the majority's consideration of policy in determining whether Art 3 had been breached.

[81] *Pretty v United Kingdom* (2002) 35 EHRR 1, para 52, referring to *Ireland v United Kingdom* (1979–80) 2 EHRR 25, para 167.

[82] *Labita v Italy* (2008) 46 EHRR 1228, para 120.

[83] *Pretty v United Kingdom* (n 81) para 52; see also *Ireland v United Kingdom* (n 81) para 167; *Moldovan and others v Romania* App nos 41138/98 and 64320/01 (ECtHR, 12 July 2005) para 101; *East African Asians* (1973) 3 EHRR 76, paras 189, 195. See also Ch 9 on dignity.

[84] *Peers v Greece* (2001) 33 EHRR 51, para 74.

The European Court of Human Rights has made clear that the assessment of this minimum level of severity is relative: 'it depends on all the circumstances of the case, such as the duration of the treatment, its physical and mental effects and, in some cases, the sex, age and state of health of the victim'.[85] Ordinarily, 'the risk which the individual runs of being subjected following expulsion to the proscribed form of treatment emanates from *intentionally inflicted acts* on the part of the public authorities in the receiving country',[86] but also where the danger emanates from non-State actors against whom 'the state has *failed to provide reasonable protection*'.[87]

C. Article 3 and socio-economic rights[88]

Human rights treaties and monitoring bodies have traditionally failed to accord the same weight to economic, social, and cultural rights as they have to civil and political rights.[89] This has led to the development of an interpretative approach known as the 'integrated' or 'holistic' approach, which has sought to show that civil and political rights have inherent socio-economic elements.[90] A treaty dealing with civil and political rights can therefore have 'its norms used as vehicles for the direct or indirect protection of norms of another treaty dealing with a different category of human rights', such as socio-economic ones.[91] For this reason, breaches of socio-economic rights are often 're-characterized' as violations of Article 3 of the ECHR, since this is an absolute right containing a clear *non-refoulement* obligation[92] whose breach cannot be justified on the grounds of insufficient resources.[93]

This section examines the potential of Article 3 to assist in climate change-related protection claims. By focusing on the underlying *human rights* which are

[85] *N v United Kingdom* (n 16) para 29.

[86] *R v Secretary of State for the Home Department, ex parte Bagdanavicius* [2005] UKHL 38, para 8 (Lord Brown) (emphasis added). The rest of the court concurred with Lord Brown's judgment: para 1 (Lord Nicholls); para 2 (Lord Hope); para 3 (Lord Walker); para 4 (Baroness Hale).

[87] ibid, para 24 (Lord Brown) (emphasis added).

[88] For a comprehensive analysis, see Foster (n 16).

[89] See, eg, Virginia Mantouvalou, 'Work and Private Life: *Sidabras and Dziautas v Lithuania*' (2005) 30 European Law Review 573; Conor Gearty and Virginia Mantouvalou, *Debating Social Rights* (Hart Publishing, 2010).

[90] Mantouvalou (n 89) 574. See especially *Airey v Ireland* (1979–80) 2 EHRR 305, para 26.

[91] Craig M Scott, 'The Interdependence and Permeability of Human Rights Norms: Towards a Partial Fusion of the International Covenants on Human Rights' (1989) 27 Osgoode Hall Law Journal 769, 771 cited in Mantouvalou (n 89) 574. See also Martin Scheinin, 'Economic and Social Rights as Legal Rights' in Allan Rosas and others (eds), *Economic, Social and Cultural Rights* (2nd edn, Kluwer, 2002). It is in part based on the idea that the enjoyment of civil and political rights is impossible if socio-economic rights are ignored, a perspective that was in fact reflected in the *travaux préparatoires* of the ECHR: Mantouvalou (n 89) 574–5. Waldron, by contrast, sees socio-economic rights as *intrinsically* valuable and as having equal weight: Jeremy Waldron, 'Liberal Rights: Two Sides of the Coin' in Jeremy Waldron, *Liberal Rights: Collected Papers 1981–1991* (Cambridge University Press, 1993) cited in Mantouvalou (n 89) 575.

[92] Den Heijer suggests that the focus on Art 3 of the ECHR as the core *non-refoulement* provision has become a 'self-fulfilling prophecy', in that it has hindered 'the development of a more profound understanding of the scope of applicability of the ECHR': Maarten den Heijer, 'Whose Rights and Which Rights? The Continuing Story of *Non-Refoulement* under the European Convention on Human Rights' (2008) 10 European Journal of Migration and Law 277, 278.

[93] See *Kalashnikov v Russia* (n 14).

compromised if the individual is removed, it avoids complex issues of causation about climate change which are not directly relevant to establishing the breach.[94]

The case of *D v United Kingdom* is often cited as evidence that complementary protection claims based on climate change impacts (lack of fresh water, food, safe shelter, etc) could succeed.[95] While it is feasible to argue that return to socio-economic deprivation could amount to inhuman or degrading treatment, on reflection I do not think that the approach in *D v United Kingdom* is necessarily the most relevant. This is because the focus of 'medical' cases is the (cessation of) treatment by the returning State, whereas in standard removal cases the focus is on the likely treatment in the country of origin.[96] One way of understanding the distinction between medical cases and standard removal cases is to use the conceptualization set out by Lord Bingham in *Ullah* of 'domestic' and 'foreign' cases.[97] This distinction does not turn on the applicant's nationality, but rather on whether the alleged breach is by the host State or the one to which return is contemplated. Thus conceived, medical cases might best be described as domestic cases.

In *D v United Kingdom*, the European Court of Human Rights noted that while Article 3 had until that point only been applied where the risk of harm emanated from intentionally inflicted acts, the provision's 'fundamental importance' meant that the court 'must reserve to itself sufficient flexibility' to address its application in other contexts.[98] Accordingly, it was not prevented from scrutinizing a claim where 'the source of the risk of proscribed treatment in the receiving country stemmed from factors which could not engage either directly or indirectly the responsibility of the public authorities of that country, or which, taken alone, did not in themselves infringe the standards of Article 3'.[99] That case concerned a man from St Kitts who had been receiving treatment for HIV while imprisoned in the UK. At the time of his application, he was in the advanced stages of an incurable illness, and it was an 'established fact that the withdrawal of his current medical treatment would hasten his death on account of the unavailability of similar

[94] See further section VII below.

[95] See, eg, Vikram Kolmannskog and Finn Myrstad, 'Environmental Displacement in European Asylum Law' (2009) 11 European Journal of Migration and Law 313. In *BB v France* App no 30930/96 (European Commission of Human Rights, 9 March 1998), the European Commission of Human Rights observed that facing AIDS alone at an advanced stage would constitute degrading treatment. Ultimately, a friendly settlement was reached in this matter so the Commission did not have to determine the issue definitively. See also *Ahmed v Austria* (1997) 24 EHRR 278; *HLR v France* (n 16); *Bensaid v United Kingdom* (2001) 33 EHRR 205; *Henao v The Netherlands* App no 13669/03 (ECtHR, 24 June 2003); *Ndangoya v Sweden* App no 17868/03 (ECtHR, 22 June 2004); *Amegnigan v The Netherlands* App no 25629/04 (ECtHR, 25 November 2004). For a successful case in the domestic context, see *CA v Secretary of State for the Home Department* [2004] EWCA Civ 1165.

[96] In the seminal removal case of *Soering v United Kingdom* (n 37) para 91 (emphasis added), the European Court of Human Rights held that: 'the decision by a Contracting State to extradite a fugitive may give rise to an issue under Article 3, and hence engage the responsibility of that State under the Convention, where substantial grounds have been shown for believing that the person concerned, if extradited, faces a real risk of being subjected to torture or to inhuman or degrading treatment or punishment in the requesting country. *The establishment of such responsibility inevitably involves an assessment of conditions in the requesting country against the standards of Article 3 of the Convention.*' 'Extradite a fugitive' should now be understood as 'remove an individual', in light of subsequent case law: see *Chahal v United Kingdom* (n 80).

[97] *Ullah* (n 12) para 9.

[98] *D v United Kingdom* (n 16) para 49.

[99] *N v United Kingdom* (n 16) para 32.

treatment in St Kitts'.[100] He argued that his removal to St Kitts would 'condemn him to spend his remaining days in pain and suffering in conditions of isolation, squalor and destitution', which would be inhuman and degrading.[101]

Even though the applicant put forward the argument that the conditions in the country of origin were themselves 'inhuman or degrading', this was not the basis on which the case was decided.[102] The court stated that the UK had 'assumed responsibility for treating the applicant's condition',[103] that he had 'become reliant on the medical and palliative care', and that the 'abrupt withdrawal of these facilities [would] entail the most dramatic consequences for him'.[104]

> Although it cannot be said that the conditions which would confront him in the receiving country are themselves a breach of the standards of Article 3 . . . his removal would expose him to a real risk of dying under most distressing circumstances and would thus amount to inhuman treatment.[105]

In other words, the UK's provision of medical treatment to him, its withdrawal of that treatment, and then (perhaps incidentally)[106] his removal to circumstances where such treatment could not be continued[107] directly engaged the UK's responsibility under Article 3 of the ECHR.

Accordingly, in the medical health cases the question asked by the European Court of Human Rights is posed differently from standard removal cases: does the cessation of medical treatment *by the sending country* amount to inhuman or degrading treatment (because such treatment cannot continue in the receiving country)? In other words, is the present state of the applicant's health such that expulsion should be precluded if the medical and social facilities that he or she needs are not available to him or her in the receiving State?[108]

The general principle is expressed as follows:

> Aliens who are subject to expulsion cannot in principle claim any entitlement to remain in the territory of a Contracting State in order *to continue to benefit from* medical, social or other forms of assistance and services provided by the expelling State. The fact that the applicant's circumstances, including his life expectancy, would be significantly reduced if he were to be removed from the Contracting State is *not sufficient in itself* to give rise to breach of Article 3. The decision to remove an alien who is suffering from a serious mental or physical illness to a country where the facilities for the treatment of that illness are inferior to those available in the Contracting State may raise an issue under Article 3, but only in a very exceptional case, where the humanitarian grounds against the removal are

[100] *D v United Kingdom* (n 16) para 40. [101] ibid.

[102] See ibid, para 49 distinguishing the reasoning in this case from other removal cases.

[103] ibid, para 53.

[104] ibid, para 52. [105] ibid, para 53.

[106] Indeed, as the dissenting judges stated in *N v United Kingdom* (n 16) dissenting judgment, para 20: 'deportation of an "applicant on his or her death bed" would *in itself* be inconsistent with the absolute provision of Article 3 of the Convention', irrespective of the conditions in the country of origin.

[107] This is why the court 'must subject all the circumstances surrounding the case to a rigorous scrutiny, especially the applicant's personal situation in the expelling State': *D v United Kingdom* (n 16) para 49.

[108] *N v Secretary of State for the Home Department* [2005] UKHL 31, para 32 (Lord Hope).

compelling. In the *D.* case the very exceptional circumstances were that the applicant was critically ill and *appeared to be close to death*, could not be guaranteed any nursing or medical care in his country of origin and had no family there willing or able to care for him or provide him with even a basic level of food, shelter or social support.[109]

(1) Protection from return to destitution

Can poor socio-economic conditions in the country of origin *themselves* amount to inhuman or degrading treatment?[110] The English Court of Appeal has stated that:

> a claim to be protected from the harsh effects of a want of resources, albeit made harsher by its contrast with the facilities available in the host country, is to my mind something else altogether.... I would hold that the application of Article 3 where the complaint in essence is of want of resources in the applicant's home country (in contrast to what has been available in the country from which he is to be removed) *is only justified where the humanitarian appeal of the case is so powerful that it could not in reason be resisted by the authorities of a civilised State.* That does not, I acknowledge, amount to a sharp legal test... an Article 3 case of this kind *must be based on facts which are not only exceptional, but extreme*; extreme, that is, judged in the context of cases all or many of which (like this one) demand one's sympathy on pressing grounds.[111]

As the court itself acknowledged, this is not a clear *legal* test, but rather an elucidation of the factual circumstances which would amount to a breach of Article 3. There is only one legal test when it comes to Article 3: whether there is a real risk of the applicant being subjected to inhuman or degrading treatment if removed.[112] Most curial analysis therefore focuses on whether the facts, in each case, satisfy that test.

The factual threshold of 'inhuman or degrading treatment' is a high one. The European Court of Human Rights has intimated that a significant decline in the applicant's circumstances, including life expectancy, will be insufficient to constitute a violation of Article 3.[113] In *D v United Kingdom*, poor medical conditions in the home country which could 'further reduce [the applicant's] already limited life expectancy and subject him to acute mental and physical suffering';[114] the lack

[109] *N v United Kingdom* (n 16) para 42 (emphasis added).

[110] The applicant in *D v United Kingdom* presented this argument, but by characterizing the breach as the direct result of treatment by the UK, this point did not have to be finally decided: see *D v United Kingdom* (n 16) paras 40–1.

[111] *N v Secretary of State for the Home Department* (n 16) paras 38–40 (emphasis added).

[112] There are cases, however, where the domestic courts have set the bar too high. For example, in *R (on the Application of 'B') v Secretary of State for the Foreign and Commonwealth Office* [2004] EWCA Civ 1344, [2005] QB 643, asylum seekers who had escaped from immigration detention in Australia to the British Consulate in Melbourne argued that returning them to the Australian authorities would breach Art 3, since it would expose them to inhuman or degrading treatment in detention. The English Court of Appeal held that the UK consular officials were only precluded from returning them to the Australian authorities if it were 'clearly necessary in order to protect them from the immediate likelihood of experiencing serious injury' (para 89). This was because the court treated the asylum seekers as 'fugitives' according to principles of extradition law. This test goes well beyond the requirements of Art 3 of the ECHR and represents an erroneous approach under that instrument.

[113] *N v United Kingdom* (n 16) para 42.

[114] *D v United Kingdom* (n 16) para 52.

of assurance that he would get a hospital bed; the absence of strong family ties or other moral or social support at home; the fact that his lack of shelter and proper diet in St Kitts could expose him to infections unable to be properly treated; and the country's generally poor health and sanitation conditions[115] were not of themselves found to breach Article 3.[116] However, the cumulative impact of these conditions in light of the applicant's *particular* circumstances, and the withdrawal of his medical treatment by the UK government, meant that his removal would violate Article 3.

In a series of domestic cases, the English courts held that the removal of subsistence support from asylum seekers living in the UK violated Article 3 of the ECHR where there was a real risk that it would result in their destitution. The House of Lords stated that treatment is inhuman or degrading 'if, *to a seriously detrimental extent*, it denies the most basic needs of any human being'.[117] While the court noted that there is no general public duty to house the homeless or provide for the destitute, the State does have such a duty if an asylum seeker 'with no means and no alternative sources of support, unable to support himself, is, by the *deliberate action of the state*, denied shelter, food or the most basic necessities of life'.[118] Relevant factors to be considered include the asylum seeker's 'age, gender, mental and physical health and condition, any facilities or sources of support available to the applicant, the weather and time of year and the period for which the applicant has already suffered or is likely to continue to suffer privation'.[119] In those cases, the threshold was met by asylum seekers who, denied State support and the right to work, were forced to sleep outdoors. Factors that contributed to this finding were 'the physical discomfort of sleeping rough, with a gradual but inexorable deterioration in their cleanliness, their appearance and their health', 'the prospect of that state of affairs continuing indefinitely', their '[g]rowing despair and a loss of self-respect', and the fact that they had 'no money of their own, no ability to seek state support and [were] barred from providing for themselves by their own labour'.[120]

More recently, the European Court of Human Rights has similarly held that 'a situation of extreme material poverty' can violate Article 3.[121] In the case of *MSS v Belgium and Greece*, the Belgian government was found to have breached its *non-refoulement* obligations under Article 3 by returning an asylum seeker to Greece, thereby knowingly exposing the applicant to conditions of detention and living conditions that amounted to degrading treatment[122]—namely:

> living in a state of the most extreme poverty, unable to cater for his most basic needs: food, hygiene and a place to live. Added to that was the ever-present fear of being attacked and

[115] ibid.

[116] ibid, para 53: 'Although it cannot be said that the conditions which would confront him in the receiving country are themselves a breach of the standards of Article 3 ... his removal would expose him to a real risk of dying under most distressing circumstances and would thus amount to inhuman treatment.'

[117] *R v Secretary of State for the Home Department, ex parte Adam* [2005] UKHL 66, para 7 (Lord Bingham) (emphasis added). See also *Secretary of State for the Home Department v Limbuela* [2004] EWCA Civ 540.

[118] *Adam* (n 117) para 7 (Lord Bingham) (emphasis added).

[119] ibid, para 8 (Lord Bingham).

[120] ibid, para 71 (Lord Scott).

[121] *MSS v Belgium and Greece* App no 30696/09 (ECtHR, Grand Chamber, 21 January 2011) para 252.

[122] ibid, para 367.

robbed and the total lack of any likelihood of his situation improving. It was to escape from that situation of insecurity and of material and psychological want that he tried several times to leave Greece.[123]

Subsequently, in *Sufi and Elmi v United Kingdom*, the European Court held that people returned to Somalia who needed to seek refuge in particular refugee camps would face a real risk of treatment contrary to Article 3 on account of the dire humanitarian conditions there:[124]

IDPs in the Afgooye Corridor have very limited access to food and water, and shelter appears to be an emerging problem as landlords seek to exploit their predicament for profit. Although humanitarian assistance is available in the Dadaab camps, due to extreme overcrowding access to shelter, water and sanitation facilities is extremely limited. The inhabitants of both camps are vulnerable to violent crime, exploitation, abuse and forcible recruitment. Moreover, the refugees living in—or, indeed, trying to get to—the Dadaab camps are also at real risk of refoulement by the Kenyan authorities. Finally, the Court notes that the inhabitants of both camps have very little prospect of their situation improving within a reasonable timeframe. The refugees in the Dadaab camps are not permitted to leave and would therefore appear to be trapped in the camps until the conflict in Somalia comes to an end. In the meantime, the camps are becoming increasingly overcrowded as refugees continue to flee the situation in Somalia. Although the IDPs in the Afgooye Corridor are permitted to leave, in reality the only place they are able to return to is Mogadishu, which the Court has found not to be a safe place for the vast majority of civilians. Consequently, there is also little prospect of their situation improving while the conflict continues.[125]

Significantly, the court noted that:

If the dire humanitarian conditions in Somalia were solely or even predominantly attributable to poverty or to the State's lack of resources to deal with a naturally occurring phenomenon, such as a drought, the test in *N. v. the United Kingdom* may well have been considered to be the appropriate one. However, it is clear that while drought has contributed to the humanitarian crisis, that crisis is predominantly due to the direct and indirect actions of the parties to the conflict.[126]

Thus, in cases concerning the deliberate action or inaction by a State, it seems that a breach of Article 3 may be more readily established. In *Sufi and Elmi v United Kingdom*, the court distinguished between its approach in *N v United Kingdom* (that '[h]umanitarian conditions would…only reach the Article 3 threshold in very exceptional cases where the grounds against removal were "compelling"')[127] and *MSS v Belgium and Greece*, to hold that in cases where ill-treatment relates

[123] ibid, para 254. See also para 253: 'The Court reiterates that it has not excluded "the possibility that the responsibility of the State may be engaged [under Art 3] in respect of treatment where an applicant, who was wholly dependent on State support, found herself faced with official indifference in a situation of serious deprivation or want incompatible with human dignity" (see *Budina v. Russia*, dec., no. 45603/05, ECHR 2009…).'

[124] *Sufi and Elmi v United Kingdom* App nos 8319/07 and 11449/07 (ECtHR, 28 June 2011) para 292.

[125] ibid, para 291.

[126] ibid, para 282.

[127] Cited in ibid, para 280.

to a deliberate act or omission by a State, the court must 'have regard to an applicant's ability to cater for his most basic needs, such as food, hygiene and shelter, his vulnerability to ill-treatment and the prospect of his situation improving within a reasonable time-frame'.[128] Whether a court would be prepared to extend this argument to find that anthropogenic climate change, even if not directly attributable to the State to which return is proposed, nonetheless required the latter test to be applied remains untested. The principal focus in every case must remain the nature of the ill-treatment a person would face if removed.

For this reason, the next section examines the situation where a State 'responsibility' link cannot be established. Whether or not the requisite factual threshold of Article 3 will be met may depend upon whether the case is argued (or resolved by the court) as a 'domestic' or a 'foreign' one. The two possible approaches are examined below.

(2) 'Domestic' case line of argument

First, following the line of reasoning in the medical cases, it could be argued that Article 3 would be breached if the act of removal constituted a withdrawal by the host State of basic socio-economic support (shelter, food, water, education, access to livelihood, etc), and there were a 'complete absence' of those resources in the home country.[129] This could presumably only be argued successfully if the applicant had resided in the host State for a particular length of time. As *D v United Kingdom* makes clear, the factual threshold is exceptionally high. That is the *only* case before the European Court of Human Rights in which non-removal has been substantiated on this basis.

The European Court of Human Rights has argued that this high threshold should be maintained in medical cases since 'the alleged future harm would emanate not from the intentional acts or omissions of public authorities or non-State bodies, but instead from a *naturally occurring illness and the lack of sufficient resources to deal with it* in the receiving country'.[130] However, as the English Court of Appeal noted in *RS (Zimbabwe)*, since there is only one legal test for Article 3, assessment

[128] ibid, para 283. This resonates with recent UK authority that has suggested a greater willingness to find a violation of Art 3 where it can be attributed to the deliberate action or inaction of a State, rather than something more 'natural', such as an illness: see *RN (Returnees) Zimbabwe CG* [2008] UKAIT 00083, para 254 citing an earlier decision of the AIT in *HS (Returning Asylum Seekers) Zimbabwe CG* [2007] UKAIT 00094, quoted in Foster (n 16) 300.

[129] See *N v Secretary of State for the Home Department* (n 108) para 50 (Lord Hope): 'What the [European] court is in effect saying is that the fact that the treatment may be beyond the reach of the applicant in the receiving State is not to be treated as an exceptional circumstance. It might be different if it could be said that it was not available there at all and that the applicant was exposed to an inevitable risk due to its complete absence. But that is increasingly unlikely to be the case in view of the amount of medical aid that is now reaching countries in the third world, especially those in sub-Saharan Africa. For the circumstances to be, as it was put in *Amegnigan v. the Netherlands*, "very exceptional" it would need to be shown that the applicant's medical condition had reached such a critical stage that there were compelling humanitarian grounds for not removing him to a place which lacked the medical and social services which he would need to prevent acute suffering while he is dying.' See also *N v United Kingdom* (n 16) para 41.

[130] *N v United Kingdom* (n 16) para 43 (emphasis added).

of harm should not depend 'on whether the "lack of sufficient resources" in the receiving State occurs as a consequence of some malign influence by that State or because of benign matters. The effect on the individual is the same in either case and it either reaches the threshold set by the ECtHR or it does not.'[131] As a matter of human rights law, this is the correct approach. However, even if the opposite view is taken, in the climate change context there may be some scope to argue that harm does emanate from acts or omissions by the authorities—either on the part of the country of origin (eg by failing to protect from foreseeable disasters),[132] or the host State (eg by failing to provide funding and technical assistance for climate change adaptation).[133]

(3) 'Foreign' case line of argument

An alternative approach would be to argue that the level of deprivation in the country of origin itself amounts to inhuman or degrading treatment. This adopts the approach in standard removal cases ('foreign' cases), where the focus is solely on the treatment in the country of origin to which the individual is likely to be subjected if removed. The advantage is that it enables a more straightforward argument to be presented, based on contemporary understandings that socio-economic deprivation can constitute inhuman or degrading treatment. In the dissenting judgment in *N v United Kingdom*, the utility of this approach was acknowledged:

> the additional grounds advanced by the Court in *D. v. the United Kingdom* and related to a lack of medical and palliative care as well as a lack of psychological support, in the home country, might be equally relevant to the finding of a *separate* potential violation of Article 3 of the Convention.[134]

Again, this is a factual rather than a legal question and would depend on whether the relevant facts—considered cumulatively—could be characterized as 'inhuman or degrading treatment'.

By way of analogy, refugee law recognizes that 'persecution' can encompass the deprivation of socio-economic rights, not just violations of civil and political rights.[135] The Australian Migration Act expressly enumerates 'significant economic hardship that threatens the person's capacity to subsist', 'denial of access to basic services, where the denial threatens the person's capacity to subsist', and 'denial of

[131] *RS (Zimbabwe) v Secretary of State for the Home Department* [2008] EWCA Civ 839, para 31. See also *ZT v Secretary of State for the Home Department* [2005] EWCA Civ 1421, para 16: '*Soering* came nowhere near to laying down any special rule about the behaviour of the receiving state, within the ambit of the single rule of article 3 in terms of inhuman and degrading treatment.' There is, however, contradictory authority in *RN (Returnees) Zimbabwe CG* (n 128), para 254; *HS (Returning Asylum Seekers) Zimbabwe CG* (n 128).

[132] See, eg, *Budayeva v Russia* (n 49).

[133] See Ch 9.

[134] *N v United Kingdom* (n 16) dissenting judgment, para 21 (fn omitted).

[135] See generally Michelle Foster, *International Refugee Law and Socio-Economic Rights: Refuge from Deprivation* (Cambridge University Press, 2007); Kate Jastram, 'Economic Harm as a Basis for Refugee Status and the Application of Human Rights Law to the Interpretation of Economic Persecution' in Simeon (n 35); Migration Act 1958 (Cth), s 91R(2).

capacity to earn a livelihood of any kind, where the denial threatens the person's capacity to subsist' as forms of persecution.[136]

How severe would the deprivation in the country of origin need to be in order to substantiate the 'inhuman or degrading treatment' threshold in a removal case? Removal is not precluded simply because conditions of treatment in the receiving State are of a lower standard than those in the host State.[137] To hold otherwise 'would be imposing an obligation on Contracting States effectively to act as indirect guarantors of [human rights] for the rest of the world'.[138] Although the courts have indicated that the Article 3 threshold is particularly high in removal cases,[139] they have (rightly) refrained from articulating a different standard on account of the 'fundamental importance' of the rights it protects.[140] As the House of Lords explained in *Januzi* in relation to the internal flight alternative, while a person may be returned to a country where the general standards of living are not as high as in the State in which protection was sought, the position would be different 'if the lack of respect for human rights posed threats to his life or exposed him to the risk of inhuman or degrading treatment or punishment'.[141]

In that context, the AIT has observed that:

> Inevitably, it will be unduly harsh if an appellant is unable for all practical purposes to survive with sufficient dignity to reflect her humanity. That is no more than saying that if survival comes at a cost of destitution, beggary, crime or prostitution, then that is a price too high.[142]

The United Nations High Commissioner for Refugees' (UNHCR) Guidelines on the internal flight alternative state that:

> It would be unreasonable, including from a human rights perspective, to expect a person to relocate to face economic destitution or existence *below at least an adequate level of subsistence*. At the other end of the spectrum, a simple lowering of living standards or worsening of economic status may not be sufficient to reject a proposed area as unreasonable. Conditions in the area must be such that *a relatively normal life can be led* in the context of the country concerned. If, for instance, an individual would be without family links and unable to benefit from an informal social safety net, relocation may not be reasonable,

[136] Migration Act 1958 (Cth), s 91R(2).

[137] *Salkic v Sweden* App no 7702/04 (ECtHR, 29 June 2004); *Amegnigan v The Netherlands* (n 95); *Januzi* (n 18) para 19 (Lord Bingham), para 45 (Lord Hope).

[138] *Z and T v United Kingdom* (n 37) 7. The House of Lords stated that in foreign cases involving rights other than Arts 2 or 3 of the ECHR, it will be necessary to establish 'at least a real risk of a flagrant violation of the very essence of the right', such that it will be 'completely denied or nullified in the destination country': *Ullah* (n 12) paras 50, 24 respectively, referring in para 24 to *Devaseelan v Secretary of State for the Home Department* [2002] IAT 702, [2003] Imm AR 1, para 111. See also *Ullah* (n 12) paras 69–70. This is a more demanding test than in a domestic case (where the Contracting State is directly implicated).

[139] 'Having regard, however, to the high threshold set by Article 3, *particularly where the case does not concern the direct responsibility of the Contracting State for the infliction of harm*': *Bensaid v United Kingdom* (n 95) para 40 (emphasis added).

[140] *F v United Kingdom* App no 17341/03 (ECtHR, 22 June 2004) 12.

[141] *Januzi* (n 18) para 19 (Lord Bingham), para 45 (Lord Hope). In *Salkic v Sweden* (n 137) the court reiterated that Art 3 will not be breached simply because the level of healthcare (including mental healthcare) in the receiving State is not of an equivalent standard to that available in the host State. See also *Amegnigan v The Netherlands* (n 95).

[142] *FB* (n 66) para 39 (emphasis removed).

unless the person would otherwise be able to sustain a relatively normal life at more than just a minimum subsistence level.[143]

Importantly, the Guidelines recognize that cultural and ethnic reasons may mean that access to land and resources will be impossible in particular areas of the country, thus precluding internal relocation. They also note that the internal flight alternative is inappropriate if it requires movement to areas such as urban slums, where conditions of severe hardship would ensue.[144] These considerations are particularly relevant in the context of climate change-related movement. The availability of an internal flight alternative will depend upon what decision-makers regard as 'reasonable', but there are necessarily limits to this. At the bare minimum, human dignity must be safeguarded.[145]

Similarly, Hugo Storey's assessment of the jurisprudence leads to the conclusion that:

> there seems to be broad agreement that if life for the individual claimant…would involve economic annihilation, utter destitution or existence below a bare subsistence level (*Existenzminimum*) or deny 'decent means of subsistence' that would be unreasonable. On the other end of the spectrum a simple lowering of living standards or worsening of economic status would not. What must be shown to be lacking is the real possibility to survive economically, given the particular circumstances of the individual concerned (language, knowledge, education, skills, previous stay or employment there, local ties, sex, civil status, age and life experience, family responsibilities, health; available or realisable assets, and so forth). Moreover, in the context of return, the possibility of avoidance of destitution by means of financial assistance from abroad, whether from relatives, friends or even governmental or non-governmental sources, cannot be excluded.[146]

Applying this reasoning, it is clear that when the (cumulative) conditions in the country of origin would result in an individual living in destitution or below a minimum subsistence level, removal to such conditions would constitute a breach of Article 3 of the ECHR.[147] This is because the conditions in the country of origin would amount to inhuman or degrading treatment. It is difficult to see how a court

[143] UNHCR, 'Guidelines on International Protection: "Internal Flight or Relocation Alternative" within the Context of Article 1A(2) of the 1951 Refugee Convention and/or 1967 Protocol relating to the Status of Refugees', UN Doc HCR/GIP/03/04 (23 July 2003) para 29 (emphasis added).

[144] ibid, para 30.

[145] See further, Ch 9.

[146] Hugo Storey, 'The Internal Flight Alternative Test: The Jurisprudence Re-examined' (1998) 10 International Journal of Refugee Law 499, 516 (fns omitted).

[147] An analogy can be found in international refugee law. 'Persecution cannot and should not be defined solely on the basis of serious human rights violations. Severe discrimination or the cumulative effect of various measures not in themselves alone amounting to persecution, as well as their combination with other adverse factors, can give rise to a well-founded fear of persecution, or, otherwise said: make life in the country of origin so insecure from many perspectives for the individual concerned, that the only way out of this predicament is to leave the country of origin': Erika Feller, 'Statement by Ms Erika Feller, Director, Department of International Protection, UNHCR' (Strategic Committee on Immigration, Frontiers and Asylum, Brussels, 6 November 2002) 3. See also UNHCR, *Handbook on Procedures and Criteria for Determining Refugee Status under the 1951 Convention and the 1967 Protocol relating to the Status of Refugees* UN Doc HCR/IP/4/Eng/REV.1 (2nd edn, Geneva, 1992) para 5.

could reach any other conclusion on a correct application of international human rights law principles.[148]

(4) Conclusion

Although the ECHR 'is essentially directed at the protection of civil and political rights',[149] this does not mean that socio-economic rights remain entirely unprotected. As the court explained in *Airey v Ireland*, although

> the Convention sets forth what are essentially civil and political rights, many of them have implications of a social or economic nature. The Court therefore considers, like the Commission, that the mere fact that an interpretation of the Convention may extend into the sphere of social and economic rights should not be a decisive factor against such an interpretation; there is no water-tight division separating that sphere from the field covered by the Convention.[150]

Nevertheless, such rights must first pass through the doorway of the protected civil and political rights to be considered. The court has consistently observed that it is not the function of the ECHR to iron out socio-economic differences between States, noting that 'the level of treatment available in the Contracting State and the country of origin may vary considerably'.[151] It has referred to the onerous burden that would otherwise be placed on Contracting States if they had to rectify global socio-economic disparities by granting a right to remain to disadvantaged people.[152] It has explained that '[o]n a purely pragmatic basis, it cannot be required that an expelling Contracting State only return an alien to a country which is in full and effective enforcement of all the rights and freedoms set out in the Convention'.[153]

Similarly, the House of Lords has referred to 'the limits that must be set on practical grounds'.[154] Lord Hope has described the jurisprudence as setting 'limits... on the extent to which [Contracting States] can be held responsible outside the areas that are prescribed by Articles 2 and 3 and by the fundamental right under Article 6 to a fair trial'. Those limits are set 'against the background of the general principle of international law that states have the right to control the entry,

[148] As Cassese has noted, 'nothing could warrant [Art 3's] possible limitation to only physical or psychological mistreatment in the area of civil rights': Antonio Cassese, 'Can the Notion of Inhuman and Degrading Treatment Be Applied to Socio-Economic Conditions?' (1991) 2 European Journal of International Law 141, 143.

[149] *N v United Kingdom* (n 16) para 44.

[150] *Airey v Ireland* (n 90) para 26.

[151] *N v United Kingdom* (n 16) para 44. The dissenting judges describe this as the majority's fear that any broader interpretation could 'make Europe vulnerable to becoming the "sick-bay" of the world': *N v United Kingdom* (n 16) dissenting judgment, para 8.

[152] *N v United Kingdom* (n 16) 44. Although, as Foster notes: '[t]his certainly raises a question as to the universality of human rights': Foster (n 16) 276.

[153] *F v United Kingdom* (n 140) 12 in relation to Art 8 of the ECHR.

[154] *EM (Lebanon) v Secretary of State for the Home Department (AF and others intervening)* [2008] UKHL 64, para 10 (Lord Hope), referring also to the European Court's decisions in *F v United Kingdom* (n 140) and *Z and T v United Kingdom* (n 37) which were not available to the House of Lords when it was considering the cases of *Ullah* (n 12) and *R (Razgar) v Secretary of State for the Home Department* [2004] 2 AC 368.

residence and expulsion of aliens'.[155] The trade-off for accepting that harm derives from a State's lack of resources to redress an applicant's predicament is that only the most exceptional cases receive international protection.[156] One can well imagine similar policy arguments being made in the context of climate change-related movement, especially in light of some of the alarmist predictions about large numbers of people who will be on the move.

It therefore seems unlikely that a lack of basic services alone would substantiate an Article 3 claim, unless they were to render survival on return impossible.[157] Something else—a distinguishing feature that makes the lack of such services particularly deleterious on the applicant—would appear to be necessary.[158] The very high threshold set in the jurisprudence means that it will likely take some decades before the negative impacts of climate change, interacting with underlying socio-economic vulnerabilities, could, in and of themselves, be regarded as constituting a violation of Article 3 giving rise to protection from removal. This is because many effects will take years to manifest at a sufficiently harmful level to engage Article 3 protection; or may be severe *temporary* effects which do not render return unlawful; or an internal flight alternative may be reasonable.

Timing will play a major role in whether claims are successful, since the imminence of harm on return, which is related to the level of deprivation in the country of origin, will be a key factor in the decision. Article 3 does not allow for pre-emptive movement where conditions are *anticipated* to become dire, and thus would not assist people trying to move before the situation became intolerable. Reliance on the remedial potential of the ECHR is therefore an imperfect way to resolve potential climate change-related displacement.

D. Domestic developments

In light of the foregoing analysis, it may come as a surprise to find that the UK has already precluded removal to socio-economic deprivation in some cases.[159] In the UK, a person who has been unsuccessful in applying for refugee status or subsidiary protection (known there as Humanitarian Leave) is automatically considered for Discretionary Leave. Discretionary Leave is not regulated by the EU

155 *EM* (n 154) para 13.

156 If anything, the anthropogenic basis of climate change should reinforce rather than undermine any claim, given that the States primarily responsible are not the ones from which movement is likely.

157 If the lack of services results from a direct act or omission by the State, then a violation of Art 3 may be more readily found: see *Sufi and Elmi v United Kingdom* (n 124).

158 On this point, see *D v United Kingdom* (n 16) para 53. Indeed, some domestic complementary protection schemes deliberately 'carve out' protection exceptions where the risk is faced generally by the population as a whole, requiring the applicant to show an individual risk: eg Immigration and Refugee Protection Act 2001 SC, c 27 (Canada), s 97(b)(ii); Qualification Directive, Art 15(c) and recital 26.

159 This section draws on cases cited in Foster (n 16) 303–7. Legislation that may assist people displaced by climate change is discussed in Ch 4. Furthermore, in a 2009 study of complementary protection in Europe, most of the countries surveyed (Austria, Belgium, Denmark, Finland, France, Germany, Ireland, Switzerland, Sweden, United Kingdom) maintained mechanisms of complementary protection based on health issues: European Council on Refugees and Exiles (ECRE), 'Complementary Protection in Europe' (July 2009) 6.

Qualification Directive[160] but is a discretionary power of the Secretary of State for the Home Department, guided by asylum policy instructions.[161] It may be granted when return could prejudice protected rights, including Article 3 of the ECHR, where the need 'does not arise from a need for protection as such, e.g. where a person's medical condition or severe humanitarian conditions in the country of return would make return contrary to Article 3'.[162]

The relevant Asylum Policy Brief notes that:

> There may be some extreme cases (although such cases are likely to be rare) where a person would face such poor conditions if returned—e.g. absence of water, food or basic shelter—that removal could be a breach of the UK's Article 3 obligations. Discretionary Leave should not be granted if the claimant could avoid the risk of suffering by leaving the UK voluntarily.[163]

This is significant because it recognizes that a violation of Article 3 based on deprivation of the basic means for survival—recognized in the domestic case of *Adam*[164]—may also apply to the non-removal context.

Even though Article 3 of the ECHR is the acknowledged source of the non-removal obligation, the UK considers these kinds of claims to fall outside the scope of Article 15(b) of the EU Qualification Directive. It seems to be trying to avoid entrenching socio-economic deprivation as an inherent aspect of 'inhuman or degrading treatment' under that instrument and instead suggests that it is purely 'humanitarian' or 'compassionate' in nature. While recital 9 of the Qualification Directive states that people who are permitted to remain 'on a discretionary basis on compassionate or humanitarian grounds' do not come within the instrument's scope, the assumption is that those people fall outside States' *non-refoulement* obligations under human rights law (eg Article 3). If Article 3 applies, then a person has a protection need and should be regulated by the Qualification Directive. The only distinction between Article 3 of the ECHR and Article 15(b) of the Qualification Directive is that the latter precludes removal to 'inhuman or degrading treatment or punishment of an applicant *in the country of origin*' (emphasis added). With respect to medical cases at least, it could be argued that the harm occurs not 'in the country of origin', but rather by the withdrawal of medical treatment in the UK (thus not directly engaging Article 15(b)).[165] However, Article 15(b) does not exclude cases of the kind envisaged by the Asylum Policy Brief above. Accordingly, such cases should be examined under that

[160] It is not incorporated in either the Immigration Rules or the Qualification Regulations of 2006, but provisions of the 1971 Immigration Act allow the Secretary of State to grant leave to a person for a reason not covered by the Immigration Rules.

[161] ECRE (n 159) 67.

[162] 'Asylum Policy Brief: Discretionary Leave' (last amended 27 October 2009).

[163] ibid.

[164] *Adam* (n 117) para 7.

[165] See also preparatory documents for the Qualification Directive: 'In the Stt. Kitts case [*D v United Kingdom*], although the lack of access to a developed health system as well as lack of a social network in itself was not considered as torture or inhuman or degrading treatment, the expulsion to this situation, which would have been life threatening to the concerned person, was described as such. Consequently, to avoid the inclusion of such compassionate grounds cases under a subsidiary protection regime, which was never the intention of this Directive, the Presidency is suggesting to limit the scope of sub-paragraph (b) by stating that the real risk of torture or inhuman or degrading treatment or punishment must prevail in his or her country of origin': Council of Europe, Presidency Note to Strategic Committee on Immigration, Frontiers and Asylum, 'Proposal for a Council Directive on Minimum Standards for the Qualification

provision and beneficiaries accorded subsidiary protection. The UK's policy reflects what Durieux describes as a shift away from a positive obligation of *protection*, in the refugee context, to *non-removability* in the human rights context.[166]

The AIT has accepted that 'poor living conditions are capable of raising an issue under Article 3 of the ECHR if they reach a minimum level of severity'.[167] Elsewhere, it has stated that it is 'uncontroversial that if as a result of a removal decision a person would be exposed to a real risk of existence below the level of bare minimum subsistence that would cross the threshold of Art 3 harm'.[168] While the use of the term 'uncontroversial' is perhaps premature,[169] the AIT has nonetheless accepted that removal would violate Article 3 ECHR where it would result in:

(a) return to 'a camp where conditions are described as "sub-human" and [the applicant would] face medical conditions described as some of the worst in the world';[170]

(b) the return of 'an amputee who had serious mental problems who would not receive either financial or medical support in the Gambia, and would only have recourse to begging for his support';[171]

(c) a 16-year-old boy's return where this would leave him destitute and without any protection;[172]

(d) the return of an applicant and his family to Kabul where they would be 'reduced either to living in a tent in a refugee camp or... in a container with holes knocked in the side to act as windows', and the applicant would be unlikely to find work and would 'be competing with others for scarce resources of food and water as well as accommodation'. Concern was also expressed for the impact of these conditions on 'five young (some of them very young) children'.[173]

In cases (b) and (c), and to some extent (d), a characteristic particular to the applicant (age, health) was pertinent to the claim's success. However, cases (a) and (d) suggest that more general country of origin conditions, if sufficiently severe, may be able to form the basis of a protection claim.[174]

and Status of Third Country Nationals and Stateless Persons as Refugees or as Persons Who Otherwise Need International Protection' Doc 12148/02 LIMITE ASILE 43 (20 September 2002) 6.

166 Jean-François Durieux, 'The Elusive "War Refugee": Violence and Discrimination in the Construct of "Refugee" as a Legal Category' (Refugee Studies Centre Lecture, University of Oxford, 10 November 2010).

167 *HS* (n 131) para 59. The cases discussed here are referred to in Foster (n 16) 303.

168 *Mandali v Secretary of State for the Home Department* [2002] UKIAT 0741, para 10.

169 Foster also sees this as potentially 'overstating the case': Foster (n 16) 303.

170 *Owen v Secretary of State for the Home Department* [2002] UKIAT 03285, para 27 cited in Foster (n 16) 304.

171 *R v Secretary of State for the Home Department, ex parte Kebbeh* [1999] EWHC Admin 388 (QB) para 58 cited in Foster (n 16) 304.

172 *Korca v Secretary of State for the Home Department* (UKIAT Appeal No HX-360001-2001, 29 May 2002) para 9 cited in Foster (n 16) 304. See also *LM (Democratic Republic of Congo) v Secretary of State for the Home Department* [2008] EWCA Civ 325.

173 *GH (Afghanistan) v Secretary of State for the Home Department* [2005] EWCA Civ 1603, para 5. The Court of Appeal rejected an appeal by the government.

174 In *NS (Relevance of Children to Removal—Art 8)* [2005] UKAIT 00081, para 55 it was held that the affect of the Asian Tsunami in Sri Lanka did not interfere with telephone and other contact

The reasoning in these cases finds closer analogies with the European Court of Human Rights' jurisprudence on general violence and Article 3 than it does with the medical cases. The court has left open the possibility that a general situation of violence could violate Article 3 if the level of violence reaches a sufficient level of intensity.[175] The applicant would need to face 'a real risk of ill-treatment *simply by virtue of an individual being exposed to such violence on return*'.[176] This is a very high threshold. Acknowledging this, the AIT held in *AM & AM* that this approach 'would be equally applicable if one substitute[d] the words "poor humanitarian conditions"' for 'situations of general violence'.[177] In that case, the AIT stated that there is nothing, in principle, that would prevent a finding that 'poor humanitarian conditions in Somalia, even if in an IDP camp'[178] violated Article 3. On the facts of that case, the AIT was not persuaded that there was 'a real risk of denial of basic food and shelter and other bare necessities of life' in Somalia. The implication, however, is that those deprivations could substantiate such a claim in the right factual circumstances.[179]

Of course, the extent to which the reasoning above can be successfully applied—and, moreover, extended—outside the jurisdictions in which it has developed remains to be seen. Other domestic provisions that may have relevance to climate change-related movement are discussed in Chapter 4.

IV. Other Rights that May Give Rise to Complementary Protection

The UN Human Rights Committee,[180] the UN Committee on the Rights of the Child,[181] the European Court of Human Rights,[182] the UN Committee on the

facilities such as to disrupt the continuation of family life, and thus removal did not breach Art 8 of the ECHR.

[175] See *HLR v France* (n 16) para 41; *NA v United Kingdom* App no 25904/07 (ECtHR, 17 July 2008) para 114. In *Ahmed v Austria* (n 95), the court found Art 3 was violated partly on account of conditions in Somalia.

[176] *NA v United Kingdom* (n 175) para 115 (emphasis added). In *AM & AM (Armed Conflict—Risk Categories) Somalia CG* [2008] UKAIT 00091, para 87, the UK Asylum and Immigration Tribunal (AIT) said there was nothing, in principle, that would prevent 'poor humanitarian conditions in Somalia, even if in an IDP camp', from violating Art 3. In that case, the evidence did not persuade the AIT that there was 'a real risk of denial of basic food and shelter and other bare necessities of life': para 157.

[177] *AM & AM* (n 176) para 87.

[178] ibid.

[179] ibid, para 157.

[180] The Human Rights Committee recognizes, at least in principle, that States' *non-refoulement* obligations may be triggered 'when considerations of non-discrimination, prohibition of inhuman treatment and respect for family life arise': Human Rights Committee, 'General Comment 15' (n 17) para 5. See also Human Rights Committee, 'General Comment No 18' (n 17). See generally Human Rights Committee, 'General Comment No 31' (n 25) para 12: 'Moreover, the article 2 obligation requiring that States Parties respect and ensure the Covenant rights for all persons in their territory and all persons under their control entails an obligation not to extradite, deport, expel or otherwise remove a person from their territory, where there are substantial grounds for believing that there is a real risk of irreparable harm, such as that contemplated by articles 6 and 7 of the Covenant, either in the country to which removal is to be effected or in any country to which the person may subsequently be removed.'

[181] The Committee on the Rights of the Child has made clear that the obligation is 'by no means limited to' those provisions (and CRC, Art 37): Committee on the Rights of the Child, 'General Comment No 6' (n 25) para 27. The language of 'irreparable harm' has been used by the Human Rights Committee to describe harm that is comparable to that contemplated by Arts 6 and 7 of the ICCPR.

[182] *N v United Kingdom* (n 16). See, eg, Art 9 cases: *Razaghi v Sweden* App no 64599/01 (ECtHR, 11 March 2003); *Gomes v Sweden* App no 34566/04 (ECtHR, 7 February 2006); *Z and T v United*

Elimination of Racial Discrimination,[183] and the House of Lords[184] have all recognized that the principle of *non-refoulement* may extend beyond protection of the right to life and the right to be free from torture or cruel, inhuman or degrading treatment or punishment. The Committee on the Rights of the Child has made clear that the *non-refoulement* obligation applies in any case where there are substantial grounds for believing that there is a real risk of 'irreparable harm' if the person is removed.[185] The language of 'irreparable harm' has been used by the Human Rights Committee to describe harm that is comparable to that contemplated by Articles 6 and 7 of the ICCPR.[186] However, so far, no other provision has independently given rise to a non-removal claim. In *Z and T v United Kingdom*, the European Court of Human Rights stated that 'it would be difficult to visualise a case in which a sufficiently flagrant violation of Article 9 would not also involve treatment in violation of Article 3 of the Convention'.[187] This is why analysis typically begins with that provision, and only if a violation is not made out are other articles even considered.[188]

Ever since *Soering v United Kingdom*, the first case in which the European Court recognized the implied principle of *non-refoulement* in Articles 2 and 3, the court has accepted that the same obligation may be implicit in other ECHR rights.[189]

Kingdom (n 37); Art 6 cases: *Drozd and Janousek v France and Spain* (1992) 14 EHRR 745; *Einhorn v France* App no 71555/01 (ECtHR, 16 October 2001); *Mamatkulov v Turkey* (n 39); *Tomic v United Kingdom* App no 17837/03 (ECtHR, 14 October 2003); *F v United Kingdom* (n 140); Art 4 case: *Ould Barar v Sweden* (1999) 28 EHRR CD 213.

[183] The Committee on the Elimination of Racial Discrimination has several times stated that: 'The Committee also urges the State party to ensure, in accordance with article 5 (b), that no person will be forcibly returned to a country where there are substantial grounds for believing that his/her life or health may be put at risk. The Committee recommends that the State party seek cooperation with UNHCR in this regard': see, eg, Committee on the Elimination of Racial Discrimination, 'Consideration of Reports Submitted by States Parties under Article 9 of the Convention: Concluding Observations of the Committee on the Elimination of Racial Discrimination: Japan', UN Doc CERD/C/JPN/CO/3-6 (16 March 2010) para 23. This seems an anomalous approach, given it is an unusually low threshold, but given its relevance it is noted here. See also Special Rapporteur on the Right to Food, 'Implementation of General Assembly Resolution 60/251 of 15 March 2006 Entitled "Human Rights Council"', UN Doc A/HRC/4/30 (19 January 2007) para 64: 'Governments have a legal obligation to help the refugees from hunger'. See also Human Rights Council Advisory Committee, 'Report to the Human Rights Council on the First Session of the Advisory Committee', UN Doc A/HRC/AC/2008/1/L.11 (15 August 2008) part 1/6.

[184] *Ullah* (n 12); *EM* (n 154); *Razgar* (n 154).

[185] Committee on the Rights of the Child, 'General Comment No 6' (n 25) para 27.

[186] Human Rights Committee, 'General Comment No 31' (n 25) para 12; *ARJ v Australia* (n 77); *Judge v Canada* Communication No 829/1998 (5 August 2002) UN Doc CCPR/C/78/D/829/1998 (20 October 2003). While neither Committee has commented further on the meaning of 'irreparable harm' in the context of removals, it has indicated in the case of interim measures that: 'what may constitute "irreparable damage" to the victim within the meaning of rule 86 [now rule 92] *cannot be determined generally*. The essential criterion is indeed the irreversibility of the consequences, in the sense of the inability of the author to secure his rights, should there later be a finding of a violation of the Covenant on the merits': *Stewart v Canada* Communication No 538/1993 (1 November 1996) UN Doc CCPR/C/58/D/538/1993 (16 December 1996) para 7.7 (emphasis added). The only dissent against the expansion of the principle of *non-refoulement* in this way was expressed in an individual opinion by Christine Chanet in *Judge v Canada* on the grounds that 'legal and practical problems would immediately arise'. See Foster (n 16) 277.

[187] *Z and T v United Kingdom* (n 37) 7.

[188] Except in *Bader v Sweden* (n 38) where the court found a violation of both Arts 2 and 3.

[189] *Soering v United Kingdom* (n 37) paras 91, 113; see also *Cruz Varas v Sweden* (1991) 14 EHRR 1, paras 69–70; *Vilvarajah v United Kingdom* (1991) 14 EHRR 248, para 103. See generally McAdam (n 3) Ch 4. For a detailed discussion of the case law on provisions other than Art 3 in this context, see also den Heijer (n 92) 280–5. In *Razgar* (n 154) para 9, Lord Bingham invoked *Bensaid United*

However, it has drawn a distinction between the 'fundamental importance' of Articles 2 and 3 of the ECHR, which are absolute and non-derogable rights, and other provisions of the ECHR, where '[s]uch compelling considerations do not automatically apply'.[190]

The most detailed analysis of the issue was undertaken by the House of Lords in the parallel cases of *Ullah* and *Razgar*, where consideration was given to whether Articles 9 or 8 (respectively) of the ECHR could found a non-removal claim.[191] The court held that, as a matter of principle, any provision of the ECHR could do so,[192] but that the threshold in such cases would be very high.[193] The applicant would need 'to establish at least a real risk of a flagrant violation of the very essence of the right'.[194] A 'flagrant denial' of a right is effectively a complete denial or nullification of the right.[195]

According to Lord Bingham, the reason why a complete nullification of the right is required is that in the case of qualified rights, the State may have a 'legitimate aim' in restricting the right, such that

> it is only in such a case—where the right will be completely denied or nullified in the destination country—that it can be said that removal will breach the treaty obligations of the signatory state however those obligations might be interpreted or whatever might be said by or on behalf of the destination state.[196]

This is said to be something that will manifest only in 'very exceptional cases'.[197] It seems that to meet the exceptionality test, the House of Lords requires 'the

Kingdom (n 95) as authority for placing reliance on Art 8 'to resist an expulsion decision, even where the main emphasis is not on the severance of family and social ties which the applicant has enjoyed in the expelling country but on the consequences for his mental health of removal to the receiving country.'

[190] *F v United Kingdom* (n 140) 12.

[191] This appeal was heard immediately following the appeal in *Ullah* (n 12), and since it was 'directly germane to the issue of principle in the present case… [it] should be read, to the extent that [it is] relevant, as incorporated in this opinion': *Razgar* (n 154) para 2. For Art 8, the possibility is also acknowledged: *Bensaid v United Kingdom* (n 95) paras 46–9; *N v United Kingdom* (n 16) dissenting judgment, para 26.

[192] *Ullah* (n 12) paras 21, 35, 39–49, 52, 53, 62, 67.

[193] ibid, para 24. See also *Razgar* (n 154) para 10: Art 8 of the ECHR could 'be engaged by the foreseeable consequences for health of removal from the United Kingdom pursuant to an immigration decision, even where such removal does not violate Article 3… an applicant could never hope to resist an expulsion decision without showing *something very much more extreme than relative disadvantage* as compared with the expelling state' (emphasis added).

[194] *Ullah* (n 12) para 50 (Lord Steyn). In *Razgar* (n 154) para 72, Lord Carswell used the term 'fundamental breach' synonymously: 'In order to bring himself within such an exceptional engagement of article 8 the applicant has to establish a very grave state of affairs, amounting to a flagrant or fundamental breach of the article, which in effect constitutes a complete denial of his rights.'

[195] *Ullah* (n 12) para 24 (Lord Bingham), adopting the test of the AIT in *Devaseelan* (n 138). This test derives from the partly dissenting opinion in *Mamatkulov v Turkey* (n 39), which said it 'is intended to convey… a breach of the [right]… which is so fundamental as to amount to a nullification, or destruction of the very essence, of the right guaranteed': Judges Bratza, Bonello, and Hedigan in *Mamatkulov v Turkey* 537–9; echoed by Lord Carswell in *EM* (n 154) para 54. They indicated that the standard of proof is the same as for an Art 3 claim, namely a 'real risk'. See also *EM* (n 154) para 4 (Lord Hope), para 34 (Lord Bingham); *Ullah* (n 12) para 50 (Lord Steyn).

[196] *Ullah* (n 12) para 24 (Lord Bingham) citing *Devaseelan* (n 138) para 111.

[197] *EM* (n 154) para 17 (Lord Hope); see also para 60 (Lord Brown): 'it is the highly exceptional facts of the case… which in combination provide utterly compelling humanitarian grounds against removal.' Here, the case was deemed 'very exceptional' for the following reasons: 'This is particularly so when the effects on the child are take into account. His mother has cared from him since his birth. He has a settled and happy relationship with her in this country. Life with his mother is the only

humanitarian grounds against…removal' to be 'compelling'.[198] The European Court of Human Rights has suggested that the applicant must demonstrate an added 'measure of persecution, prosecution, deprivation of liberty or ill treatment' beyond a 'mere' violation of the right.[199] Again, the justification for this appears to be a policy one: the ECHR does not make Contracting States the 'indirect guarantors of freedom of worship for the rest of the world',[200] and thus a higher threshold of harm (beyond the absolute, non-derogable rights of Articles 2 and 3) must be met.[201]

Applying this to the climate change context, it may be possible to show—in exceptional circumstances—that a right is violated if the very essence of the right is destroyed or nullified. In time, this could well become the case. To what extent it needs to remain 'exceptional' to qualify is unclear: as the House of Lords observed in *N's case*, the sad irony was that the applicant's circumstances were no longer exceptional, even though their impacts were very severe.[202] As in that case, advances in medical treatment mean that people will rarely now be close to death at the time of the hearing, even though once treatment ceases they may rapidly become so. As Foster points out, 'requiring the person to be effectively dying seems to ignore the fact that "degrading treatment" does not need to amount to a loss of life—otherwise Art 3 would have no independent operation'.[203] Furthermore, it ignores the fact that the test relates to *foreseeability* of harm.[204]

V. The Role of the ICESCR

The UN Committee on Economic, Social and Cultural Rights has not yet considered whether any rights in the International Covenant on Economic, Social and Cultural Rights (ICESCR) contain a *non-refoulement* obligation. This may be facilitated by the envisaged creation of an individual and group communications procedure, similar to that of the Committee against Torture and the Human Rights Committee, but this is not yet operational.[205]

Foster, who has written extensively on socio-economic deprivation as a basis for international protection,[206] rejects the common assumption that the ICESCR is inapplicable in the asylum context. This assumption is based on the idea that the ICESCR rights are subject to progressive implementation according to the resources of indi-

family life he knows. Life with his father or any other member of his family in Lebanon, with whom he has never had any contact, would be totally alien to him', paras 17 and 18 (Lord Hope); see also para 47 (Baroness Hale).

198 ibid, para 17 (Lord Hope); see also para 60 (Lord Brown).

199 *Z and T v United Kingdom* (n 37) 7.

200 ibid.

201 As Baroness Hale acknowledged in the House of Lords: 'There clearly is some additional threshold test indicating the enormity of the violation to which the person is likely to be exposed if returned': *Razgar* (n 154) para 42. See also *F v United Kingdom* (n 140). Den Heijer argues, though, that the Art 3 threshold 'is a jurisprudential construction which can very well be extrapolated to other provisions if considered necessary': den Heijer (n 92) 294.

202 *N v Secretary of State for the Home Department* (n 108) paras 13 (Lord Nicholls), 50 (Lord Hope), 93 (Lord Brown).

203 Foster (n 16) 294.

204 ibid.

205 See Optional Protocol to the ICESCR (opened for signature 24 September 2009, not yet in force) UN Doc A/RES/63/117 (5 March 2009).

206 See, eg, Foster (n 135); Foster (n 16).

vidual State parties. Foster argues that this approach is flawed for two reasons. First, some ICESCR rights are immediately binding.[207] Secondly, even where they are not, Article 2 of that treaty imposes 'two key duties of an immediate nature': (a) the obligation to 'take steps' to realize rights, which includes a 'core obligation to ensure the satisfaction of, at the very least, minimum essential levels of each of the rights enunciated in the Covenant';[208] and (b) the duty to permit rights to be exercised without discrimination on specified grounds.[209] This means that 'a State party in which any significant number of individuals is deprived of essential foodstuffs, of essential primary healthcare, of basic shelter and housing, or of the most basic forms of education is, *prima facie*, failing to discharge its obligations under the Covenant'.[210]

Accordingly, she suggests that:

> Where the person fears a violation based on the receiving state's failure to respect rights (by withdrawing or preventing access to rights or actively denying them to a particular segment of the population) or failure to protect rights (by being unable or unwilling to protect against violation by non-state actors), the assessment is arguably no more complicated than where a civil and political right is at issue.[211]

The focus is again on the *active* withholding of a right, rather than a general lack of resources. Even if decision-makers were prepared to examine ICESCR rights directly in the *non-refoulement* context—and 'there is insufficient authority at present for holding states accountable for *refoulement* on the basis of the ICESCR'[212]—it seems that the element of differential, individual harm would pose an obstacle in climate change-related claims.[213]

VI. Analytical Assessment

In light of the analysis above, this section raises some thematic issues highlighting gaps in the existing normative framework, particularly in relation to climate change impacts resulting from slower-onset processes as opposed to rapid-onset disasters.

[207] Art 3 (equality between men and women); Art 7(a)(i) (equal pay); Art 8 (right to form trade unions and to strike); Art 10(3) (protection of children from exploitation); Art 13(2)(a) (free primary education); Art 13(3) (freedom of parents to choose the type of education for their children).

[208] Committee on Economic, Social and Cultural Rights, 'General Comment No 3: The Nature of States Parties' Obligations (Art 2(1))' UN Doc E/1991/23 (14 December 1990) para 10.

[209] Foster (n 16) 279.

[210] Committee on Economic, Social and Cultural Rights, 'General Comment No 3' (n 208) para 10. See respectively Committee on Economic, Social and Cultural Rights, 'General Comment 12: The Right to Adequate Food (Art 11)' UN Doc E/C.12/1999/5 (12 May 1999); 'General Comment No 13: The Right to Education (Art 13)' UN Doc E/C.12/1999/10 (8 December 1999); 'General Comment No 14: The Right to the Highest Attainable Standard of Health (Art 12)' UN Doc E/C.12/2000/4 (11 August 2000); 'General Comment No 15 (2002): The Right to Water (Arts 11 and 12)' UN Doc E/C.12/2002/11 (20 January 2003); 'General Comment No 18 (2005): The Right to Work (Art 18)' UN Doc E/C.12/GC/18 (6 February 2006).

[211] Foster (n 16) 281.

[212] ibid, 284 (fn omitted).

[213] On which, see section VI.B. below.

A. Timing

The protection possibilities discussed above may assist in cases of sudden movement in response to a disaster or emergency, but are a very uneasy fit for the slow-onset impacts of climate change. Even those who have called for a considerable widening of the Refugee Convention framework seem to assume that protection must be linked to 'flight', rather than to departure 'before the circumstances degenerate to life-threatening proportions'.[214]

Existing international refugee and complementary protection frameworks do not adequately address the time dimension of pre-emptive and staggered movement. Even though it is the severity of harm,[215] and not the timing of it, which determines a protection need, the two are necessarily interrelated. Since the impacts of slow-onset climate change processes may take some time before they amount to sufficiently serious harm, the timing of a protection claim is crucial.[216] The ability of existing legal mechanisms to respond to climate-related movement—through complementary protection in particular—would depend on the point in time at which protection were sought, based on the severity of the immediate impacts on return. These are matters that any new protection or migration agreement, whatever its form, would need to address.[217]

In a case concerning the right to life and the potential use of nuclear weapons, the UN Human Rights Committee held that for a person to be considered a 'victim' of a violation of the ICCPR,[218] and thus eligible to bring an individual complaint, 'he or she must show either that an act or an omission of a State party has already adversely affected his or her enjoyment of such right, *or that such an effect is imminent*'.[219] This was so despite the very strong statement by the Committee in General Comment 14 that 'the designing, testing, manufacture, possession and deployment of nuclear weapons are among the greatest threats to the right to life which confront mankind today' and that 'the production, testing, possession, deployment and use of nuclear weapons should be prohibited and recognized as crimes against humanity'.[220] The refusal of the Committee to find a claim admissible on the basis of a potential threat to life, despite its recognition of the very serious threat that nuclear weapons pose to it, does not augur well for a successful claim on the basis of potential, slow-onset climate change impacts, especially given

[214] Gaim Kibreab, 'Climate Change and Human Migration: A Tenuous Relationship?' (2009) 20 Fordham Environmental Law Review 357, 398 referring to Aristide R Zolberg and others, *Escape from Violence: Conflict and the Refugee Crisis in the Developing World* (Oxford University Press, 1989) 31, who advocate that the term 'refugee' should encompass any flight from life-threatening violence.

[215] In refugee law, to constitute 'persecution' acts must be sufficiently serious by their nature or repetition as to constitute a severe violation of basic human rights, or they must amount to an accumulation of measures of equivalent severity: Goodwin-Gill and McAdam (n 16) 91.

[216] This is apparent in some of the cases that have already been brought before the Refugee Review Tribunal in Australia and the Refugee Status Appeals Authority in New Zealand: see, eg, *0907346* [2009] RRTA 1168 (10 December 2009). For other cases, see Ch 2.

[217] See Ch 9.

[218] Optional Protocol to the ICCPR, UNGA res 2200A (XXI) of 16 December 1966 (entered into force 23 March 1976) 999 UNTS 302, Art 1.

[219] *Aalbersberg v The Netherlands* Communication No 1440/2005 (12 July 2006) UN Doc CCPR/C/87/D/1440/2005 (14 August 2006) para 6.3 (emphasis added).

[220] Human Rights Committee, 'General Comment No 14' (n 24).

the far less forceful comments of the Committee about the links between climate change and the right to life.[221]

The European Court of Human Rights has also insisted on a requirement of imminence and directness of a threat. In *Gounaridis v Greece*, a case relating to potential environmental damage arising from the construction of a new road, the European Commission of Human Rights held that the applicants needed to show in a defensible and detailed way that the probability of the potential harm would directly affect them such that it amounted to a violation, not just a general risk.[222] Likewise, in *Tauira v France*, the Commission found that there was insufficient evidence to show that French nuclear testing in the Pacific would *directly* affect the applicants' right to life, private life, and property.[223] In a case concerning the impact of a nuclear power plant on the right to life, the court said that the applicants were alleging 'not so much a specific and imminent danger in their personal regard as a general danger in relation to all nuclear power plants'.[224] The dissenting judges noted the irony that 'it is virtually impossible to prove imminent danger in the case of inherently dangerous installations'.[225] Elsewhere, the European Court of Human Rights noted that the applicants 'failed to show that the operation of the power station exposed them personally to a danger that was not only serious but also specific and, above all, imminent'.[226]

Significantly, though, in *EHP v Canada*,[227] the UN Human Rights Committee accepted in principle that dumping nuclear waste could violate the right to life—not only for present residents, but also for future generations. This is a rare example of future harm being recognized as a potential violation. However, the precedential value of the matter is limited, not only because of the non-binding nature of the Committee's views, but also because the matter was deemed inadmissible as local remedies had not been exhausted, and it preceded the more recent nuclear weapons cases in which the Committee has delimited the scope of a 'victim'.

The requirement of imminence has also been a hallmark of cases on Article 3 of the ECHR. The relevant question is 'whether the applicant's illness has reached such a critical stage (ie *he is dying*) that it would be inhuman treatment to deprive him of the care which he is currently receiving and send him home to an early death unless there is care available there to enable him to meet that fate with dignity'.[228] A key reason why almost all Article 3 medical cases have failed is that at the time of the application, the applicant's condition was not sufficiently 'advanced' or

[221] See Office of the High Commissioner for Human Rights (n 29) paras 21–4.

[222] *Gounaridis and others v Greece* App no 41207/98 (European Commission of Human Rights, 21 October 1998) in Loucaides (n 46) 169.

[223] *Tauira and others v France* App no 28204/95 (European Commission of Human Rights, 4 December 1995). See also *LCB v United Kingdom* App no 23413/94 (ECtHR, 9 June 1998). These cases are cited in Loucaides (n 46) 169.

[224] *Athanassoglou and others v Switzerland* App no 27644/95 (ECtHR, 6 April 2000).

[225] ibid (joint dissenting opinion).

[226] *Balmer-Schafroth v Switzerland* App no 22110/93 (ECtHR, 26 August 1999).

[227] *EHP v Canada* Communication No 67/1980 (27 October 1982) UN Doc CCPR/C/OP/1 (1984) 20.

[228] *N v Secretary of State for the Home Department* (n 108) para 69 (Baroness Hale) (emphasis added).

'terminal', and the cases were thus declared inadmissible.[229] It was the already terminal condition of the applicant in *D v United Kingdom* which made the case 'exceptional'.[230]

In *N's case*, the court stated:

> The rapidity of the deterioration which she would suffer and the extent to which she would be able to obtain access to medical treatment, support, and care, including help from relatives, *must involve a certain degree of speculation*, particularly in view of the *constantly evolving situation* as regards the treatment of HIV and AIDS worldwide.[231]

One could imagine a similar rationale being used in the climate change context, given uncertain timescales about when climate change impacts will be most severe, and the 'constantly evolving situation' with respect to human adaptation and resilience. This is where the standard of proof from refugee law may be instructive. A 'well-founded fear of persecution' can be less than a 50 per cent chance of harm.[232] Any refugee determination is necessarily an 'essay in hypothesis, an attempt to prophesy what might happen to the applicant in the future, if returned to his or her country of origin'.[233] A degree of speculation about future risk does not preclude a protection need from being recognized. Thus, the assessment of the intensity, severity, and nature of *future harm*, based on the individual's circumstances, is the key factor that leads to refugee status being granted. That assessment is not a prediction, but rather a *supposition*, based on the available evidence. Foster has argued that the approach in *N v United Kingdom* ignores the fact that the legal test relates to *foreseeability* of harm.[234]

The precautionary principle in international environmental law provides an interesting parallel here. As expressed in the 1992 Rio Declaration, 'full scientific certainty shall not be used as a reason for postponing cost-effective measures to prevent environmental degradation' where 'threats of serious or irreversible damage' exist.[235] Thus, although climate science indicates with great certainty that anthropogenic greenhouse gas emissions are changing the climate, uncertainties remain as to precisely when particular impacts will be felt, how severe they will be, and the extent to which natural and social systems will be able to adapt. Like

[229] See, eg, *Karara v Finland* App no 40900/98 (European Commission of Human Rights, 29 May 1998); *SCC v Sweden* App no 46553/99 (ECtHR, 15 February 2000); *Henao v The Netherlands* (n 95); *Ndangoya v Sweden* (n 95). In *N v United Kingdom*, the applicant was not expected to survive for more than one to two years if removed. Although Lord Nicholls pondered 'why is it unacceptable to expel a person whose illness is irreversible and whose death is near, but acceptable to expel a person whose illness is under control but whose death will occur once treatment ceases (as well may happen on deportation)?', he still refused the appeal: *N v Secretary of State for the Home Department* (n 108) para 13.

[230] *N v Secretary of State for the Home Department* (n 108) para 36 (Lord Hope).

[231] *N v United Kingdom* (n 16) para 50 (Lord Hope) (emphasis added).

[232] *INS v Cardoza-Fonseca*, 480 US 421 (1987) 431; *Chan v Minister for Immigration and Ethnic Affairs* (1989) 169 CLR 379. The UNHCR *Handbook* (n 147) para 42, states: 'In general, the applicant's fear should be considered well-founded if he can establish, to a reasonable degree, that his continued stay in his country of origin has become intolerable to him for the reasons stated in the definition, or would for the same reasons be intolerable if he returned there.'

[233] Goodwin-Gill and McAdam (n 16) 54.

[234] Foster (n 16) 294.

[235] Declaration of the United Nations Conference on the Human Environment (adopted 14 June 1992) UN Doc A/CONF.151/26, vol I (12 August 1992) Annex I ('Rio Declaration') Principle 15.

the 'well-founded fear' of persecution test in international refugee law, the precautionary principle operates, in effect, as an insurance policy against potential harm. Definitive proof of harm is not necessary: rather, the possibility of a risk that serious harm may ensue is sufficient to warrant protection (in the former case) and due diligence (in the latter). A precautionary approach appropriately acknowledges the high stakes involved in inaction.[236]

B. The individual nature of the harm

The traditional western approach of individualized decision-making about protection on technical legal grounds seems highly inappropriate to the situation of climate change-related displacement, in which the responsibility for displacement is highly diffuse (attributable to a large number of polluting States over many years, rather than to direct ill-treatment of a particular person by a certain government) and the numbers of those displaced may require group-based rather than individualized solutions. If and when States recognize that it is no longer possible for people to continue to live in their traditional homes, then it would be misplaced, in my view, to require individuals to reach a destination country and show that they meet a particular definition. Rather, as has been the case with schemes such as temporary protection in the EU, group determination in the Netherlands, Temporary Protected Status in the US, and ad hoc visa regimes in Australia responding to particular crises (East Timor, Kosovo, China),[237] it would seem more appropriate for States to designate certain countries as demonstrating sufficient, objective characteristics that 'justify' movement, thereby obviating the need for people wishing to leave them to show specific reasons why climate change is personally affecting them. Prima facie refugee status is similarly predicated on the fact that a person has fled a particular country (generally in conflict), and is deemed on that purely objective evidence to have a protection need. Any such schemes should not preclude the possibility of protection for people leaving countries that have not been designated, however.

Existing jurisprudence relating to socio-economic-based protection and environmental claims requires some *individual* factor that makes the situation intolerable for the particular applicant.[238] A considerable relaxation of this requirement would be needed if human rights law is to protect against return to climate change-related harms.

That said, special characteristics of the applicant may improve the possibility of protection being granted.[239] For example, the child-centric approach advocated by

[236] See Boyle (n 55) 16–17, referring to the environment/right to life jurisprudence of the European Court of Human Rights.

[237] For details on these schemes, see Ch 4.

[238] See section VI.A. on 'timing'.

[239] *N v United Kingdom* (n 16) para 29. See also *Adam* (n 117); *Fadele v United Kingdom* (1990) HRCD vol 1(1) 15, cited in Nicholas Blake and Raza Hussain, *Immigration, Asylum and Human Rights* (Oxford University Press, 2003) 100; *Taspinar v The Netherlands* (1984) 8 EHRR 47; *Mayeka v Belgium* (2008) 46 EHRR 23, cited in Foster (n 16) 306. See also Committee on the Rights of the Child, 'General Comment No 6' (n 25) para 27.

Baroness Hale in *EM (Lebanon)*,[240] based on the fact that the 'best interests of the child' must be a 'primary consideration' in any decision affecting a child, means that children may have a higher chance of being granted protection under Article 3. This is apparent in some of the cases already cited above.[241] As the Scottish Inner House (court of final instance) recently affirmed:

> Best interests [of the child] are not merely relevant. They are given a hierarchical importance. The decision maker is being told by Article 3 [of the CRC] that they are not just something to be taken into account but something to be afforded a grander status. They are to be regarded as a matter of importance. That having been said, the measure of that importance in the final balance will depend upon the facts and circumstances of the particular case.[242]

The Canadian courts have found that in the consideration of a humanitarian and compassionate claim, an immigration official erred by failing to give 'due consideration to the best interests of the applicants' three young Canadian-born children', including by failing 'to mention the serious issues facing children in Bangladesh, such as poor educational opportunities, diseases and natural disasters'.[243]

However, a word of caution is needed when it comes to special characteristics generally. There is an important difference between assessing risk on the basis of the applicant's particular circumstances, and requiring an applicant to show additional 'special distinguishing features'.[244] Just as in cases of generalized violence it is wrong in principle to limit the concept of 'persecution' to measures immediately identifiable as direct and individual,[245] so in the case of broad-ranging climate impacts, the relevant question is whether the applicant faces a real risk of serious harm if removed, not whether the applicant is at greater risk than others.[246]

Decision-makers have been very reluctant to hold that generalized violence in the country of origin is sufficient to preclude return. However, recent case law from the Court of Justice of the EU has acknowledged that under human rights law, 'the more the applicant is able to show that he is specifically affected by reason of factors particular to his personal circumstances, the lower the level of indiscriminate violence required for him to be eligible for subsidiary protection'.[247] Similarly, the European

[240] *EM* (n 154) para 28.

[241] For example *ZK (Afghanistan) v Secretary of State for the Home Department* [2010] All ER (D) 265 (Jun) paras 25–6.

[242] *HS v Secretary of State for the Home Department* [2010] CSIH 97 (Scottish Extra Division, Inner House, Court of Session) para 15.

[243] *Mazharul Hasan v Canada (Public Safety and Emergency Preparedness)* 2008 FC 1100, para 5.

[244] *Salah Sheekh v The Netherlands* App no 1948/04 (ECtHR, 11 January 2007) para 148. See also UNHCR, *Asylum in the European Union: A Study of the Implementation of the Qualification Directive* (UNHCR, 2007) 74.

[245] Goodwin-Gill and McAdam (n 16) 129, fn 364 citing *R v Secretary of State for the Home Department, ex parte Jeyakumaran* (No CO/290/84, QBD, 28 June 1985).

[246] See also Case C-465/07 *Elgafaji v Staatssecretaris van Justitie* [2009] OJ C90/4, para 39: 'the more the applicant is able to show that he is specifically affected by reason of factors particular to his personal circumstances, the lower the level of indiscriminate violence required for him to be eligible for subsidiary protection.' See also James C Hathaway, *The Law of Refugee Status* (Butterworths, 1991) 97.

[247] *Elgafaji* (n 246) para 39.

Court of Human Rights has observed in the context of Article 3 of the ECHR that a 'real risk' of inhuman or degrading treatment or punishment does not require an applicant to establish 'further special distinguishing features concerning him personally in order to show that he was, and continues to be, personally at risk', since to require this 'might render the protection offered by that provision illusory'.[248] Recalling the statements of the New Zealand and Australian decision-makers in Chapter 2 in relation to Pacific Islanders seeking protection on account of climate impacts, it is possible that a similar rationale could be brought to bear there.

C. The role of climate change in the legal analysis of the 'harm' feared

Finally, it is important to examine the relevance of 'climate change' in any legal analysis of harm. Kälin suggests that 'it is conceptually sounder to look at sudden-onset disasters as a cause of displacement, and not to limit the focus to those triggered by global warming'.[249] Similarly, in the slow-onset context, Kolmannskog and Trebbi argue that the focus should not be on why someone left their home initially, 'but rather whether the gradual degradation has reached a critical point where they cannot be expected to return now'.[250] Thus, in a complementary protection claim, the focus is the nature of potential *harm*, not its cause. In a human rights analysis, whether the source of that harm is attributable to climate change or other socio-economic or environmental pressures is immaterial (and misplaces the focus of the inquiry);[251] what matters is the harm likely to be faced by the individual if removed. This is what Kälin describes as the 'returnability test', which emphasizes the 'prognosis'—whether it is safe to return—rather than the underlying motivations for movement. Such a test would be based on the 'permissibility, feasibility (factual possibility) and reasonableness of return'.[252]

Thus, a decision-maker's task is to determine whether returning the particular individual to the conditions overall in the country of origin will amount to a breach of a protected right, *not* the precise cause of that harm.[253] Focusing on the latter may complicate and narrow climate change-related claims. By contrast, a human rights-based approach which documents all the rights at risk, based on general country conditions as well as the personal circumstances of the applicant, will better substantiate the protection claim.

[248] *Salah Sheekh* (n 244) para 148.

[249] Walter Kälin, 'Conceptualising Climate-Induced Displacement' in Jane McAdam (ed), *Climate Change and Displacement: Multidisciplinary Perspectives* (Hart Publishing, 2010) 85.

[250] Vikram Kolmannskog and Lisetta Trebbi, 'Climate Change, Displacement and Protection: A Multi-Track Approach' (2010) 92 International Review of the Red Cross 713.

[251] An exception would be if the harm could be directly linked back to the action or negligence of the home State, in which case it could be a juridically relevant fact. See, eg, *Budajeva v Russia* (n 49).

[252] Kälin (n 249) 98.

[253] Although note recent UK authority which suggests a varying threshold in Art 3 removal cases 'dependent upon the responsibility of the receiving state for the circumstances complained of': *RN* (n 131) para 254 citing an earlier decision of the AIT in *HS* (n 128) quoted in Foster (n 16) 300.

VII. Do States Have Obligations Towards the Displaced under International Environmental Law?

Although there may be strong ethical claims, which suggest that a focus on climate change is warranted,[254] it presently adds little to a protection claim. The final section of this chapter explains the reasons for this conclusion by examining State responsibility for climate change under international environmental law.

Whether States are *legally* responsible for climate change is 'one of the most complicated and debated issues', which needs to be examined within the broader context of State responsibility for environmental damage.[255]

Although it has been argued that States' obligations under the Kyoto Protocol are of an *erga omnes* character—that is, they are obligations owed to the whole international community which can be invoked by one State on behalf of all—the orthodox view is that international environmental law obligations have not yet achieved this status.[256] We therefore need to turn to consider the specific obligations which States have concluded under treaty law, and principles of customary international law on State responsibility, to determine the extent of liability.

In international environmental law, the global atmosphere and climate are considered a 'common resource' of vital interest to humanity.[257] Under treaty law, States have obligations to implement programmes for mitigating greenhouse gas (GHGs) emissions,[258] to prevent, reduce, and control pollution of the atmosphere and the marine environment,[259] and to conserve biodiversity,[260] among many others. The latter are relevant where displacement is due to a loss of livelihood or resources resulting from disappearing plant and animal species. The regulation of climate change sits within the context of the protection of the atmosphere. However, whereas there are mechanisms in international and regional human

[254] See, eg, Peter Penz, 'International Ethical Responsibilities to "Climate Change Refugees"' in McAdam (n 249); Simon Caney, 'Climate Change, Human Rights and Moral Thresholds' in Humphreys (n 2); Denis G Arnold (ed), *The Ethics of Global Climate Change* (Cambridge University Press, 2011).

[255] Malgosia Fitzmaurice, 'Responsibility and Climate Change' (2010) 53 German Yearbook of International Law 89, 89.

[256] See ibid, 97, referring to the views of Joy-Dee Davis, 'State Responsibility for Global Climate Change: The Case of the Maldives' (Master of Arts in Law and Diplomacy Thesis, The Fletcher School, Tufts University, 2005) 37–8.

[257] See Recommendations of the International Meeting of Legal and Policy Experts, Ottawa (1989) 19 Environmental Policy and Law 78.

[258] For example, Stockholm Declaration, Principle 21; Convention on Long-Range Transboundary Air Pollution (adopted 13 November 1979, entered into force 16 March 1983) 1302 UNTS 217; Vienna Convention for the Protection of the Ozone Layer (adopted 22 March 1985, entered into force 22 September 1988) 1513 UNTS 293; UN Framework Convention on Climate Change (adopted 9 May 1992, entered into force 21 March 1993) 1771 UNTS 107 ('UNFCCC'); Kyoto Protocol to the Framework Convention on Climate Change (adopted 11 December 1997, entered into force 16 February 2005) 2303 UNTS 148 ('Kyoto Protocol').

[259] For example, Stockholm Declaration, Principle 21; Rio Declaration, Principle 2; UNFCCC; Kyoto Protocol; United Nations Law of the Sea Convention (adopted 10 December 1982, entered into force 16 November 1994) 1833 UNTS 396, Arts 192–95.

[260] For example, Convention on Biological Diversity (adopted 5 June 1992, entered into force 29 December 1993) 1760 UNTS 79.

rights law for individual complaints to be made against State conduct, such avenues are far less developed in the international environmental law field, where the plane of legal responsibility is primarily between States alone.

In recognizing that climate change is of 'common concern' to the international community, the UN has effectively acknowledged the artificiality of spatial boundaries in this context.[261] However, there seems to be little political appetite among developed States, at least,[262] to take this to the (perhaps logical) conclusion that the human rights impacts of climate change must therefore also be of international concern, demanding responses that go beyond the traditional State territory/jurisdiction link.

Under customary international law, every State has an obligation not to knowingly allow its territory to be used for acts that are contrary to the rights of other States.[263] This principle is understood in the field of environmental law as requiring States to refrain from using their territory in a way that causes environmental harm beyond their borders. The customary law principle of responsibility for transboundary environmental harm is well established:

> no State has the right to use or permit the use of its territory in such a manner as to cause injury by fumes in or to the territory of another or the properties or persons therein, when the case is of serious consequence and the injury is established by clear and convincing evidence.[264]

The concept of sustainable development adds additional specificity to this concept by limiting the manner in which States may seek to realize the (nascent) 'right to development'.[265] Principle 3 of the Rio Declaration states that it 'must be fulfilled so as to equitably meet developmental and environmental needs of future generations', implying limits on emissions that may jeopardize the ability of future generations to live and develop in a healthy environment. Principle 2 requires States to ensure that their exploitation of resources does 'not cause damage to the environment of other States or of areas beyond the limits of national jurisdiction', again suggesting limitations on carbon emissions because of their potential damage to vulnerable populations.[266] Although the Rio Declaration is not formally binding on States, it represents a set of principles that States have agreed should guide their conduct.

[261] Birnie, Boyle, and Redgwell (n 44) 130.

[262] See Marc Limon, 'Human Rights Obligations and Accountability in the Face of Climate Change' (2010) 38 Georgia Journal of International and Comparative Law 543.

[263] *Corfu Channel case (United Kingdom v Albania)* [1949] ICJ Rep 4, 22.

[264] *Trail Smelter Arbitration (United States v Canada)* (1938–41) 3 RIAA 1905, 1965; see also Stockholm Declaration, Principle 21; Rio Declaration, Principle 2; see also ILC, 'Prevention of Transboundary Harm from Hazardous Activities' in 'Report of the International Law Commission on the Work of its Fifty-Third Session', UN Doc A/56/10 (2001); *Advisory Opinion on the Legality of the Threat or Use of Nuclear Weapons* [1996] ICJ Rep 226, 241–2.

[265] See *Centre for Minority Rights Development (Kenya)* (n 58) paras 228, 277ff, where the court found a violation of the right to development.

[266] Rio Declaration, Principle 2. See also Ilona Millar, 'There's No Place Like Home: Human Displacement and Climate Change' (2007) 14 Australian International Law Journal 71, 86ff.

A. Liability for climate-related damage

The legal analysis in this book has so far concentrated on whether, and how, international law can assist those displaced by climate change-related factors. An underlying question is whether States bear legal responsibility for 'causing' such harm and have duties to remedy it.

In legal terms, it is very difficult to establish causation for climate change. Even though the science can explain that certain environmental impacts cannot be attributed to any other cause—in other words, there is no plausible explanation for certain patterns other than changes to the climate—the law demands a more precise test. To establish legal liability, it is not enough simply to show that 'industrialized countries' have largely caused climate change through their historical and present carbon emissions.

The primary problem is that to establish responsibility for a wrong, it has to be shown that the alleged conduct was illegal or otherwise wrongful at the time it was carried out. This immediately presents difficulties, because the law of most States, and international law itself, did not contain any rules prohibiting certain levels of carbon emissions until very recently. Consequently, it is very difficult to establish legal liability for historical emissions which have predominantly contributed to anthropogenic climate change, because such emissions by companies or governments during those periods were not unlawful. The case for liability only becomes stronger following the adoption of emissions reductions targets since the 1990s, emissions in excess of which might well entail legal liability.[267]

Even if an underlying wrong can be identified, there are further legal difficulties in suing and the narrow focus of most litigation is not really an effective way of addressing the issue. Ordinarily, applicants must show that particular acts by particular actors (such as companies) at particular times have led to particular consequences for particular groups. For example, using a domestic law framework of negligence law, experts

[267] Tol and Verheyen suggest that, at the very least, States were aware of the impacts of unchecked greenhouse gas emissions by 1992, when the UNFCCC was adopted: see Richard SJ Tol and Roda Verheyen, 'State Responsibility and Compensation for Climate Change Damages: A Legal and Economic Assessment' (2004) 32 Energy Policy 1109, 1118. Other relevant literature they refer to includes: Claudia Kemfert and Richard SJ Tol, 'Equity, International Trade and Climate Policy' (2002) 2 International Environmental Agreements 23; Richard SJ Tol, 'Welfare Specification and Optimal Control of Climate Change: An Application of FUND' (2002) 24 Energy Economics 367; Theodore Panayotou, Jeffrey D Sachs, and Alix P Zwane, 'Compensation for "Meaningful Participation" in Climate Change Control: A Modest Proposal and Empirical Analysis' (2002) 43 Journal of Environmental Economics and Management 437; James Cameron and Durwood Zaelke, 'Global Warming and Climate Change: An Overview of the International Legal Process' (1990) 5 American University Journal of International Law & Policy 249; Roda Verheyen, 'The Legal Framework of Adaptation and Adaptive Capacity' in Richard JT Klein, Saleemal Huq, and Joel B Smith (eds), *Climate Change, Adaptive Capacity and Development* (Imperial College Press, 2003). See further Roda Verheyen, *Climate Change Damage and International Law: Prevention Duties and State Responsibility* (Martinus Nijhoff, 2005); Rebecca M Bratspies and Russell A Miller (eds), *Transboundary Harm in International Law: Lessons from the Trail Smelter Arbitration* (Cambridge University Press, 2006) Chs 8–9; Christina Voigt, 'State Responsibility for Climate Change Damages' (2008) 77 Nordic Journal of International Law 1.

have suggested that anyone seeking to sue for damage arising from climate change 'would face formidable legal, logistical, evidentiary and financial obstacles'.[268]

First, there are significant difficulties in establishing that the resulting harm was a reasonably foreseeable consequence of a particular defendant's activities. The global nature of climate change, and the fact that alleged harms may be 'geographically and temporally divorced from the adverse consequences', mean that it will be difficult for an applicant to show a sufficiently direct or specific relationship such as to establish a duty of care.[269] To put it starkly, does a polluting power station operator in Australia owe a duty of care to the people of Tuvalu or Kiribati, in the event that its emissions affect livelihoods there? The law does not usually stretch so far. There is a question whether courts would be prepared to recognize such a duty as a matter of legal policy, given its potentially far-reaching consequences. Doing so would amount to a 'revolutionary rather than an evolutionary development of the law'.[270]

Secondly, even if a duty were recognized, it would be difficult to establish that it had been breached. This is because a court would have to decide whether the particular harm was reasonably foreseeable and the scientific risks of carbon emissions have only become well known relatively recently. That said, since knowledge about the impacts of climate change has developed over time, it is arguable that the requisite standard of care should increase over time as well.[271]

Thirdly, establishing causation is likely to 'pose the greatest obstacle'[272] because of the difficulty in establishing an evidentiary link between the act/omission and the harm, and attributing responsibility according to legal principles. Eliciting expert scientific evidence to prove this is likely to be fraught with difficulty. This is because 'the impacts involve intensification of existing climatic phenomena, such as more frequent storms, rather than the creation of unique "signature diseases" such as asbestosis, which is caused only by exposure to asbestos'. Further, 'because climate is affected by several factors interacting in complex ways, it is difficult for scientists to tease out what percentage of any climate change is affected by GHGs, and it is even more difficult to determine what percentage is affected by a specific polluter or group of polluters'.[273]

Similar problems arise when approaching the issue from a wider international law perspective.[274] In most scenarios, the State itself is not the primary emitter. Rather,

[268] Peter Cashman and Ross Abbs, 'Liability in Tort for Damage arising from Human-Induced Climate Change' in Rosemary Lyster (ed), *In the Wilds of Climate Law* (Australian Academic Press, 2010) 245.

[269] ibid, 245 citing Nicola Durrant, 'Tortious Liability for Greenhouse Gas Emissions? Climate Change, Causation and Public Policy Consideration' (2007) 7 Queensland University of Technology Law and Justice Journal 404, 406. See also *Agar v Hyde* (2000) 201 CLR 552, paras 66–7 (Gaudron, Gummow, and Hayne JJ).

[270] Cashman and Abbs (n 268) 245.

[271] ibid, 248 referring to Durrant (n 269) 411.

[272] ibid, 249 referring to Joseph Smith and David Shearman, *Climate Change Litigation: Analysis of the Law, Scientific Evidence and Impacts on the Environment, Health and Property* (Presidian Legal Publications, 2006) 107.

[273] Cashman and Abbs (n 268) 250 referring to Bradford Mank, 'Civil Remedies' in Michael B Gerrard (ed), *Global Climate Change and US Law* (American Bar Association, 2008) 201.

[274] On the question of causation and State responsibility under international law, see Verheyen, *Climate Change Damage and International Law* (n 267) Ch 5.

its responsibility will arise via its due diligence function—that is, the regulation and control of carbon emissions by companies within its territory or jurisdiction.[275] It is not only necessary to establish that a particular State is responsible for harm and to disentangle it from the responsibility of other emitting States (including as regards historical and contemporary emissions), but also to identify what specific legal rule has been violated. As noted above, international law did not contain any rules prohibiting carbon emissions until very recently. Jurisprudence of the ICJ suggests that States have a duty to prevent and control foreseeable risks where there is an actual and serious harm that is likely to recur, or where there is a known risk to other States.[276] But even if foreseeability is established, it is unclear whether the State must have foreseen the general, or the precise, nature of the damage that could be caused.[277] For example, is it sufficient to foresee that climate change may damage the environment, or must it be foreseen that it will require particular communities to move from particular areas? Tol and Verheyen suggest that the severity of the risk should determine the threshold for establishing a duty of care: the more obvious the risk, the more States must demonstrate that they have taken all necessary measures to prevent it from materializing.[278]

Furthermore, although international law recognizes that more than one State may be responsible for the same wrongful act,[279] it is unclear whether one State would have to provide reparation for the whole injury, or only the part it caused.[280] In any case, in the climate change context, it is doubtful that emissions caused by a number of States would count as 'the same internationally wrongful act' (and thus Article 47 of the Articles on State Responsibility would not apply).[281] As the Commentary to those Articles notes, situations can arise:

> where several States by separate internationally wrongful conduct have contributed to cause the same damage. For example, several States might contribute to polluting a river by

[275] See n 264 above. For analysis, see Fitzmaurice (n 255) 121–8.

[276] See *Trail Smelter Arbitration* (n 264); *Corfu Channel case* (n 263) 18–22 and analysis in Birnie, Boyle, and Redgwell (n 44) 153. The ILC Articles on Transboundary Harm (n 264) Art 2 indicate that a risk can include 'a low probability of causing disastrous harm' and a 'high probability of causing significant harm'. Fitzmaurice (n 255) 130 argues that the Articles on Transboundary Harm 'are of limited use for climate change.'

[277] Tol and Verheyen (n 267) 1117.

[278] ibid, 1118.

[279] ILC Articles on Responsibility of States for Internationally Wrongful Acts in 'Report of the International Law Commission on the Work of its Fifty-Third Session', UN Doc A/56/10 (2001) 313, Art 47: 'Where several States are responsible for the same internationally wrongful act, the responsibility of each State may be invoked in relation to that act.'

[280] This issue was not settled in *Case concerning Phosphate Lands in Nauru (Nauru v Australia)* (Preliminary Objections of the Government of Australia) [1992] ICJ Rep 240, 262, where the ICJ emphasized that its decision on jurisdiction was separate from the question whether Australia, if found responsible, would be obliged to provide reparation for all or only part of the damage Nauru had suffered, given the role of the British and New Zealand governments, which were not parties to the matter. The case was ultimately withdrawn by agreement, and Australia agreed to pay by instalments the full amount of Nauru's claim. The two other governments subsequently agreed to contribute to the payments: ILC Articles on Responsibility of States for Internationally Wrongful Acts (n 279) 315, Commentary to Art 47, para 4. Art 47(2)(b) 'recognizes that where there is more than one responsible State in respect of the same injury, questions of contribution may arise between them.... On the other hand, there may be cases where recourse by one responsible State against another should not be allowed': ibid, 318, Commentary to Art 47, para 10.

[281] Fitzmaurice (n 255) 129.

the separate discharge of pollutants.... In such cases, the responsibility of each participating State is determined individually, on the basis of its own conduct and by reference to its own international obligations.[282]

Accordingly, as Kälin and Schrepfer observe, this means that a State could only be held responsible for harm caused by its own emissions, which would make it necessary for the injured State to establish causation between the specific harm suffered and the conduct of the emitting State.[283] As the preceding discussion has highlighted, this would be very difficult to do.

Further, certain areas of international law contain other limitations. Under international human rights and refugee law, for example, countries generally only owe obligations to people already in their territory or jurisdiction. This means that while the US, for instance, might be considered responsible for its own emissions, which breach the human rights of people in its territory, it would be far more difficult to demonstrate that these same emissions, drifting ethereally and randomly as carbon particles around the world, constitute violations of the rights of distant populations.

This was at issue in a petition lodged by the Inuit indigenous peoples of the Arctic regions of the US and Canada before the Inter-American Commission on Human Rights. They argued that the effects of climate change on their environment, allegedly caused by the acts and omissions of the US (in particular, its failure to ratify the Kyoto Protocol), violated certain fundamental rights, including the right to life, health, physical integrity, security, a means of subsistence, property, inviolability of the home, and the benefits of culture.[284]

In particular, the Inuit explained that animals on which they rely for food were disappearing, damaging their subsistence harvest and health. Thawing permafrost was causing landslides and complicating food storage. Travel was becoming increasingly dangerous and difficult due to unpredictable weather, with the warmer climate making traditional knowledge about the safety of sea ice unreliable (leading to increased drownings each year).

Although the petition was ultimately dismissed, and instead evidence was presented by way of a general hearing on human rights issues rather than as contentious litigation, the case raised yet unanswered questions about whether a State's failure to ratify key international instruments aiming to combat climate change could *inter alia* constitute a breach of human rights law, the principle of good faith in upholding international obligations, and the duty to cooperate in international environmental law.[285]

Litigation can serve a political function, in addition to asserting legal rights. It may help to raise awareness of the impacts of climate change on the on-going

[282] ILC Articles on Responsibility of States for Internationally Wrongful Acts (n 279) 317–18, Commentary to Art 47, para 8.

[283] Walter Kälin and Nina Schrepfer, 'Protecting People Crossing Borders in the Context of Climate Change: Normative Gaps and Possible Approaches' (Study for the Swiss Ministry of Foreign Affairs, April 2011) 7, forthcoming as a paper in the UNHCR Legal and Protection Policy Research Series.

[284] Petition to the Inter American Commission on Human Rights (n 28); Earthjustice, 'Global Warming and Human Rights: Testimony of Martin Wagner before the Inter-American Commission on Human Rights' (1 March 2007) 3.

[285] See, eg, Stockholm Declaration, Principle 24; Rio Declaration, Principles 7 and 27.

viability of settlement in particular areas, and the attendant human rights consequences.[286] In this way, it may also precipitate legal and/or policy change. For example, the 400-inhabitant Eskimo village of Kivalina, on a small island north of the Arctic Circle, is seeking up to $400 million compensation to cover relocation to the mainland. Even if the litigation does not proceed or is unsuccessful, it could drive industry to negotiate, and, as with tobacco litigation, lead to settlements and regulation.[287] However, litigation is a very expensive and slow process, and using the courts to seek general legal remedies for climate-related displacement is unlikely to yield the desired results.[288] I agree with Kälin and Schrepfer's conclusion that 'a responsibility based approach to population movements in the context of climate change is not fruitful',[289] largely because it can focus attention on very difficult-to-establish matters of causation and divert attention away from the protection needs of the displaced.[290]

B. Conclusion

Addressing climate-related displacement from an international environmental law perspective has a number of limitations. First, the plane of legal responsibility is primarily between States. Individuals therefore have a lesser legal capacity than they

[286] See, eg, the Tuvaluan threat to sue Australia and the US in the ICJ for damages: Robin Pomeroy, 'US Faces Legal Battles as Climate Bogeyman' (*World Environment News*, 30 August 2002) <http://www.planetark.org/dailynewsstory.cfm/newsid/17512/story.htm> accessed 16 March 2010; Reuters, 'Tuvalu Seeks Help in US Global Warming Lawsuit' (*World Environment News*, 30 August 2002) <http://www.planetark.com/dailynewsstory.cfm/newsid/17514/story.htm> accessed 16 March 2010; Rebecca E Jacobs, 'Treading Deep Waters: Substantive Law Issues in Tuvalu's Threat to Sue the United States in the International Court of Justice' (2005) 14 Pacific Rim Law and Policy Journal 103. See also Palau's reported plans to seek an Advisory Opinion from the ICJ as to whether States have a legal obligation to ensure that their greenhouse gas emissions do not harm other States: 'Palau Seeks UN World Court Opinion on Damage Caused by Greenhouse Gases' (*UN News Service*, 22 September 2011) <http://www.un.org/apps/news/printnews.asp?nid=39710> accessed 5 October 2011.

[287] See John Schwartz, 'Courts as Battlefields in Climate Fights', *the New York Times* (New York, 27 January 2010) A1.

[288] Arguing against human rights litigation in the climate change context, see Eric A Posner, 'Climate Change and International Human Rights Litigation: A Critical Appraisal' (2007) 155 University of Pennsylvania Law Review 1925. For literature generally, see further, Owen Cordes-Holland, 'The Sinking of the Strait: The Implications of Climate Change for Torres Strait Islanders' Human Rights Protected by the *ICCPR*' (2008) 9 Melbourne Journal of International Law 405; RoseMary Reed, 'Rising Seas and Disappearing Islands: Can Island Inhabitants Seek Redress under the Alien Tort Claims Act?' (2002) 11 Pacific Rim Law and Policy Journal 399; Donna Green and Kirsty Ruddock, 'Could Litigation Help Torres Strait Islanders Deal with Climate Impacts?' (2009) 9 Sustainable Development Law and Policy 23; Natalie L Bridgeman, 'Human Rights Litigation under the ATCA as a Proxy for Environmental Claims' (2003) 6 Yale Human Rights and Development Law Journal 1; Dinah Shelton, 'The Environmental Jurisprudence of International Human Rights Tribunals' in Romina Picolotti and Jorge Daniel Taillant (eds), *Linking Human Rights and the Environment* (University of Arizona Press, 2003); Felicity Millner and Kirsty Ruddock, 'Climate Litigation: Lessons Learned and Future Opportunities' (2011) 36 Alternative Law Journal 27; Tim Stephens, *International Courts and Environmental Protection* (Cambridge University Press, 2009) Ch 10.

[289] Kälin and Schrepfer (n 283) 1; see also 64. They suggest that '[e]lements of responsibility may be accounted for by other means, e.g. through fair burden-sharing agreements or respective contributions to the adaptation fund established under the UNFCCC': 8.

[290] See further, Ch 7.

enjoy under human rights law. Secondly, as discussed above, there is considerable difficulty in quantifying the harm caused by the carbon emissions of any particular State, and identifying causation between emissions and detrimental effects, when all States have contributed to emissions at some point. Thirdly, there remain difficulties in establishing the accountability of corporations for carbon emissions in a legal system in which States remain the primary duty-bearers.

Creative arguments have, however, been made on the causation issue. Some scholars have attempted to apportion responsibility for climate-related displacement according to the State's share of responsibility for carbon emissions. Byravan and Chella Rajan, for instance, have suggested that people living in areas that are likely to be rendered uninhabitable on account of climate change impacts should have the early option of migrating to other countries, in numbers roughly proportionate to the host countries' cumulative greenhouse gas emissions.[291] According to their calculations, this would mean that, per year, the US (as the highest emitter) would take in 866,000 people, while Italy (as the 10th highest emitter) would take 8,600. These figures are comparable to the actual numbers of immigrants already absorbed by those States each year. Their argument is reminiscent of Peter Schuck's proposed quota system for protecting refugees, which went even further by suggesting that States could 'trade their quotas by paying others to fulfil their obligations'.[292] However, as Kälin and Schrepfer rightly argue, 'directly linking obligations to admit, protect and assist persons moving across borders to the issue of legal responsibility under international law based on the "polluter pays" principle would be detrimental to the interests of both the countries of origins and possible countries of admission and refuge'.[293] It treats humans like carbon particles that can be traded, which is one of the main reasons why States like the Federated States of Micronesia resisted the inclusion of 'relocation' in the United Nations Framework Convention on Climate Change (UNFCCC). Furthermore, it hardly accounts for the complexity of contributory causes in any given displacement situation, and overlooks the role of intervening factors and other human actors in determining how the effects of climate change manifest themselves in a particular place.[294] Such a blunt approach is also unlikely to garner genuine political support.

[291] Sujatha Byravan and Sudir Chella Rajan, 'Providing New Homes for Climate Change Exiles' (2006) 6 Climate Policy 247, 249.

[292] Peter H Schuck, 'Refugee Burden-Sharing: A Modest Proposal' (1997) 22 Yale Journal of International Law 243, 248. See also James C Hathaway and R Alexander Neve, 'Making International Refugee Law Relevant Again: A Proposal for Collectivized and Solution-Oriented Protection' (1997) 10 Harvard Human Rights Journal 115.

[293] Kälin and Schrepfer (n 283) 64 (emphasis removed). In the context of refugee burden-sharing, UNHCR has observed that 'simple readimission or redistribution of persons between States is often not enough': UNHCR, 'International Cooperation to Share Burden and Responsibilities', Discussion Paper (Expert Meeting in Amman, Jordan, 27–28 June 2011) para 24.

[294] It 'is not a promising approach and does not provide a solution to questions arising in this context': Kälin and Schrepfer (n 283) 8.

VIII. Conclusion

This chapter has demonstrated the difficulties—at this point in time—of successfully demonstrating that climate change impacts on vulnerable communities substantiate an international protection claim under human rights law. While it seems that Articles 2 and 3 of the ECHR, and Articles 6 and 7 of the ICCPR, remain the strongest sources of protection for climate change-related claims, there will likely be difficulties in establishing the requisite degree of imminence of harm in many cases. It will be important to point to the cumulative impact of deprivations 'in the light of all the circumstances'.[295] Just as 'persecution' in refugee law may be demonstrated either by a single severe act or by a series of less severe acts which, by virtue of their nature or repetition, cumulatively reach that threshold, breaches of a number of individual human rights might collectively found a protection claim.

While protection gaps exist both for people fleeing sudden-onset disasters and those seeking to escape the effects of slower-onset impacts of climate change, there is a more obvious lacuna for the latter group. In part, this is because existing protection principles are more analogous to situations of sudden flight than pre-emptive movement. Kälin and Schrepfer argue that the distinction comes down to whether movement is understood as 'forced' or 'voluntary',[296] which in turn influences the degree to which those moving will be regarded as being in need of international protection (an obligatory response) as opposed to 'economic migrants' (a discretionary response). Whether this distinction provides a sound basis for determining protection needs is a question underlying the remaining chapters, since it does not properly capture 'the reality of migration as an adaptation strategy which lies somewhere in between these two categories'.[297]

[295] *BB v France* (n 95) para 53.

[296] On which, see Chs 2 and 8.

[297] Kälin and Schrepfer (n 283) 35. See also discussion in Ch 9 about the applicability of principles relating to distress.

4

State Practice on Protection from Disasters and Related Harms

I. Introduction

Whereas the last chapter examined the extent to which human rights-based jurisprudence may preclude the return of people to climate-related harms, the present chapter examines particular legislative and ad hoc schemes developed by States which may assist people fleeing the impacts of disasters and other serious harm.[1] To date, most responses to cross-border climate change-related or environmental displacement have been domestic ones rather than international agreements.[2] This chapter accordingly catalogues temporary protection responses, asylum-type mechanisms, and ad hoc humanitarian schemes (group and individual) to elucidate the potential scope of existing domestic and regional frameworks in responding to climate change-related movement. Overall, it shows how varied and unpredictable these mechanisms are.

[1] As UNHCR's Assistant High Commissioner (Protection) notes: '[A]sylum is also one of the responses suitable to situations which do not fit the classical refugee paradigm, and which involve a temporary protection need. This has been partially recognised, particularly in Europe, through subsidiary protection arrangements, but also through discretionary provisions of various sorts in the immigration laws of a number of countries outside this region': Erika Feller (Assistant High Commissioner (Protection)), 'The Refugee Convention at 60: Still Fit for Its Purpose?' (Workshop on Refugees and the Refugee Convention 60 Years On: Protection and Identity, Prato, 2 May 2011) 8.

[2] An exception is the 2009 Kampala Convention adopted by the African Union, which includes an obligation to 'take measures to protect and assist persons who have been internally displaced due to natural or human made disasters, including climate change': African Union Convention for the Protection and Assistance of Internally Displaced Persons in Africa (adopted 22 October 2009, not yet in force) ('Kampala Convention') Art 5(4). While this is the first treaty to expressly recognize climate change as a form of 'natural or human made' disaster, it essentially just elaborates on the description of an 'internally displaced person' in the Guiding Principles on Internal Displacement, UN Doc E/CN.4/1998/53/Add.2 (11 February 1998) which refers to 'natural or human-made disasters', and which only binds States with respect to people moving within their own borders, rather than across international frontiers.

II. Legislative Protection Responses

Legislative responses include temporary humanitarian assistance, through schemes such as Temporary Protected Status in the United States (US); (potentially) temporary protection in the European Union (EU); and longer-lasting refugee-like protection in countries such as Sweden and Finland. These are examined in turn below.

A. Temporary protection

A number of countries have mechanisms for providing temporary protection to people displaced by sudden disasters. The scope of the protection is set out in law, but often, as in the case of the EU and the US, an executive decision is required before the protection can be accessed.[3]

(1) United States

Temporary Protected Status (TPS) is a discretionary status in the US designed to provide safe haven for people who are reluctant to return to potentially dangerous situations in their home country. Protection is not automatic: the Attorney General must first 'designate' a country before its nationals are eligible.

A country may be 'designated' where there is an on-going armed conflict threatening people's personal safety, or where:

- (i) there has been an earthquake, flood, drought, epidemic, or other environmental disaster in the state resulting in a substantial, but temporary, disruption of living conditions in the area affected,
- (ii) the foreign state is unable, temporarily, to handle adequately the return to the state of aliens who are nationals of the state, and
- (iii) the foreign state officially has requested designation under this subparagraph.[4]

TPS is thus a blanket form of relief granted on the basis of objective country of origin conditions, rather than circumstances particular to the individual.[5] Crucially, it only benefits people already in the US at the time of a disaster, *and* whose government requests assistance under this mechanism. A grant of TPS enables beneficiaries to work and precludes deportation for the period of the designation.[6] TPS

[3] By the Council of the European Union and the Attorney General respectively.

[4] Immigration and Nationality Act, INA § 244(b), 8 USC § 1254a(b).

[5] See generally, Ruth E Wasem and Karma Ester, 'Temporary Protected Status: Current Immigration Policy and Issues' (Congressional Research Service, 9 September 2010).

[6] US Citizenship and Immigration Services, 'Temporary Protected Status—Haiti' (last updated 1 February 2010) <http://www.uscis.gov/portal/site/uscis/menuitem.eb1d4c2a3e5b9ac89243c6a7543f6d1a/?vgnextchannel=e54e60f64f336210VgnVCM100000082ca60aRCRD&vgnextoid=e54e60f64f336210VgnVCM100000082ca60aRCRD> accessed 3 January 2011. It has been reported that France, Canada, and the Dominican Republic also eased their immigration rules in light of the earthquake: see Michael B Farrell, 'TPS: Haiti's Illegal Immigrants Given Temporary Protection in

can be granted for periods between six and 18 months, and it can be extended if country conditions do not change. However, as its name implies, it is a temporary status, and people on TPS are not eligible to become legal permanent residents (LPRs) in the US without a special act of Congress.[7]

In January 2010, as a result of the earthquake in Haiti, the Department of Homeland Security determined an 18-month designation for Haitians who had continuously resided in the US since 12 January 2010. Haitians in the country unlawfully, as well as those living there on another visa, could apply for TPS. Following Hurricane Mitch in 1998, the Attorney General indicated that the deportation of people from El Salvador, Guatemala, Honduras, and Nicaragua would be temporarily suspended, and TPS was granted a month later to people from Honduras and Nicaragua on the grounds of the extraordinary degree of displacement and damage there.[8] In 2001, TPS was granted to people from El Salvador on account of two earthquakes there.[9] There have been other calls for TPS to be granted on account of natural disasters in Peru, Pakistan, Sri Lanka, India, Indonesia, Thailand, Somalia, Myanmar, Malaysia, the Maldives, Tanzania, Seychelles, Bangladesh, and Kenya, but the US government did not take a formal position on those countries.[10]

There is, accordingly, nothing in principle which would prevent TPS from being granted to people affected by a climate change-related disaster. However, it is unlikely to assist people facing slow-onset impacts of climate change, given the time they take to manifest and their 'creeping' effect, rather than their sudden nature. Furthermore, given that TPS is only available to designated nationals already in the US at the time of the disaster, not to those who flee *after* an event, it may have little relevance to citizens of many affected countries, such as Kiribati, Tuvalu, and Bangladesh. The US is not a common destination country for these communities and many of the worst affected would lack the means to travel there in the first place. Nonetheless, opponents of TPS see it as an immigration amnesty for unauthorized migrants already in the US and as a magnet for further unauthorized movement.[11]

US, *The Christian Science Monitor* (15 January 2010) <http://www.csmonitor.com/USA/2010/0115/TPS-Haiti-s-illegal-immigrants-given-temporary-protection-in-US> accessed 16 February 2010.

[7] Wasem and Ester (n 5) 6. This occurred in 1992 when Chinese people on Deferred Enforced Departure (DED) were permitted to adjust to LPR status under the Chinese Student Protection Act of 1992: PL 102–404, 106 Stat 1969 (9 October 1992) and for Haitians under the FY 1999 Omnibus Consolidated and Emergency Supplemental Appropriations Act: PL 105–277, 112 Stat 2681, 105th Congress 2nd Session (21 October 1998). Legislation for some other TPS groups has been introduced but not enacted, and included in comprehensive immigration reform legislation. In the 111th Congress, HR 264 would enable some current TPS holders to convert to LPR status if they have lived in the US for five years or more; are of good moral character; have no criminal convictions; have successfully completed a course on reading, writing, and speaking in English (with exceptions on account of disability); have accepted the values and cultural life of the US; and have completed at least 40 hours of community service: see Wasem and Ester (n 5) 7.

[8] ibid.

[9] ibid.

[10] Wasem and Ester (n 5) 5.

[11] ibid, 5–6, referring to Ruth E Wasem, 'US Immigration Policy on Haitian Migrants' (Congressional Research Service Report RS 21349, 31 March 2010).

Apart from TPS, the Attorney General may provide discretionary relief from deportation. The policy is that 'all blanket relief decisions require a balance of judgment regarding foreign policy, humanitarian, and immigration concerns'.[12] Work authorization is not automatic and must be applied for separately. It has at times been granted to people whose TPS has not been renewed.[13]

(2) European Union

The EU Temporary Protection Directive was designed as an exceptional mechanism[14] to respond to mass influx on account of armed conflict, endemic violence, or generalized human rights violations.[15] It could potentially be activated to respond to a sudden influx of people on account of environmental or climate change impacts, since Article 2(c), which sets out the Directive's scope of application, is not exhaustive.[16]

The drafting history reveals that Finland sought to have included in the definition recognition of 'persons who have had to flee as a result of natural disasters', but this was not supported by other Member States, with Belgium and Spain noting that 'such situations were not mentioned in any international legal instrument on refugees'.[17] Curiously, in 2004, when the Temporary Protection Directive was concluded, the United Kingdom (UK) stated that the instrument would 'ensure that each European Member State plays its part in providing humanitarian assistance to people forced from their homes by war and natural disasters'.[18]

Despite considerable speculation about the possible reach of the Temporary Protection Directive in the context of climate-related movement, there does not appear to be any discussion about formally expanding the instrument.[19] Kälin and Schrepfer suggest that the 'solidarity clause' of the Lisbon Treaty, which provides that '[t]he Union and its Member States shall act jointly in a spirit of solidarity if a Member State is... the victim of a natural or man-made disaster',[20] could

[12] This most commonly occurs as DED or Extended Voluntary Departure (EVD). See Wasem and Ester (n 5) 3.

[13] For example, to 190,000 Salvadorans in 1992; to 3,600 Liberians (whose present grant of DED runs until 30 September 2011): ibid, 4.

[14] Council Directive (EC) 2001/55 on Minimum Standards for Giving Temporary Protection in the Event of a Mass Influx of Displaced Persons and on Measures Promoting a Balance of Efforts between Member States in Receiving Such Persons and Bearing the Consequences thereof [2001] OJ L212/12, Art 2(a).

[15] ibid, Art 2(c).

[16] Vikram Kolmannskog and Finn Myrstad, 'Environmental Displacement in European Asylum Law' (2009) 11 European Journal of Migration and Law 313, 316ff; UK Home Office, 'UK Plans in Place to Protect Victims of Humanitarian Disasters' (20 December 2004) Press Release; Walter Kälin, 'Conceptualising Climate-Induced Displacement' in Jane McAdam (ed), *Climate Change and Displacement: Multidisciplinary Perspectives* (Hart Publishing, 2010).

[17] Council of the European Union, 'Outcome of Proceedings of Working Party on Asylum', Doc 6128/01 LIMITE ASILE 15 (16 February 2011) 4.

[18] UK Home Office (n 16) 1.

[19] Walter Kälin and Nina Schrepfer, 'Protecting People Crossing Borders in the Context of Climate Change: Normative Gaps and Possible Approaches' (Study for the Swiss Ministry of Foreign Affairs, April 2011) 39, forthcoming as a paper in the UNHCR Legal and Protection Policy Research Series.

[20] Treaty of Lisbon amending the Treaty on European Union and the Treaty establishing the European Community (signed 13 December 2007) OJ C306/1 ('Lisbon Treaty') Art 188R.

provide a basis for action with respect to natural disasters *within* an EU Member State. However, as they observe, it is limited to assistance to a Member State in its own territory (so does not cover cross-border movement), and it does not extend to non-Member States. Article 10A of the Lisbon Treaty may provide more traction in its requirement that the EU 'work for a high degree of cooperation in all fields of international relations', including to 'assist populations, countries and regions confronting natural or man-made disasters', although it does not expressly refer to the admission of those displaced by such disasters[21] (and this is not ordinarily understood as a component of the duty to cooperate).[22]

Finally, given the empirical evidence on the likely nature of climate change-related movement, it remains uncertain whether the EU would ever be faced by a 'mass influx' from a climate-affected country sufficient to overwhelm the regular asylum processing procedures and warrant the exceptional grant of temporary protection.

B. Asylum-type mechanisms

(1) European Union

The EU Qualification Directive, which provides the framework for individual protection in the EU, does not contain an express provision on protection from environmental or climate change-related impacts, although the potential for such movement to be covered under 'inhuman or degrading treatment' (based on Article 3 of the European Convention on Human Rights[23]) has been discussed extensively in Chapter 3 above. Although the possibility of including 'environmental disasters' as a ground of subsidiary protection was raised during the drafting process,[24] this does not seem to have been entertained seriously in deliberations and, given the nature of the negotiations, it was very unlikely ever to be adopted.[25]

Early drafts of the Qualification Directive contained an additional ground on which subsidiary protection could be granted, namely where a person was at risk of a 'violation of a human right, sufficiently severe to engage the Member State's international obligations'.[26] This was subsequently modified to prevent return to:

> acts or treatment outside the scope of subparagraphs (a) to (c) in an applicants country of origin, or in the case of a stateless person, his or her country of former habitual residence,

[21] Kälin and Schrepfer (n 19) 39. [22] See Ch 9 on the duty to cooperate.

[23] European Convention on Human Rights (formally the Convention for the Protection of Human Rights and Fundamental Freedoms) (drafted 4 November 1950, entered into force 3 September 1953) ETS No 5 ('ECHR').

[24] Council of the European Union, 'Note from President to Asylum Working Party: Discussion Paper on Subsidiary Protection', Doc 13167/99 LIMITE ASILE 41 (19 November 1999) 6.

[25] See Jane McAdam, 'The European Union Qualification Directive: The Creation of a Subsidiary Protection Regime' (2005) 17 International Journal of Refugee Law 461.

[26] Commission of the European Communities, 'Proposal for a Council Directive on Minimum Standards for the Qualification and Status of Third Country Nationals and Stateless Persons as Refugees or as Persons Who Otherwise Need International Protection', COM (2001) 510 final (12 September 2001) Art 15(b). See discussion in Jane McAdam, *Complementary Protection in International Refugee Law* (Oxford University Press, 2007) 81–4.

when such acts or treatment are sufficiently severe to entitle the applicant to protection against refoulement in accordance with the international obligations of Member States.[27]

Commentary on this provision, which was deleted from the final text, made clear that '[b]y using the wording "acts or treatment" it is ensured that only man-made situations, and not for instance situations arising from natural disasters or situations of famine, will lead to the granting of subsidiary protection'.[28] By extension, and from the context of the deliberations, it is apparent that Member States did not consider the other provisions on subsidiary protection to cover such flight either. That does not foreclose the possibility that treatment resulting from such situations could amount to inhuman or degrading treatment, however.

(2) National laws

At the national level, Swedish asylum law contains a provision extending protection to people who are 'unable to return to the[ir] country of origin because of an environmental disaster'.[29] To date, however, it has not been used. In any case, it is unclear if this would extend to people displaced for climate change-related reasons, since it seems that it was only ever intended to cover people fleeing specific environmental disasters such as Chernobyl, rather than more generally. Kolmannskog and Myrstad note that the drafting history reveals discussion of the idea that environmental displacement could include so-called 'sinking' island States and other longer-term scenarios,[30] but commentary on the provision as adopted clarified that it was only intended to apply in cases of sudden disasters, and would only be available if there were no internal flight alternative.[31]

Finnish asylum law also provides that a person may be granted humanitarian protection if 'he or she cannot return to his or her country of origin or country of former habitual residence as a result of an environmental catastrophe or a bad security situation which may be due to an international or internal armed conflict or a poor human rights situation'.[32] Again, the law is untested in relation to climate change-related displacement.

27 Council of the European Union, 'Outcome of Proceedings', Doc 12199/02 LIMITE ASILE 45 (25 September 2002) Art 15, as suggested in Council of the European Union, 'Presidency Note to Strategic Committee on Immigration, Frontiers and Asylum on 25 September 2002: Proposal for a Council Directive on Minimum Standards for the Qualification and Status of Third Country Nationals and Stateless Persons as Refugees or as Persons Who Otherwise Need International Protection', Doc 12148/02 LIMITE ASILE 43 (20 September 2002) 4.

28 Council of the European Union, 'Presidency Note' (n 27) 7. This interpretation was confirmed by Greens MEP Jean Lambert, who was the European Parliament's Rapporteur on the topic in 2002. In an interview with Finn Myrstad, she stated: 'It became crystal clear that it was no point of pushing the agenda for (environmental refugees), there was no majority.... We had enough with keeping humanitarian protection in there along with asylum. That was the real battle': Finn Myrstad, 'Not Much Protection for Climate Refugees in the EU' (29 July 2008) <http://www.myrstad.eu/not-much-protection-for-climate-refugees-in-the-eu/> accessed 13 July 2011.

29 Aliens Act 2005: 716, Ch 4, s 2(3).

30 Swedish *travaux préparatoires* (SOU) (1995) 75, 147 in Kolmannskog and Myrstad (n 16) 323.

31 Proposition 1996/97: 25, 100–1 in Kolmannskog and Myrstad (n 16) 323. See Ch 3.

32 Aliens Act (301/2004, amendments up to 1152/2010 included) (Finland) ('Aliens Act (Finland)') s 88a.

According to Kälin, even though Swiss asylum law does not expressly mention natural or environmental disasters, legislation on temporary admission and subsidiary protection could be interpreted so as to accommodate climate-related scenarios.[33] Again, however, the focus is on *disasters* rather than slow-onset changes, and this creates an obvious protection gap.

In 2010, Argentina adopted new immigration legislation providing access to provisional residence permits for individuals who cannot return to their country of origin because of a natural or environmental disaster.[34] Additionally, international protection (through permanent residence) is available there to people who are not refugees but who are protected by the principle of *non-refoulement* where their human rights would be at risk in the country of origin.[35] This also has the potential to apply to individuals affected by climate impacts.

(3) Legislative proposals

In 2006, the Belgian Senate adopted a resolution (introduced by Philippe Mahoux of the socialist party) calling for Belgium to agitate in the United Nations (UN) for the recognition of an international 'environmental refugee' status.[36] During debate, some senators opposed the resolution because it did not sufficiently address the root causes of the problem, although none raised any technical or political difficulties with respect to amending the Refugee Convention.[37] In 2008, two further resolutions were introduced in the lower house—one expressly calling for a Protocol to the Refugee Convention, the other not expressly mentioning this—but

[33] Kälin (n 16) 100. He notes that this was the conclusion of an inter-departmental roundtable discussion by the Swiss Ministry of Foreign Affairs on 13 January 2009, at which he was present. See further, Kälin and Schrepfer (n 19) 38, referring to Asylum Act of 26 June 1998 (as at 1 April 2011) Switzerland, Arts 44, 66–79; Bundesgesetz über die Ausländerinnen und Ausländer vom 16 Dezember 2005 (as at 24 January 2011) Arts 83–8; Motion Zisyadis, 'Internationaler Status für Umweltflüchtlinge' (19 December 2007), Antwort des Bundesrates vom 13.2.08 und Interpellation Grüne Fraktion, 'Flüchtlingsstatus für Umweltflüchtlinge' (4 October 2007), Antwort des Bundesrates vom 28.11.2007.

[34] Decree No 616/2010 Official Bulletin No 31.898 (6 May 2010) (regulating immigration law 25.871) Art 24(h). The provision delimits its scope of application to foreigners who enter the country as 'transit residents'.

[35] ibid, Art 23(m).

[36] Sénat de Belgique, 'Proposition de résolution visant à la reconnaissance dans les conventions internationales du statut de réfugié environnemental', Doc 3-1556/3 (21 March 2006) Belgian Senate in François Gemenne, 'Environmental Changes and Migration Flows: Normative Frameworks and Policy Responses' (PhD thesis, Institut d'Etudes Politiques de Paris and University of Liège, 2009) 300, 445–7 (for copy of text adopted by the Commission des relations extérieures et de la défence, 'Proposition de résolution visant à promouvoir la reconnaissance dans les conventions internationales du statut de réfugié environnemental').

[37] Convention relating to the Status of Refugees (adopted 28 July 1951, entered into force 22 April 1954) 189 UNTS 137, Art 1A(2), read in conjunction with Protocol relating to the Status of Refugees (adopted 31 January 1967, entered into force 4 October 1967) 606 UNTS 267 (together 'Refugee Convention').

a vote on both is still pending.[38] Even if they were adopted, they would be non-binding on the Parliament.[39]

Prior to being elected as the new Australian government in late 2007, the Australian Labor Party proposed the creation of a Pacific Rim coalition to accept climate change 'refugees', and encouraged the Australian government to lobby the UN to 'ensure appropriate recognition of climate change refugees in existing conventions, or through the establishment of a new convention on climate change refugees'.[40] However, when a Senator from the Greens Party proposed the Migration (Climate Refugees) Amendment Bill 2007, calling for protection visas to be granted to people fleeing 'a disaster that results from both incremental and rapid ecological and climatic change and disruption', the Labor party was quick to note that its idea of an international response required a collaborative approach with other countries, rather than unilateral action by Australia.[41] In government, Labor has not formulated any policies on this issue. A Senate inquiry in 2009 revealed that '[w]hen asked about the possibility of forced re-location from Pacific island countries such as Kiribati and Tuvalu, DFAT [the Department of Foreign Affairs and Trade] informed the committee that it was not aware of any government consideration of this matter. Invited to comment again on whether these two islands were under consideration, DFAT replied no.'[42]

III. Ad Hoc Humanitarian Schemes

A. Group-based schemes

Even in the absence of specific legislation, a number of countries provide some form of protection to people fleeing natural disasters.[43] For various reasons, States may prefer ad hoc humanitarian responses that permit them to determine on a

[38] Chambre des Représentants de Belgique, 'Proposition de résolution relative à la prise en considération et à la création d'un statut de réfugié environnemental par les Nations-Unies et l'Union européenne' (déposée par M Jean Cornil et consorts of the socialist party) Doc 52–1451/001 (3 October 2008) (calling for a Protocol to the Refugee Convention) in Gemenne (n 36) 300, 469–78 (for text); Chambre des Représentants de Belgique, 'Proposition de résolution visant à la reconnaissance d'un statut spécifique pour les réfugiés climatiques' (déposée par Mme Juliet Boulet, M Wouter Devriendt et Mme Zoé Genot of the Greens) Doc 52–1478/001 (14 October 2008) in Gemenne (n 36) 300, 460–8 (for text).

[39] Email from François Gemenne to author (4 January 2011).

[40] Australian Labor Party (ALP), *Our Drowning Neighbours: Labor's Policy Discussion Paper on Climate Change in the Pacific* (ALP Policy Discussion Paper, 2006) 10.

[41] Commonwealth of Australia, *Senate: Official Hansard* (9 August 2007) 100ff.

[42] Senate Foreign Affairs, Defence and Trade References Committee, *Economic Challenges facing Papua New Guinea and the Island States of the Southwest Pacific* (Commonwealth of Australia, 2009) para 6.60 (making reference to *Committee Hansard* (21 November 2008) 28). On 21 November 2009, a spokesperson for the then Climate Change Minister, Penny Wong, was reported as acknowledging that permanent migration may eventually be the only option for some people, which will need to be dealt with by governments in the region: Adam Morton, 'Land of the Rising Sea', *Sydney Morning Herald* (Sydney, 21 November 2009) 5 <http://www.smh.com.au/environment/land-of-the-rising-sea-20091120-iqub.html> accessed 27 November 2009.

[43] Sometimes this has been as a result of calls from UNHCR: see UNHCR, 'High Commissioner's Dialogue on Protection Challenges: Breakout Session 1: Gaps in the International Protection Framework and its implementation: Report by the Co-Chairs' (8–9 December 2010) 4.

situation-by-situation basis whether they wish to provide 'protection', for what duration, and in what form. Typically, though, this is emergency protection after a particular event, rather than pre-emptive protection for projected longer-term impacts. Further, it is unpredictable as a protection tool and 'dominated by humanitarian considerations rather than a rights-based approach'.[44]

Sometimes, special historical or cultural links may foster humanitarian goodwill towards people displaced by a sudden disaster. For example, the African community offered special protection to Haitians following the earthquake there in 2010.[45] The African Union reportedly considered a proposal to create a new country for them in Africa, citing 'a sense of duty and memory and solidarity'.[46] Historical ties and symbolism of 'home' are of paramount importance here.[47] Botswana and Tanzania have admitted people from neighbouring States escaping natural disasters, such as floods, though only on a temporary, ad hoc basis and for humanitarian reasons.[48] Latin American countries offer asylum to extended categories of refugees on a regional basis. Caribbean countries provided temporary asylum to Montserratians fleeing volcanic eruptions in the 1990s (see below). While not every country has the capacity to absorb large numbers of migrants, localized solutions may provide more culturally appropriate responses than a universal burden-sharing mechanism.

[44] Kälin and Schrepfer (n 19) 38, referring also to Roberta Cohen and Megan Bradley, 'Disasters and Displacement: Gaps in Protection' (2010) 1 Journal of International Humanitarian Legal Studies 95, 131.

[45] The President of Senegal offered to resettle displaced Haitians as descendants of African slaves: 'They have a right to return to Africa, their original land. They were colonized by the Americans. We will find them land.' See Scott Baldauf, 'Haitians to Africa? Senegal Resettlement Plans Gain Steam', *Christian Science Monitor* (2 February 2010) <http://www.csmonitor.com/World/Africa/2010/0202/Haitians-to-Africa-Senegal-resettlement-plans-gain-steam> accessed 3 August 2011.

[46] The African Union is reportedly considering a proposal to create a new country for them in Africa, citing 'a sense of duty and memory and solidarity'. See ibid.

[47] Indeed, in the context of the proposed relocation of Nauru in the 1960s, discussed in Ch 5, Tabucanon and Opeskin argue: 'The Nauruans had a strong moral and legal claim against three developed states that had the means to rehabilitate their island or accept them as new settlers [Australia, New Zealand, and the United Kingdom]. The Nauruans were the beneficiaries of a "sacred trust", and Australia, the United Kingdom and New Zealand were obliged, at least in principle, to advance their interests under the mandate and the trusteeship. In correspondence quoted above, the Australian Prime Minister acknowledged the "clear obligation" of the three governments to provide a satisfactory future for the Nauruans, and this, more than anything else, may explain the depth of the resettlement negotiations': Gil Marvel Tabucanon and Brian Opeskin, 'The Resettlement of Nauruans in Australia: An Early Case of Failed Environmental Migration' (2011) 46 Journal of Pacific History (forthcoming), text available online <http://papers.ssrn.com/sol3/papers.cfm?abstract_id=1851910> accessed 17 July 2011.

[48] Kälin and Schrepfer (n 19) 37. See also Ch 2.

(1) Montserrat and the United Kingdom

Following volcanic eruptions in Montserrat (a UK overseas territory[49]) from July 1995, the UK provided a voluntary evacuation scheme whereby citizens[50] of Montserrat could obtain two years' exceptional leave[51] to enter/remain in the UK (provided they had a UK sponsor). Until August 1997, they had to fund their own airfare, which few could afford; after that time, some assistance was made available. They were granted permission to remain for an initial two year period, with full access to employment, income support, housing benefits, education, and healthcare as UK citizens.[52] By 1997, almost 3,500 Montserratians had relocated to the UK.[53]

In 1998, as their two year period of leave began to expire and return to Montserrat remained impossible, the Montserratians were eventually granted indefinite leave to remain in the UK.[54] On 21 May 2002, all British overseas territories citizens from Montserrat were automatically granted British citizenship (the main form of British nationality) when the principal provisions of the British Overseas Territories Act 2002 commenced. It is fair to say that this development had nothing to do with the Montserrat crisis. Extending British citizenship to Montserratians may have been good timing from the recipients' perspective, and may have appeared as a humanitarian gesture from the perspective of the international community, but it was part of a restructuring in British nationality law announced in 1999 (which extended British citizenship to virtually all the remaining British overseas territories) and implemented irrespective of the volcanic eruptions.[55]

49 From 1 January 1983, the remaining British colonies were termed 'Dependent Territories' under the British Nationality Act 1983. In practice, they were renamed 'British overseas territories' on 16 March 1999 when the then Foreign Secretary, Robin Cook, presented to Parliament a White Paper entitled *Partnership for Progress and Prosperity: Britain and the Overseas Territories* (White Paper, Cm 4264, 1999). The change to 'British overseas terrtitory' was made *de jure* as of 26 February 2002 by the British Overseas Territories Act 2002.

50 This is a term of convenience. In nationality terms, those with a sufficiently close connection to Montserrat were 'citizens of the UK & Colonies' prior to 1983 (see the British Nationality Act 1948), then 'British Dependent Territories citizens' (under the British Nationality Act 1981), then 'British overseas territories citizens' from 16 March 1999/26 February 2002 (as explained in n 49). In immigration terms, it was a matter for the immigration laws of Montserrat to provide an immigration status for those entitled or allowed to reside there. I am grateful to Laurie Fransman for clarifying this.

51 'Leave' is the term used by the Immigration Act 1971 for 'immigration permission'. 'Exceptional leave' is a term of convenience referring to leave granted outside the Immigration Rules (which are made under the Immigration Act 1971 and list the categories of persons to whom a grant of leave will normally be confined).

52 Select Committee on International Development (UK), *First Report, Montserrat* (HC 267) 18 November 1997, para 87. For details of services, see Select Committee on International Development, *First Special Report: Government Response to the First Report from the International Development Committee, Session 1997–98: Montserrat* (HC 532) 16 February 1998, Annex 3, 'Moving to the UK: Your Questions Answered'.

53 Select Committee on International Development, *Sixth Report: Montserrat: Further Developments* (HC 726) 4 August 1998, para 22.

54 Leave without time restrictions or other conditions.

55 I am grateful to Laurie Fransman for providing me with this information. There are parallels here with New Zealand's Pacific Access Category, discussed in section IV. below.

Additionally, on 21 August 1997 an Assisted Regional Voluntary Relocation Scheme was established by the UK government to assist Montserratians relocating within the Caribbean region or to the US or Canada. It has been described as little more than a six-month holiday package,[56] since residents with less than £10,000 per head of savings were granted £2,400 per adult over six months plus £600 for each child under 18 years of age.[57] Some assistance for travel to the US and Canada was also made available, since the US granted TPS to Montserratians in August 1997 and Canada granted landed settlement rights.[58]

While this may sound like a relatively orderly process, it was in fact very protracted. The Prime Minister of Antigua-Barbuda expressed concern that the UK's reluctance to facilitate movement meant that countries in the region were effectively being forced to accommodate large numbers of Montserratians.[59] While he committed his government to continue to 'open its doors to Montserratians in the spirit of good Caribbean neighbourliness that has always marked our relationship with the island' during 'their greatest hour of need', he argued that 'appropriate arrangements must be made primarily by the Government of the United Kingdom whose responsibility Montserrat remains'.[60]

The apparent reluctance by the UK to take responsibility for its own nationals displaced by the Montserrat volcanoes does not bode well for the potential relocation of whole communities affected by climate change, especially where there are no strong ties with other States. It also shows that while temporary schemes may provide initial assistance, in some cases long-term solutions will need to be found. As has been discussed at length in the refugee context, leaving people in limbo for protracted periods is inconsistent with international human rights law and can cause considerable psychological and social harm.

(2) Natural disasters

In the absence of an international protection framework for displacement by disasters,[61] the United Nations High Commissioner for Refugees (UNHCR) has called on States to provide discretionary responses in situations of natural disaster. Following the 2004 Asian Tsunami, it called for a halt on returns to areas affected by the devastation.[62] The UK suspended involuntary returns of failed

[56] Claude ES Hogan, 'Seeking Refuge in the Mother Country: UK Maintained Strict Exclusion Policy Despite Erupting Volcano' (Montserrat Country Conference, Montserrat, 13–14 November 2002) <http://www.cavehill.uwi.edu/BNCCde/montserrat/conference/papers/hogan.html> accessed 2 August 2011.

[57] Select Committee on International Development (n 52) para 81.

[58] Hogan (n 56).

[59] Prime Minister Bird, 'Address' (18th Meeting of the Conference of Heads of Government of the Caribbean Community, Jamaica, 30 June–4 July 1997) cited in Hogan (n 56).

[60] A Joint CARICOM/OECS mission to Montserrat in April 1996 reiterated the 'easing of immigration and work permit requirements for migrating Montserratians' relocating to CARICOM Member States: Ann Dummett and Andrew Nicol, *Subjects, Citizens and Others: Nationality and Immigration Law* (Weidenfeld and Nicolson, 1990) 203, cited in Hogan (n 56).

[61] On which, see Ch 9.

[62] Kolmannskog and Myrstad (n 16) 323 citing UNHCR Executive Committee Conclusion No 103 (LVI) 'The Provision of International Protection including through Complementary Forms of Protection' (2005).

asylum seekers to India, Sri Lanka, Thailand, and Indonesia, while Canada and Australia fast-tracked permanent and temporary visa applications for people coming from tsunami-affected regions, and offered permanent residents from these regions the opportunity to expedite the procedure for sponsoring family members.[63] In Germany, the Federal Ministry of the Interior indicated that returns should be stalled, although only some of the Bundesländer implemented this. The Netherlands halted deportations until March 2005.[64] France, Canada, and the Dominican Republic also reportedly eased their immigration rules in light of the 2010 Haitian earthquake.[65] Sometimes, however, discretionary halts to removal may mean that people are able to remain, but do not have many substantive rights. These kinds of responses are not sustainable in the long term.

(3) Other domestic examples

There are other examples of State practice extending special protection to particular groups, including for environmental or socio-economic reasons (which are pertinent to the climate change context). For example, between 2001 and 2006, it was Danish practice not to return young children to Afghanistan because of drought. This was subsequently extended to landless people from areas where there was a lack of food and who would be especially vulnerable on return.[66] Denmark has also provided humanitarian asylum to single women and families with young children who would otherwise be returned to areas where living conditions are very harsh, such as on account of famine.[67] Unaccompanied minors may be granted complementary protection there if they will be placed in 'an emergency situation' if returned.[68] Australia has responded to particular crises (eg East Timor, Kosovo, China) by creating ad hoc visa categories.[69] Belgium created ad hoc temporary protection schemes during the crises in the Former Yugoslavia and Rwanda.[70]

Canada's Immigration and Protection Regulations provide that:

> The Minister may impose a stay on removal orders with respect to a country or a place if the circumstances in that country or place pose a generalized risk to the entire civilian popu-

[63] Benjamin Glahn, '"Climate Refugees"? Addressing the International Legal Gaps—Part II', *International Bar Association* (undated) <http://www.ibanet.org/Article/Detail.aspx?ArticleUid=3E9DB1B0-659E-432B-8EB9-C9AEEA53E4F6> accessed 13 January 2011.

[64] Kolmannskog and Myrstad (n 16) 324.

[65] Farrell (n 6). See also *Shpati v Canada (Minister of Public Safety and Emergency Preparedness)* [2010] FC 367 (Canada) para 47, where the court stated: 'Nor do I rule out the possibility that an enforcement officer may defer removal in circumstances in which new events have occurred after the negative PRRA [Pre-Removal Risk Assessment] decision, such as natural disasters in the form of tsunamis or earthquakes or political upheavals such as "coup d'états."'

[66] Kolmannskog and Myrstad (n 16) 324.

[67] ibid, citing Aliens (Consolidation) Act No 826 of 24 August 2005 (Denmark) Art 9b(1).

[68] Aliens (Consolidation) Act (Denmark) Art 9c(3)(ii), cited in European Council on Refugees and Exiles (ECRE), 'Complementary Protection in Europe' (July 2009) 7.

[69] See, eg, Migration Regulations 1994 (Cth) Sch 2 (Australia), Subclass 448 Kosovar Safe Haven (Temporary).

[70] Dirk Vanheule, 'The Qualification Directive: A Milestone in Belgian Asylum Law' in Karin Zwaan (ed), *The Qualification Directive: Central Themes, Problem Issues, and Implementation in Selected Member States* (Wolf Legal Publishers, 2007) 71.

lation as a result of... (b) an environmental disaster resulting in a substantial temporary disruption of living conditions.[71]

The stay can be cancelled if the circumstances no longer pose 'a generalized risk to the entire civilian population'.[72]

Finnish law makes provision for immigration for humanitarian reasons, where, on the basis of a joint proposal from the Ministry of the Interior and the Ministry for Foreign Affairs, the government decides in a plenary session to admit aliens to Finland on special humanitarian grounds.[73] While initial admission is on a temporary basis, a continuous residence permit may be granted after three years if the grounds for issuing the temporary permit still exist.[74] A permanent residence permit may be granted four years after that.[75]

In Germany, the local state (*Land*) authorities may authorize a stay on removal on humanitarian or international law grounds for particular groups of people.[76] Swiss law provides provisional protection to people exposed to a serious general danger, especially during internal armed conflict or situations of generalized violence.[77] Although climate change is not expressly mentioned, the provision may cover this.[78] Provisional protection is granted for up to five years, after which time a resident permit is granted. This expires as soon as protection is withdrawn, but it can be challenged on a case-by-case basis.[79] If protection has not been withdrawn after 10 years, an establishment permit may be granted.[80]

What most of these ad hoc schemes have in common is the designation of *particular* countries as demonstrating sufficient, objective characteristics that 'justify' movement,[81] thereby obviating the need for people wishing to leave them to show specific reasons why—in the climate change context—climate change is personally affecting them. This has parallels with prima facie refugee status in refugee law, and even with the development of the international refugee protection regime, in which a series of international agreements based on the refugees' country

[71] Immigration and Refugee Protection Regulations, SOE/2002-227 (Canada) s 230(1).

[72] ibid, s 230(2).

[73] Aliens Act (301/2004, amendments up to 1152/2010 included) (Finland) s 93.

[74] ibid, s 113(2); ECRE (n 68) 37.

[75] ibid, s 56(1).

[76] Residence Act of 30 July 2004 (Federal Law Gazette I, p 1950), last amended by the Act on Implementation of Residence- and Asylum-Related Directives of the European Union of 19 August 2007 (Federal Law Gazette I, p 1970) (Germany) ('Residence Act (Germany)'), s 60a(1).

[77] 'Asylum Act (Switzerland)' Ch 1, Art 4.

[78] Kälin (n 16) 100. He notes that this was the conclusion of an inter-departmental roundtable discussion by the Swiss Ministry of Foreign Affairs on 13 January 2009, at which he was present.

[79] Asylum Act (Switzerland) Ch 4, Art 74(2); ECRE (n 68) 62.

[80] Asylum Act (Switzerland) Ch 4, Art 74(3).

[81] The Migration (Climate Refugees) Amendment Bill 2007, proposed by the Australian Greens, suggested a mechanism whereby an individual application for a 'climate change refugee visa' would trigger a requirement for the Minister for Immigration to make a declaration about the 'climate change circumstance' on which the application was based, thus creating a visa pathway for others similarly affected. Similar suggestions have been made in some of the treaty proposals: see Ch 7.

of origin was created in response to particular (albeit, politically selected) crises, prior to the establishment of a universal refugee instrument.[82]

B. Discretionary grounds for individual claimants[83]

Many States have some form of discretionary leave for non-citizens to remain on humanitarian or compassionate grounds.[84] Their applicability to a person seeking protection on the basis of climate change-related displacement will vary from jurisdiction to jurisdiction, since each has different requirements as to eligibility for humanitarian protection and it remains to be seen whether decision-makers would be prepared to construe their circumstances as being of an exceptional humanitarian nature. Some humanitarian mechanisms can only be triggered once a failed asylum application has been made. Others take into account the length of time a person has already spent in the country in which they are seeking to remain (and thus the level of integration there). Some statuses do not provide very extensive rights, or may be temporary. A temporary status does not solve the problem for those who are permanently displaced. In some cases, political pressure has meant that States have had to convert people from a temporary to permanent status after a certain period of time.

The following provides a brief descriptive overview of some domestic humanitarian schemes which may be relevant in the context of climate change-related displacement.

In Austria, two kinds of humanitarian residence permits may be granted.[85] An *Aufenthaltsbewilligung* (residence permit) is for a temporary, time-bound stay, whereas a *Niederlassungsbewilligung* (establishment permit) may be permanent, since the beneficiary has already started to integrate into society.[86] Humanitarian grounds are automatically assessed for any applicant seeking to remain in Austria.[87] In Belgium, leave of stay due to exceptional circumstances may be granted for reasons such as the practical or legal impossibility of return (eg non-readmission or no means of transport), or if the applicant has developed close links to Belgium. It results in a one year extendable and renewable permit.[88] In Denmark, humanitarian status may be granted if essential considerations of a humanitarian nature conclusively warrant it. It may be granted on a

[82] See generally, Guy S Goodwin-Gill and Jane McAdam, *The Refugee in International Law* (3rd edn, Oxford University Press, 2007) 16–20.

[83] See also discussion of UK Discretionary Leave in Ch 3.

[84] For an analysis of State practice in 10 European countries, see ECRE (n 68). Of the 10 States surveyed by ECRE, only Switzerland did not have such a provision: 7.

[85] Federal Act concerning Settlement and Residence in Austria, Federal Law Gazette No 100/2005 in version Federal Law Gazette No 31/2006 (Austria) (The Settlement and Residence Act) Ch 2, s 7, Arts 72–4.

[86] For the rights granted, see ECRE (n 68) 18.

[87] Ministry of the Interior, 'Ministerrat beschließt Neuregelung des Humanitären Aufenthalts', Press Release No 5318 (24 February 2009) <http://www.bmi.gv.at/> accessed 3 August 2011, cited in ECRE (n 68) 18.

[88] Intégration Sociale, 'Circulaire relative à la régularisation de séjour pour raisons médicales et son impact sur le droit à l'aide sociale', *Intégration Sociale* (20 February 2008) 3 <http://www.mi-is.be/sites/default/files/doc/OB%202008-02-20%20FR.pdf> accessed 3 August 2011, cited in ECRE (n 68) 25.

temporary or a permanent basis.[89] Importantly, humanitarian protection is only available to registered asylum seekers (seeking Refugee Convention protection). Additionally, Denmark precludes removal on the basis of exceptional reasons or hindrances to deportation, which again may lead to either temporary or permanent stay.[90] This may be granted if exceptional reasons make it appropriate (including respect for family unity); if a person who does not qualify for international protection has not been able to be removed for at least 18 months; or for certain unaccompanied minors, including—relevantly—if there is reason to assume that the unaccompanied minor will be placed in an emergency situation upon return.

As already discussed, Finland's complementary protection regime provides that a residence permit may be granted if a person cannot return because of an environmental disaster.[91] Discretionary temporary residence permits are also available if a person cannot be removed from the country (including for temporary health reasons).[92] A continuous residence permit may be granted if refusing a residence permit would be manifestly unreasonable with regard to the applicant's health, ties to Finland, or on other compassionate grounds, particularly given the circumstances they would face in their home country and/or their vulnerable position.[93]

Similarly, as discussed above, Swedish law protects as 'persons otherwise in need of protection' those who are unable to return to their country of origin because of an environmental disaster.[94] This sits within Sweden's complementary protection regime, which typically results in permanent residence. Additionally, but outside the complementary protection framework, a residence permit (normally permanent) may be granted on grounds of an exceptionally distressing situation.[95] This is determined by a holistic assessment of the applicant's circumstances, with particular attention paid to his or her state of health, integration in Sweden, and situation in the country of origin.[96] The criteria may be applied less stringently to children.

Germany provides for temporary suspension of deportation (*Duldung*).[97] Unsuccessful asylum seekers may be permitted to remain in Germany if there are practical or legal obstacles to their removal, or if humanitarian, personal considerations, or important public interests require it.[98] This is little more than a temporary stay on deportation in six-month increments. The permit can be extended if removing the individual would constitute an extraordinary hardship for him or her,[99] and permanent residence may only be granted after an 18-month suspension

[89] Aliens (Consolidation) Act No 785 of 10 August 2009 (Denmark) ss 9b, 11.
[90] ibid, ss 9c(1), 9c(2), 9c(3).
[91] Aliens Act (301/2004, amendments up to 1152/2010 included) (Finland) s 88a.
[92] ibid, s 51.
[93] ibid, s 52 in ECRE (n 68) 34.
[94] Aliens Act 2005:716 (Sweden) Ch 4, s 2. [95] ibid, Ch 5, s 6. [96] ibid.
[97] Residence Act (Germany) s 60a.
[98] ibid, s 60a(2) in ECRE (n 68) 47. [99] ibid, s 25(4).

of deportation.[100] The rights entitlements under *Duldung* status are very limited and do not include the right to work.[101]

In Switzerland, a residence permit may be granted on exceptional grounds.[102] Its aim is to regularize the legal position of people who have been in Switzerland for at least five years, and whose deportation would be particularly harsh because of his or her integration there.[103] Provisional admission may be authorized where a deportation order cannot be reasonably carried out, which would be the case if it would place a person in concrete danger, such as in a situation of generalized violence or medical necessity.[104]

In Ireland, leave to remain may be granted on a discretionary basis to an individual whose asylum claim has been rejected. The Minister for Justice, Equality and Law Reform has the discretion on humanitarian or other grounds not to issue a deportation order.[105] Australian law contains a similar discretionary provision.[106] In the Netherlands, a temporary residence permit may be granted to an individual 'for whom return to his country of origin would, in the opinion of the Minister, constitute an exceptional hardship in the context of the overall situation there'.[107] In South Africa, the Minister may grant an individual or a group 'the rights of permanent residence for a specified or unspecified period when special circumstances exist which justify such a decision'.[108] The operation of Discretionary Leave in the UK has already been discussed at length in Chapter 3.

In addition to complementary protection grounds,[109] the New Zealand Immigration Act provides that a person may be permitted to remain in New Zealand if 'there are exceptional circumstances of a humanitarian nature that would make it unjust or unduly harsh for the person to be deported from New Zealand', and it would not be contrary to the public interest to permit the person to remain.[110]

In Canada, people with a removal order against them may apply for Pre-Removal Risk Assessment (PRRA). A favourable PRRA decision normally leads to eligibility for permanent residence.[111] However, since this involves an assessment of the refugee and complementary protection grounds in Canadian law, its usefulness for people fleeing the effects of climate change may be limited. Canada also considers Humanitarian and Compassionate (H&C) claims.[112] H&C is usually a measure of last resort for people seeking to remain in Canada. H&C decisions take

[100] ibid, s 26(1). [101] See ECRE (n 68) 48–9.

[102] Asylum Act (Switzerland) Ch 2, art 14.

[103] See Federal Office for Migration, 'Loi sur l'asile (LAsi): questions et réponses' <http://www.bfm.admin.ch/content/dam/data/migration/asyl_schutz_vor_verfolgung/asylgesetz/abstimmungen/faq_asylg_f_neu.pdf> accessed 3 August 2011, cited in ECRE (n 68) 64.

[104] Aliens Act 2005 (as amended until January 2009) (Switzerland) Ch 11, Art 83(6), cited in ECRE (n 68) 64.

[105] Immigration Act 1999, No 22 of 1999 (Ireland) ss 3(3)(b), 3(6).

[106] Migration Act 1958 (Cth) (Australia) s 417.

[107] Vreemdelingenwet 2000 (Aliens Act 2000) (Netherlands) Art 29(1)(d).

[108] Immigration Act 2002, Act No 13 of 2002 (South Africa) Art 31(2).

[109] Immigration Act 2009, No 51 (NZ) ss 130, 131. [110] ibid, s 207.

[111] Immigration and Refugee Protection Act 2001, SC 2001, c 27 (Canada) s 112.

[112] ibid, s 25.

into account considerations such as establishment and relationships in Canada, the best interests of the child, hardships affecting the applicant, and the country of origin's ability to provide medical treatment.[113] A successful claim results in permanent residence.

IV. Migration Responses

Finally, New Zealand's Pacific Access Category (PAC) deserves a brief mention here given widespread misunderstandings about its purpose. The PAC visa was created in 2002. It was based on an existing scheme for Samoans and replaced previous work schemes and visa waiver schemes for people from Tuvalu, Kiribati, and Tonga.[114] This visa has mistakenly been hailed as a 'protection' response to people at risk of climate change-related displacement in the Pacific, both in media and academic circles.[115] Although the scheme was extended to citizens of Tuvalu after a plea from that country's government for special immigration assistance to enable some of its citizens to move, it is a traditional migration programme rather than one framed with international protection needs in mind.[116]

The scheme permits an annual quota of 75 citizens each from Tuvalu and Kiribati and 250 each from Tonga (and previously Fiji), plus their partners and dependent children, to settle in New Zealand.[117] Eligibility is restricted to applicants between the ages of 18 and 45, who have a job offer in New Zealand, meet a minimum income requirement, and have a minimum level of English. Selection is by ballot.

Contrary to some reports that the visas were not being taken up, an analysis of the statistics since the scheme's commencement in July 2002 until August 2011 reveals that on average, 1,897 applications have been made by people from Kiribati and 419 from Tuvalu each year, with 3,226 being lodged from Kiribati in 2011

[113] Citizenship and Immigration Canada, 'Frequently Asked Questions: The Refugee System: Balanced Refugee Reform' <http://www.cic.gc.ca/english/information/faq/refugees/ref-reform-faq.asp> accessed 3 August 2011.

[114] Lianne Dalziel, 'Government Announces Pacific Access Scheme' (*Beehive*, 20 December 2001) <http://www.beehive.govt.nz/node/12740> accessed 8 December 2008.

[115] For example, it is relied upon in Camillo Boano, Roger Zetter, and Tim Morris, 'Environmentally Displaced People: Understanding the Linkages between Environmental Change, Livelihoods and Forced Migration' (2008) Forced Migration Policy Briefing 1 (Refugee Studies Centre, University of Oxford), and in François Gemenne, 'Climate Change and Forced Displacements: Towards a Global Environmental Responsibility? The Case of the Small Island Developing States (SIDS) in the South Pacific Ocean' (2006) Les Cahiers du CEDEM, Université de Liège <http://www.cedem.ulg.ac.be/m/cdc/12.pdf> accessed 8 December 2008. See also Corlett's critique : David Corlett, 'Tuvalunacy, or the Real Thing?', *Inside Story* (27 November 2008) <http://inside.org.au/tuvalunacy-or-the-real-thing/> accessed 28 January 2011. It appears that the misunderstanding was perpetuated by Al Gore's film, *An Inconvenient Truth* (2006).

[116] Interestingly, programmes like this may ultimately be the basis on which veiled assistance is afforded to those at risk of climate-induced displacement, since this may be politically more palatable than an explicit scheme to address the issue.

[117] 'Pacific Access Category' (*Immigration New Zealand*, last updated 29 November 2010) <http://www.immigration.govt.nz/migrant/stream/live/pacificaccess/> accessed 14 January 2011.

and 456 from Tuvalu.[118] Only in the first year of the scheme was the quota for Kiribati unmet, with 51 visas granted, and in 2004 only 64 visas were granted to Tuvaluans.[119] In most years, the 75 places have been exceeded by at least 10–15 additional visa grants; in 2010, a record 125 visas were granted to I-Kiribati and 120 to Tuvaluans.[120] The programme is well known in Tuvalu and Kiribati: almost every person interviewed referred to and welcomed it, although noted that some improvements could be made.[121] Conversations with locals in Tuvalu and Kiribati revealed that many older people embraced the opportunities that their younger relatives would have by migrating to New Zealand under the scheme. While they did not want to have to move themselves, they talked about the need to plan and adapt for the future. They regarded education and training at home as a sensible way for young people to make themselves attractive migrants for developed countries, in particular by training in areas where they would be likely to get employment (such as cooking).[122] Younger people spoke positively about the scheme and the opportunities it would afford them, although they noted that life in New Zealand is expensive and one has to work very hard to bring in an income.[123]

Although New Zealand does not formally have any humanitarian visas relating to climate change and displacement, it has expressed its commitment to 'respond to climatic disasters in the Pacific and manage changes as they arise'.[124] In 2008, in

[118] See Immigration New Zealand Statistics: Ballot System <http://www.immigration.govt.nz/NR/rdonlyres/28F41153-42CF-402B-B507-7BF31FC4C37C/0/B1BallotReportReceivedandDecided06JUL2011.pdf> accessed 3 August 2011. These figures include both initial registration for the ballot, and re-registration (for people who have applied in the ballot on a previous occasion).

[119] See ibid. Figures as at August 2011 show that only 41 visas have been granted to I-Kiribati in 2011. It is unclear whether further grants will be made later in the year.

[120] See ibid.

[121] People interviewed commented on difficulties in securing a job offer in New Zealand. The job issue has partly been addressed by NZ companies visiting the islands and agreeing to provide a job to anyone selected. There is no requirement that the particular job be taken up once the person arrives in NZ. Some find the NZ$60 registration fee expensive (it is NZ$25 for re-registrations). Although I did not encounter this view in my own interviews, one community leader reportedly condemned the scheme as a new type of 'slavery immigration', whereby educated Tuvaluans renounce stable, white-collar government employment at home to end up as cleaners or fruit-pickers in New Zealand: quoted in Shawn Shen, 'Noah's Ark to Save Drowning Tuvalu' (2007) 10 Just Change: Critical Thinking on Global Issues 18, 19.

[122] 'I know the older generation like myself, I don't want to move to anywhere. But the younger generation—maybe they are lucky if they want to move, there's also higher educations. They want to build up their life. They don't want to come here and live here. Those are the people who would be able to migrate and travel and stay overseas': Interview with Sir Kamuta Latasi, Speaker of the Tuvaluan Parliament (and former Prime Minister) (Funafuti, Tuvalu, 27 May 2009).

[123] 'But those who are probably younger, they are excited, because of course they're earning money more than they earn here. But they realize too that life is expensive in NZ and Australia. But they like to go because of their kids—for education and the opportunities that they don't have here': Interview with an official from the Ministry of Environment, Lands and Agricultural Development (Tarawa, Kiribati, 14 May 2009); 'Those who have gone [for seasonal work] have thought it would be an easy nine months or seven months and then return with a lot of money and in fact that's not true. You have to go there to work. And it's those people that think it's easy money who have come back with negative comments on what has happened in NZ': Interview with Tebao Awerika, Deputy Secretary, Ministry of Foreign Affairs and Immigration (Tarawa, Kiribati, 12 May 2009).

[124] 'Background: Environmental Migrants/Relocation/Displacement', New Zealand Government Poznan Delegation Brief for UNFCCC COP14, 343 (released pursuant to an Official Information Act request). The President of Kiribati has noted that so far, the country most receptive

response to 'the perceived need for action' on 'environmental migrants' in a number of Pacific countries,[125] the New Zealand government adopted a more proactive approach to the issue. Rather than simply 'correct[ing] misperceptions about New Zealand's position on the environmental migrants issue [ie no agreement to resettle people from Tuvalu]' and 'outlining New Zealand's current commitment to climate change adaptation efforts in the Pacific region',[126] it decided instead to:

a) acknowledge the concerns of Pacific Island countries in relation to this issue;
b) stress that current climate change efforts in the Pacific should continue to focus on adaptation, and should be underpinned by the desire of Pacific peoples to continue to live in their own countries; and
c) reaffirm that New Zealand has a proven history of providing assistance where needed in the Pacific, and that our approach to environmentally displaced persons would be consistent with this.[127]

Nevertheless, it remains without a specific visa category for people displaced by, or seeking to migrate away from, the impacts of climate change.

V. Conclusion

This chapter has shown that although many States have ad hoc or discretionary schemes in place for people fleeing natural disasters, they are inconsistent and unpredictable—both in terms of when they take effect and the status they accord.

Recently, UNHCR has advocated the reinforcement and development of temporary protection schemes as a means of eliciting initial international support for managing climate change-related displacement within a rights-based

to his plea for more migration has been East Timor: see remarks quoted in Morton (n 42). This accords with comments made by the President of East Timor, Dr José Ramos-Horta, Diplomacy Training Programme 20th Anniversary Public Lecture (Faculty of Law, University of New South Wales, 23 July 2009) <http://tv.unsw.edu.au/video/dr-jose-ramos-horta-dtp-20th-anniversary-public-lecture> accessed 14 December 2009. In 2009, Fiji's interim Minister of Foreign Affairs stated that his country would consider taking 'climate change refugees from Tuvalu and Kiribati in the future' on the basis of 'historical ties with both these two countries' and the fact that there are already Tuvaluans and I-Kiribati living in Fiji: Radio New Zealand, 2009, cited in Richard Bedford and Charlotte Bedford, 'International Migration and Climate Change: A Post-Copenhagen Perspective on Options for Kiribati and Tuvalu' in Bruce Burson (ed), *Climate Change and Migration: South Pacific Perspectives* (Institute of Policy Studies, Victoria University of Wellington, 2010) 90. See also Kirsty Graham, Deputy Permanent Representative for New Zealand, 'Statement' (United Nations Permanent Forum on Indigenous Issues (Seventh Session): Discussions on the Pacific, New York, 23 April 2008), noting that '[t]he Pacific Islands are extremely vulnerable to climate change and its effects. As a result, we actively support the South Pacific Regional Environment Programme, the development of the Pacific Framework and Action Plan on Climate Change and community-level projects on adaptation to climate change'.

[125] New Zealand Ministry of Foreign Affairs and Trade, 'Climate Change and the Issue of Environmental Migrants: A Proposed Revised Approach' (8 August 2008) (document circulated to the Prime Minister, the Minister Responsible for Climate Change Issues, the Minister for the Environment, and the Minister of Immigration, released pursuant to an Official Information Act request) 5.

[126] 'Background: Environmental Migrants/Relocation/Displacement' (n 124) 343.

[127] ibid, 344.

framework.[128] According to the Assistant High Commissioner (Protection), '[i]n our assessment the time has come to work with states to develop an internationally agreed doctrine of temporary protection, which would ensure the availability of interim protection to people in temporary need'.[129] This may appeal to host States because it does not require them to permanently resettle people. It may appeal to those at risk of climate change-related displacement because it secures interim assistance and a temporary right of stay elsewhere. Having a foothold via a temporary protection mechanism may also help to shore up a subsequent, more permanent status if it becomes necessary. An instrument modelled on the Guiding Principles on Internal Displacement, for example, could identify the specific needs of the displaced within the framework of States' existing international human rights obligations and help to formalize long-standing ad hoc schemes of temporary protection.[130] This is discussed further in Chapter 9.

In some situations, temporary protection will be sufficient. However, it is likely that some people fleeing climate change-related impacts will need permanent solutions (such as the inhabitants of some small island States).[131] In such cases, it would be preferable to circumvent the political charade of temporary protection and instead develop policies that can facilitate more permanent movement for those who need and desire it. This is why principled advocacy for durable migration options remains very important, especially in contexts such as the Pacific where relatively small numbers of people may need assurances that they can ultimately settle elsewhere. This does *not* mean that top-down relocation policies should be imposed. Rather, on-going governmental dialogue about migration as an adaptation response, informed by in-depth community consultation and participation, must be part of the climate change agenda. The next chapter examines this in more detail.

[128] See, eg, António Guterres, 'High Commissioner's Closing Remarks: 2010 Dialogue on Protection Gaps and Responses' (UNHCR, Geneva, 9 December 2010) 2; António Guterres, 'Statement by António Guterres, United Nations High Commissioner for Refugees' (Nansen Conference on Climate Change and Displacement in the 21st Century, Oslo, 6 June 2011) 4, 5.

[129] Feller (n 1) 9.

[130] See, eg, Guy S Goodwin-Gill, '*Non-Refoulement* and the New Asylum Seekers' (1986) 26 Virginia Journal of International Law 897; Deborah Perluss and Joan F Hartman, 'Temporary Refuge: Emergence of a Customary Norm' (1986) 26 Virginia Journal of International Law 551.

[131] See Ch 5. This was also noted in 2008 by the UN High Commissioner for Refugees: António Guterres, 'Keynote Address' (Ten Years of Guiding Principles on Internal Displacement: Achievements and Future Challenges, Oslo, 16 October 2008).

5

'Disappearing States', Statelessness, and Relocation

It is for the people to determine the destiny of the territory
and not the territory the destiny of the people.[1]

I. Introduction

The 'disappearing States' or 'sinking islands' phenomenon has become the 'canary in the coalmine'[2]—the litmus test for the dramatic impacts of climate change on human society. Atlantis-style predictions of whole countries disappearing beneath the waves raise fascinating legal issues. As a purely academic exercise, pondering the dissolution of a State because of climate change rather than conflict, cession, merger, or succession entails novel questions that go to the heart of legal rules on the creation and extinction of States. Certainly, the potential loss of territory or the total displacement of a population and/or government for environmental or climatic reasons is novel.[3] However, much of this deliberation is taking place in the abstract, such that the premises for why, when, whether, and how States might 'disappear', and the consequences of this, do not always sit comfortably with the empirical evidence.[4] There is a risk that however academically stimulating and challenging these questions of extinction are, their practical relevance is undermined by some of the assumptions on which they are based. This, in turn, could lead to the adoption of well-intentioned but ultimately misguided policies.[5]

Though this chapter avoids alarmist discourses of 'drowning' States, it recognizes that in the absence of adaptation and migration strategies, there could be widespread population displacement from small island nations rendered uninhabitable

[1] *Western Sahara Case* (Advisory Opinion) [1975] ICJ Rep 12, 122 (Separate Opinion of Judge Dillard).

[2] See President Mohamed Nasheed (Maldives), 'Address to the United Nations General Assembly' (21 September 2009) <http://www.unmultimedia.org/tv/unifeed/d/13548.html> accessed 1 October 2009.

[3] Susin Park, 'Climate Change and the Risk of Statelessness: The Situation of Low-Lying Island States', UNHCR Legal and Protection Policy Research Series, PPLA/2011/04 (May 2011) 3.

[4] Graeme Hugo, 'Climate Change-Induced Mobility and the Existing Migration Regime in Asia and the Pacific' in Jane McAdam (ed), *Climate Change and Displacement: Multidisciplinary Perspectives* (Hart Publishing, 2010).

[5] See Ch 7.

as a result of climate change impacts on their already fragile ecosystems. There is, accordingly, a risk of '*de facto* statelessness, which could turn into *de jure* statelessness should the affected States be considered to have ceased existence'.[6]

This chapter is anchored in a case study of two small Pacific island nations, Kiribati and Tuvalu, which have become emblematic of the so-called 'sinking States' and 'climate refugee' phenomenon. It is a response to the observation that a lack of specificity in climate migration research means that many of the normative and policy recommendations being made at the macro level are divorced from context.[7] It draws extensively on my own fieldwork in these two small island States in 2009, as well as on the work of geographers over the past few decades who have documented the socio-economic conditions that bear on patterns of movement in the Pacific. The international law analysis of statehood and statelessness is thus set against this particular empirical backdrop.

The chapter is underscored by two premises. First, the disappearance of the international legal personality of small island States is not inevitable. Even if territory is rendered uninhabitable for the bulk of the population, this will not automatically trigger the extinction of the State itself.[8] Secondly, it may be counter-productive to perpetuate the idea of the 'disappearing State'. As Barnett and Adger have argued:

> the result of lost confidence in atoll-futures may be the end of the habitability of atolls. But this could, we argue, be brought about less by the physical impacts of climate change per se, and more by a common expectation of serious climate impacts leading to changes in domestic resource use and decreased assistance from abroad.[9]

While vulnerability narratives may have a certain political potency, they may also undermine capacity and contribute to a sense of fatalism that accelerates the demise.[10] This is why a number of affected States have strongly countered the discourse of 'climate refugees' and proposals to include relocation obligations within multilateral treaties.[11] Similarly, a key message from a United Nations High Commissioner for Refugees-convened (UNHCR) international expert meeting

[6] Park (n 3) 23. See also UNHCR, supported by the International Organization for Migration and the Norwegian Refugee Council, 'Climate Change and Statelessness: An Overview' (Submission to the 6th Session of the Ad Hoc Working Group on Long-Term Cooperative Action (AWG-LCA 6) under the UN Framework Convention on Climate Change (UNFCCC), 1–12 June 2009) <http://unfccc.int/resource/docs/2009/smsn/igo/048.pdf> accessed 11 August 2011; 'Submission by the High Commissioner for Refugees to the Office of the High Commissioner for Human Rights Compilation Report Universal Periodic Review: The Republic of the Marshall Islands' (April 2010) 2: 'The worst case scenario, involving complete submersion under rising sea levels, may cause widespread "external displacement" and a *de facto* or *de jure* loss of the sovereign State itself. UNHCR's mandate and experience in the field of statelessness may be relevant in that situation'.

[7] Jon Barnett and Michael Webber, 'Migration as Adaptation: Opportunities and Limits' in McAdam (n 4).

[8] See also 'Summary of Deliberations on Climate Change and Displacement' (UNHCR Expert Roundtable on Climate Change and Displacement, Bellagio, 22–25 February 2011) (April 2011) ('Bellagio Deliberations') para 30.

[9] Jon Barnett and W Neil Adger, 'Climate Dangers and Atoll Countries' (2003) 61 Climatic Change 321, 330.

[10] Orientalist-like vulnerability narratives are constructed around assumptions about power, knowledge, and 'solutions': Jon Barnett and John Campbell, *Climate Change and Small Island States: Power, Knowledge and the South Pacific* (Earthscan, 2010) 165.

[11] See Ch 7.

on climate change and displacement was that 'the legal presumption of continuity of statehood needs to be emphasized and the notion and language that such states will "disappear" (*i.e.*, lose their international legal personality) or "sink" ought to be avoided'.[12]

However, as that meeting rightly noted, small island developing States face profound humanitarian and protection issues as a consequence of climate change.[13] If people seek to move to or remain in another country without permission to do so, they may find themselves in a very precarious legal position, potentially without work rights, basic health care, or social services.[14] It is not clear that they would be recognized as *de jure* stateless as a matter of international law, and even if they were, few States have statelessness determination procedures in place to recognize this.[15]

This chapter therefore provides a measured response to some of the ill-informed notions circulating in the literature by: (a) examining empirical evidence about projected climate change impacts on Kiribati and Tuvalu, in conjunction with pre-existing environmental and socio-economic stressors; (b) analysing the position in international law with respect to State continuity and extinction; (c) demonstrating why people who may move from affected small island States would be unlikely to be regarded as 'stateless persons' as a matter of international law, and where the protection gaps arise; (d) examining why proposals for *en masse* relocation of national groups to other States is problematic from the perspective of human rights law; and (e) examining alternative constructs for the maintenance of nationhood, such as self-governance in free association with another State, as a means of preserving culture, identity, and community. The chapter posits that considering these issues now, while there is time to enhance mechanisms for planned, pre-emptive movement, could provide greater certainty and predictability for the future and prevent displacement.

II. Conceptual Problems: Macro versus Micro

The 'sinking island State' phenomenon is frequently raised in the media and scholarly literature,[16] but rarely analysed.[17] As 'a synecdoche: a representation of all

[12] Bellagio Deliberations (n 8) 'Main Messages' 2. [13] ibid, para 31.

[14] See further, Park (n 3) 14.

[15] See The Equal Rights Trust, *Unravelling Anomaly: Detention, Discrimination and the Protection Needs of Stateless Persons* (The Equal Rights Trust, 2010).

[16] See, eg, James Bone and Robin Pagnamenta, 'We Are Sinking, Say Maldive Islanders, But There Is Still Time to Save the World', *The Times* (London, 23 September 2009) <http://www.timesonline.co.uk/tol/news/environment/article6845261.ece> accessed 11 August 2011; Rowan Callick, 'Don't Desert Us, Say Sinking Pacific Islands', *The Australian* (Australia, 30 July 2009) 6; Rupert Colville, 'Vanishing Homelands: New Twist to a Sad Tale' *Bangkok Post* (Bangkok, 7 February 2008); Chloe Lambert, 'That Sinking Feeling: What Would You Do If Your Country Was Disappearing under the Sea?', *The Times* (London, 18 March 2009); Jo Lateu, 'That Sinking Feeling: Climate Refugees Receive Funds to Leave Islands', *New Internationalist* (Oxford, March 2008) 26; Nicholas Schmidle, 'Wanted: A New Home for My Country', *The New York Times* (New York, 10 May 2009) 38; Craig Sherborne, 'Sinking Sandbanks', *The Monthly* (Melbourne, March 2009) <http://www.themonthly.com.au/node/1472> accessed 12 September 2009.

[17] Some exceptions include Park (n 3); Carol Farbotko, 'Representing Climate Change Space: Islographs of Tuvalu' (PhD thesis, University of Tasmania, School of Geography and Environmental

threatened islands and greenhouse disasters',[18] it has become emblematic of the most extreme impacts of climate change on human society, but is used haphazardly even by experts in the field.[19] The 'sinking islands' are often portrayed as the epitome of climate-related movement, connoting a straightforward cause (climate change) and effect (displacement). However, this is not borne out by the empirical evidence.

In part, such assumptions may be fuelled by the way some small island States themselves have used the imagery of the drowning homeland to emphasize the impacts of climate change. Certainly, some Pacific leaders have invoked climate change as an existential threat,[20] and as a form of 'eco-terrorism'.[21] Perhaps the most arresting example of this to date was an underwater Cabinet meeting held by the government of the Maldives in September 2009 to highlight its concerns about rising sea levels.[22]

At a more formal level, in June 2009, the Pacific Island States, with the support of a number of other countries, sponsored a United Nations (UN) General Assembly resolution on 'Climate Change and Its Possible Security Implications'.[23] During debate, delegates referred to the unprecedented 'real possibility' of 'the disappearance of whole nations',[24] and the resolution's 'pursuit of greater guarantees of our territorial integrity'.[25] The President of the Federated States of Micronesia placed climate change in a human rights context by stressing its impacts on 'our own security and territorial integrity, and on our very existence as inhabitants of very small and vulnerable island nations'. In this regard, he said that climate

Studies 2008); Carol Farbotko, 'Tuvalu and Climate Change: Constructions of Environmental Displacement in the *Sydney Morning Herald*' (2005) 87B(4) Geografiska Annaler 279. For some helpful international law scholarship, focusing in particular on law of the sea issues, see Rosemary Rayfuse, 'W(h)ither Tuvalu? International Law and Disappearing States' (2009) University of New South Wales Faculty of Law Research Series Working Paper No 9, 8 <http://law.bepress.com/unsw-wps/flrps09/art9/> accessed 11 August 2011; Alfred HA Soons, 'The Effects of a Rising Sea Level on Maritime Limits and Boundaries' (1990) 37 Netherlands International Law Review 207; David Freestone and John Pethick, 'Sea Level Rise and Maritime Boundaries: International Implications of Impacts and Responses' in Gerald H Blake (ed), *Maritime Boundaries* (Routledge, 1994).

[18] John Connell, 'Losing Ground? Tuvalu, the Greenhouse Effect and the Garbage Can' (2003) 44 Asia Pacific Viewpoint 89, 104.

[19] At a recent conference in Geneva, one distinguished academic referred to the 'tens of millions of people who will sink from their islands'. The remark went uncontested despite the fact that the island States at risk do not have combined populations of anything near this magnitude.

[20] As early as 1992 the South Pacific Forum 'reaffirmed that global warming and sea level rise are the most serious threats to the Pacific region and the survival of some island states': Forum Communiqué (23rd South Pacific Forum, Honiara, 8–9 July 1992) Doc SPFS(92)18, para 7. See further, Lorraine Elliott, 'Climate Migration and Climate Migrants: What Threat, Whose Security?' in McAdam (n 4); Ch 1.

[21] President Anote Tong (Kiribati) stated at a UN conference on small island States that 'deliberate acts by some [polluting countries] to secure benefits at the cost of others could only be compared to an act of terrorism—eco-terrorism if you wish': 'Kiribati Accuses Polluting Countries of "Eco-Terrorism"', *Agence France-Presse* (Port Louis, 13 January 2005) <http://www.spacedaily.com/2005/050113083110.lc5jiprp.html> accessed 6 May 2011.

[22] See, eg, 'Maldives Cabinet Makes a Splash', *BBC News* (London, 17 October 2009) <http://news.bbc.co.uk/2/hi/8311838.stm> accessed 20 October 2009.

[23] UNGA res 63/281 (3 June 2009). On climate change as an existential threat, see Elliott (n 20).

[24] UNGAOR 63rd Session, 9th plenary meeting (25 September 2008) UN Doc A/63/PV.9, Mr Chin (Palau).

[25] ibid, Mr Litokwa Tomeing (President of the Marshall Islands).

change must be approached 'from a holistic perspective rather than limiting it to the dimensions of sustainable development, to humanitarian or technical issues, or to economic or environmental issues'.[26] The President of Vanuatu noted the risk that 'some of our Pacific colleague nations will be submerged. If such a tragedy should happen, then the United Nations and its members will have failed in their first and most basic duty to a Member and its innocent people, as stated in Article 1 of the Charter of the United Nations'.[27] Arguing along the same lines, the President of Nauru expressed the expectation that 'the Security Council will review particularly sensitive issues such as the implications of the loss of land and resources and the displacement of people for sovereignty and international legal rights'.[28]

However, while the image of an island disappearing beneath the rising sea provides a potent, frightening basis from which to lobby for global reductions in carbon emissions, it is not necessarily as useful in getting the international community to develop normative frameworks to respond to climate change-related movement. Indeed, it may contribute to misunderstandings about the likely patterns, timescale, and nature of such movement. That is not to say that this approach is disingenuous, but rather that it is important to be alert to the particular objectives it may promote: raising awareness of climate impacts on small island States, providing pressure for political outcomes in climate negotiations, and making the international community aware that a failure to act on global emissions may ultimately lead to serious destruction of human society and structures. Often this sort of advocacy involves simplifying the issues,[29] and partially because of this, Pacific governments cannot agree among themselves on a common approach to the issue of movement.

III. The Nature of 'Disappearance'

Though some States themselves use the 'disappearing islands' imagery to dramatic effect, the empirical evidence suggests that a simple 'climate change' cause and effect is not so straightforward, and motivations for movement even less so. Though climate change is having real impacts on small island States, the Atlantis-style predictions that have captivated the imagination of some are unlikely to materialize as the means by which States cease to exist.

The following section provides some general background about Kiribati and Tuvalu, and the impact that climate change is predicted to have there (in

[26] UNGAOR 63rd Session, 10th plenary meeting (25 September 2008) UN Doc A/63/PV.10, Mr Emanuel Mori (President of the Federated States of Micronesia).

[27] UNGAOR 63rd Session, 11th plenary meeting (26 September 2008) UN Doc A/63/PV.11, Mr Kalkot Matas Kelekele (President of the Republic of Vanuatu).

[28] UN Doc A/63/PV.9 (n 24) Mr Marcus Stephen (President of the Republic of Nauru); Marcus Stephen (President of the Republic of Nauru), 'On Nauru, A Sinking Feeling', *The New York Times* (New York, 18 July 2011) <http://www.nytimes.com/2011/07/19/opinion/19stephen.html?_r=1> accessed 22 July 2011.

[29] See Ch 1.

conjunction with existing environmental and socio-economic pressures). It concludes that the focus on loss of *territory* as the indicator of a State's 'disappearance' may be misplaced, since these countries are likely to become uninhabitable as a result of diminished water supplies long before they physically disappear.[30] In international law terms, the absence of population, rather than territory, may provide the first signal that an entity no longer displays the full indicia of statehood (namely, a defined territory, a permanent population, an effective government, and the capacity to enter into relations with other States).

Kiribati and Tuvalu are independent small island States in the South Pacific. They are both Least Developed Countries (LDCs) on account of their relatively low national income, weak human assets, and high economic vulnerability.[31] Despite belonging to different ethnic groups (Polynesian and Micronesian respectively), they were claimed by the British in 1892 as a single protectorate—the Gilbert and Ellice Islands—which became a Crown colony in 1916 until independence was achieved some 30 years ago (Tuvalu in 1978; Kiribati in 1979).

Kiribati has a population of around 95,000,[32] while Tuvalu is the world's smallest State (apart from the Vatican), with only 10,000 people.[33] Though relatively tiny in population terms, both countries suffer from significant overpopulation and crowding.[34] This has attendant problems for employment, food and water security, and environmental sustainability. Of the States threatened by eventual annihilation, Kiribati has the largest population (especially in light of future population growth), and virtually no capacity for long-term internal migration because of an absence of high land.[35] Half of Kiribati's population lives on the main island of Tarawa, and the population is increasing rapidly, particularly as people move from outer islands in search of work in the urban centre.[36] On its

[30] As Crawford observes, 'the substrate of the State is not property, it is the people of the State seen as a collective': James Crawford, *The Creation of States in International Law* (2nd edn, Oxford University Press, 2006) 717. Note that in October 2011, Tuvalu and Tokelau each declared a state of emergency because of severe water shortages: Kirsty Needham, 'Australia Helps as Tuvalu Runs Dry', *Sydney Morning Herald* (Sydney, 7 October 2011) 3; 'Tokelau, Tuvalu Declare State of Emergency', *ABC News*, (5 October 2011); Jen Lacey, 'Need Water Now: Australia Sends Desal Plant to Tuvalu' (*ABC News*, 10 October 2011) <http://www.abc.net.au/local/stories/2011/10/10/3335802.htm?site=sydney> accessed 10 October 2011; 'Water Situation Getting More Desperate on Tuvalu', *Pacific Beat* (ABC Radio National, Australia, 10 October 2011) <http://www.radioaustralia.net.au/pacbeat/stories/201110/s3335901.htm> accessed 10 October 2011.

[31] See <http://unfccc.int/cooperation_support/least_developed_countries_portal/frequently_asked_questions/items/4743.php> accessed 6 July 2011.

[32] Republic of Kiribati, *National Adaptation Program of Action* (NAPA) (January 2007) 1, estimated in 2005.

[33] The 2002 census stated the population as 9,359: Government of Tuvalu, *Tuvalu's National Adaptation Programme of Action* (May 2007) 6.

[34] ibid; Kiribati NAPA (n 32).

[35] The average height of land in Kiribati and Tuvalu is three to four metres above sea-level: Kiribati NAPA (n 32) iii; Tuvalu NAPA (n 33) 13. Banaba, which is the only high land, was all but depopulated in the mid-1940s when its population was relocated to Fiji to facilitate phosphate mining. The President of Kiribati has mentioned the possibility of eventually relocating the government there as a means of maintaining a presence on the territory for as long as possible: interview with President Anote Tong (Tarawa, Kiribati, 12 May 2009).

[36] A 2005 census put South Tarawa's population at 40,300, an increase of almost 43 per cent over a decade: Sherborne (n 16). This represents a population density of 2,558 people per square

southern tip, the population density of the 1.2 square kilometre islet of Betio is greater than that of Hong Kong, but without the high-rise apartments to house it.[37]

Similarly, in Tuvalu, about 42 per cent of the population resides on Funafuti, which is the capital and only urban centre.[38] Internal migration to the centre is high due to increased dependence on imported foods as domestic subsistence agricultural production has declined.[39] Like Kiribati, there is no high ground for long-term internal resettlement.

In both countries, sanitation is poor and pollution is high, with beach toileting and washing very common. In Kiribati, only 20 per cent of households have access to a sewerage system, and 64 per cent do not use toilets.[40] Septic tanks seep into the groundwater supply, which is often brackish, and the tank infrastructure is too rudimentary to keep up with population growth.

The vast majority of people are unemployed. In Kiribati, less than a quarter of people have regular jobs, and of them, half work in government administration.[41] The average annual wage on Tarawa in Kiribati is 3,000 Australian dollars.[42] Most people maintain subsistence livelihoods, which are highly susceptible to climate change impacts on water and land availability, including through coastal erosion and salination of the soil.[43] In Tuvalu, only around 30 per cent of people are employed, and two thirds of them work for the government.[44] Those without jobs

kilometre: quoting Dr Gerald Haberkorn, head of the population division in the Secretariat of the Pacific Community in Samisoni Pareti, 'Urban Crisis: Towns Bursting with Growing Population and Limited Resources' <http://www.islandsbusiness.com/islands_business/index_dynamic/containerNameToReplace=MiddleMiddle/focusModuleID=17439/overideSkinName=issueArticle-full.tpl.> accessed 6 July 2011. See generally Charles W Stahl and Reginald T Appleyard, *Migration and Development in the Pacific Islands: Lessons from the New Zealand Experience* (AusAID, April 2007).

37 Betio's population density is 10,400 people per square kilometre: Pareti (n 36). By contrast, in Hong Kong it was 6,426 persons per square kilometre in 2010: US Department of State, 'Background Note: Hong Kong' (15 March 2011) <http://www.state.gov/r/pa/ei/bgn/2747.htm> accessed 7 May 2011.

38 It has a population density of 1,610 people per square kilometre, compared to the outer islands where it is only 222 people per square kilometre: Tuvalu NAPA (n 33) 6. As Connell observed in 2003, the concentration of public sector employment on the main island of Funafuti 'has resulted in rapid internal migration and urbanisation', and it accommodates almost half of Tuvalu's population: Connell (n 18) 93. See further, John Connell and John P Lea, *Urbanisation in the Island Pacific: Towards Sustainable Development* (Routledge, 2002).

39 For example, it is reported that 60 per cent of pulaka pit plantations in Tuvalu have already been destroyed by salt-water intrusion, and the remaining 40 per cent remain highly sensitive to it: Tuvalu NAPA (n 33) 28.

40 Sherborne (n 16) citing a Kiribati government report, *The Challenge: Things (Beginning to) Fall Apart.*

41 In 2000, only 11 per cent of I-Kiribati had a regular job: Kiribati NAPA (n 32) 1. By 2009, the figure was reported to be 25 per cent: Sherborne (n 16).

42 Sherborne (n 16).

43 Kiribati NAPA (n 32) 4.

44 See Connell (n 18) 92; Asian Development Bank, *Country Partnership Strategy Tuvalu 2008–2012* (Asian Development Bank, 2008) 41 <http://www.adb.org/documents/CPSs/TUV/2008/CPS-TUV-2008-2012.pdf> accessed 14 October 2011. Employment rate figures are based on the 2002 census for Tuvalu. Censuses of Population and Housing are held every 10 years in Tuvalu, with the next census scheduled for 2012: Government of Tuvalu Central Statistics division <http://www.spc.int/prism/country/tv/stats/Census%20&%20Surveys/Census_index.htm> accessed 21 July 2011.

rely heavily on their employed relatives (including via remittances from overseas) and foreign aid.[45] As Connell observes, the social and resourcing problems 'are largely a function of overcrowding', and 'the conventional disadvantages of small island states [are] further accentuated by [Tuvalu's] very small size, extreme isolation and fragmentation'.[46]

In the Otin Taai Declaration of 2004, the Pacific Council of Churches acknowledged the ways in which human-induced climate change will significantly affect Pacific island countries. Likely impacts include loss of coastal land and infrastructure due to erosion, inundation, sea-level rise, and storm surges; an increase in the frequency and severity of cyclones, creating risks to life, health, and homes; loss of coral reefs, with attendant implications for the ecosystems on which many islanders' livelihoods depend; changing rainfall patterns, leading to flooding in some areas, drought in others, and threats to fresh water supplies; salt-water intrusion into agricultural land; and extreme temperatures.[47]

There are also underlying environmental and socio-economic conditions which enhance the islands' vulnerability. For example, Kiribati's National Adaptation Programme of Action (NAPA) identifies the following issues impacting on its ability to cope with further changes to the climate:

- emerging 'unacceptable level of inequity';
- increasing population;
- deteriorating states of coastal zones, coral reefs, fisheries, fresh ground water, health and biodiversity;
- inadequate urban services such as water supply and sanitation;
- overexploitation of natural resources in urban Tarawa; and
- difficultly [sic] in enforcing land use management strategies and controls.[48]

Tuvalu's NAPA also acknowledges the increasing environmental issues 'arising from the growing urbanization of Funafuti and decreasing labour force on the outer islands',[49] which place further stress on 'sensitive environments and major sources of food security and livelihood'.[50] Further, coastal erosion from sea-level rise, overland flooding, storm surges, tropical cyclones, and major hurricanes is exacerbated by anthropogenic causes including building, excavation, and coastal development activities.[51] The risk of displacement is therefore likely to result from

[45] Remittances constitute 7 per cent of GDP in Tuvalu: *Asian Development Outlook 2011: South–South Economic Links* (Asian Development Bank, 2011) 234. Primarily drawn from seafarers and Funafati-based public servants, remittance incomes account for 50 to 60 per cent of family income on some islands: Asian Development Bank (n 44) 2, 21. The Kiribati NAPA notes that Kiribati 'will continue to depend on assistance from abroad for socio economic development. This dependency is likely to increase into the foreseeable future because of the additional need of protection from climate variability and climate change and the adverse affects on key infrastructure and environmental resources': Kiribati NAPA (n 32) 7.

[46] Connell (n 18) 92.

[47] The Pacific Churches' Consultation on Climate Change—A Statement and Recommendations from the World Council of Churches and WCC Member Churches in the Pacific (Tarawa, March 2004) ('Otin Taai Declaration').

[48] Kiribati NAPA (n 32) 1 (fn omitted).

[49] Tuvalu NAPA (n 33) 18.

[50] ibid, 24.

[51] ibid, 12.

a combination of these factors overlaid by climate change, which may provide a tipping point.

Acknowledging these underlying problems does not undermine the devastating impacts that climate change may have on these countries. In fact, it helps to emphasize the importance of implementing context-appropriate adaptation strategies. Physical adaptation will be inadequate if underlying socio-economic limitations, which also require funding and management, remain unaddressed. In the absence of a holistic approach, responses will be piecemeal and inadequate.[52] International responses to climate change—including funding for adaptation—must be attuned to this so that funds, technical expertise, and policy responses can be directed in a suitable and sustainable manner.

A. Climate change, human movement, and the State

The discussion about 'sinking islands' is premised on the assumption that at some point, the territories of States such as Kiribati and Tuvalu will disappear on account of rising sea levels—either completely, or to the point that they can no longer sustain permanent populations. Yet, many false assumptions are made about the role of climate change in forcing movement away from small Pacific island countries, which in turn impact on debates about their on-going statehood. Though it might seem logical to assume that criteria conditioning a State's creation should also apply with respect to its 'legal demise', such that 'where the territory of a state becomes submerged by the sea, or where the population of a state evacuates *en masse* to other territories... it should be possible to conclude that the state has ceased to exist',[53] the law is not so clear-cut.

Although international law contemplates the disappearance of States, it does so within the context of State succession. The conventional ways in which a State can become extinct—through voluntary absorption by another State, merger with another State, or extinction by dissolution (voluntary or involuntary)[54]—all presuppose that a successor State begins to exist on, or assumes control over, the territory of the previous State. Indeed, the two treaties on State succession define this as 'the replacement of one State by another in the responsibility for the international relations of territory'.[55] There is never simply a void. As Marek observed in her leading work on the identity and continuity of States, a State's extinction entails a

[52] See Robin Bronen, 'Climate-Induced Community Relocations: Creating an Adaptive Governance Framework based in Human Rights Doctrine' (2011) 35 New York University Review of Law and Social Change 356.

[53] Matthew CR Craven, 'The Problem of State Succession and the Identity of States under International Law' (1998) 9 European Journal of International Law 142, 142 159 (citations omitted).

[54] See generally, Crawford (n 30) ch 17.

[55] Vienna Convention on Succession of States in Respect of Treaties (adopted 23 August 1978, entered into force 6 November 1996) 1946 UNTS 3, Art 2(1)(b); Vienna Convention on Succession of States in Respect of State Property, Archives and Debts (adopted 8 April 1983, not yet in force) UN Doc A/CONF.117/14 (1983) Art 2(1)(a).

succession and prevents any further continuity of that State; a 'miraculous resurrection' is impossible.[56]

By contrast, the potential extinction of a State because of climate change would be markedly distinct. The presumption would be that the abandoned territory could no longer sustain human life and thus could not be assumed by any other State. Unless the territory were ceded to another State, the normal rules on State succession would not apply. Accordingly, the conventional international law rules on extinction are largely inapplicable to the climate change context. For this reason, this chapter turns its attention to the *creation* of States, to determine at what point the absence of certain criteria of statehood might lead other States (and the international community, through international organizations) to conclude that a State no longer exists.

IV. What is a 'State'?

Whether or not a State exists is a 'mixed question of law and fact'.[57] The absence of a formal international law definition of a 'State' might be explained by the fact that (a) questions about an entity's nature only tend to arise in borderline cases, and (b) States like to preserve as much freedom of action as possible with respect to new States.[58] Logically, this might also be said to apply in reverse—to enable States themselves to determine when an entity's loss of the indicia of statehood should indicate the end of that State. Crawford queries whether the rules determining statehood 'have been kept so uncertain or open to manipulation as not to provide any standards at all'.[59]

The classic formulation of statehood is contained in Article 1 of the 1933 Montevideo Convention on the Rights and Duties of States,[60] which is generally regarded as reflecting customary international law. The four elements of statehood are: a defined territory, a permanent population, an effective government, and the capacity to enter into relations with other States. While all four criteria would seemingly need to be present for a State to come into existence, the lack of all four may not mean the end of a State. This is because of the strong presumption of continuity of existing States,[61] which may account for the fact that since

[56] Krystyna Marek, *Identity and Continuity of States in Public International Law* (Librairie E Droz, 1954) 5–6 referred to in Crawford (n 30) 669. But see the case of Syria: 690.

[57] Humphrey Waldock, 'General Course on Public International Law' (1962) 106 Recueil des Cours de l'Académie de Droit International 5, 146 cited in Crawford (n 30) 5. For an overview of the competing theories of statehood, see Crawford (n 30) Ch 1.

[58] Crawford (n 30) 45; see also 40.

[59] ibid, 45. Crawford notes that, at times, States have treated as States entities that do not come within the accepted definition of the term, such as the Holy See (1870–1929), British India (1919–47), and the United Nations (UN) membership of Byelorussia and Ukraine.

[60] Montevideo Convention on the Rights and Duties of States (adopted 26 December 1933, entered into force 26 December 1934) 165 LNTS 19.

[61] Crawford says international law is 'based on this assumption': Crawford (n 30) 715, 701. He says 'there is a strong presumption against the extinction of States once firmly established': 715. See generally references referred to there: Marek (n 56) 548; Oscar Schachter, 'State Succession: The

the establishment of the UN Charter in 1945,[62] there have been very few cases of State extinction and virtually none of involuntary extinction.[63] It is also significant that so-called 'failed States' have continued to be recognized as States even during the period when they were objectively failing.[64] As Craven observes, an analysis of State practice reveals that 'in many cases the issue is not simply one of determining the existence of the state, but rather the degree of identity and extent of continuity'.[65]

The next section briefly highlights the key elements of each criterion of statehood to tease out possible implications for the 'sinking island' scenario.

A. Defined territory

Crawford writes: 'Evidently, States are territorial entities'.[66] But do they need to remain so in order to preserve their legal status? Certainly, there is no minimum amount of territory that needs to be held, and the loss of some territory at least should not affect the legal status of the entity, since it is not necessary for a State to have precisely defined boundaries.[67] The requirement is only that 'the right to be a State is dependent at least in the first instance upon the exercise of full governmental powers with respect to some area of territory'.[68] Such territory does not have to be contiguous, and '[l]ittle bits of States can be enclaved within other States'.[69]

Once and Future Law' (1993) 33 Virginia Journal of International Law 253, 258–60; Roda Mushkat, 'Hong Kong and Succession of Treaties' (1997) 46 International and Comparative Law Quarterly 181, 183–7; Martti Koskenniemi, 'The Wonderful Artificiality of States' (1994) 88 Proceedings of the American Society of International Law 22.

[62] Charter of the United Nations (adopted 26 June 1945, entered into force 24 October 1945) 1 UNTS XVI.

[63] Crawford (n 30) 715.

[64] Daniel Thürer, 'The "Failed State" and International Law' (1999) 81 International Review of the Red Cross 731. Crawford argues, however, that the notion of a 'failed State' involves some conceptual confusion, and that many cited cases of 'failed States' are in fact crises of government or governance, rather than about the extinction of the State in question: Crawford (n 30) 721–2. If a 'failed State' describes 'a situation where the structure, authority (legitimate power), law, and political order have fallen apart and must be reconstituted in some form, old or new', the very notion of 'reconstitution' suggests that a reformulation of the State is possible, *qua* State, rather than as some other kind of entity: I William Zartman, 'Introduction: Posing the Problem of State Collapse' in I William Zartman (ed), *Collapsed States: The Disintegration and Restoration of Legitimate Authority* (Lynne Rienner Publishers, 1995) 1 cited in Crawford (n 30) 720.

[65] Craven (n 53) 160. He goes on to state: 'What this means is that emphasis should not be so much upon the existence of "external" rules of succession that allow for the "transference" of rights and duties from one subject to another, but rather upon determining the extent to which legal continuity should follow from elements of material (social, cultural or political) identity'.

[66] Crawford (n 30) 46.

[67] See, eg, *Deutsche Continental Gas-Gesellschaft v Polish State* (1929) 5 AD 11, 15 in which the German–Polish Mixed Arbitral Tribunal observed that it is sufficient that 'territory has a sufficient consistency, even though its boundaries have not yet been accurately delimited'.

[68] Crawford (n 30) 46. The field of critical geopolitics has critiqued the international law position as a statist 'territorial trap': see, eg, John Agnew, 'The Territorial Trap: The Geographical Assumptions of International Relations Theory' (1994) 1 Review of International Political Economy 53; Cornelius Castoriadis, *The Imaginary Institution of Society* (Polity Press, 1987).

[69] ibid, 47 referring to *Case concerning Sovereignty over Certain Frontier Land (Belgium/Netherlands)* [1959] ICJ Rep 209, 212–13, 229; *Case concerning Right of Passage over Indian Territory*

While 'territory which was once connected to land and then submerged by the sea can continue to be regarded as a connected part of State territory', a wholly artificial construction cannot.[70]

The link between statehood and territory is crucial. Inherent in the possession of territory (as an indicator of statehood) is exclusive control over it.[71] Crawford therefore frames the territorial requirement of statehood as 'a constituent of government and independence', rather than as a separate criterion.[72] While Lowe argues that the concept of a State 'is rooted in the concept of control of territory', he says this is arguably more about ensuring that the criterion of independence is met rather than about the territory per se, since such control is 'to ensure that activities within its borders are not regulated by any other State'.[73] Jessup argued that the rationale for a State needing to possess territory was that 'one cannot contemplate a State as a kind of disembodied spirit... [T]here must be some portion of the earth's surface which its people inhabit and over which its Government exercises authority'.[74]

And yet, as will be examined below, States can continue to function even when their governments operate from outside their territory. The mechanism of the 'government in exile' has enabled governments to function extraterritorially, although this has always been contemplated as temporary and exceptional. Furthermore, it is premised on the continued existence of a permanent population on the State's territory (although the government in exile also retains jurisdiction over nationals abroad). Indeed, the general requirement that States have 'a certain coherent territory effectively governed'[75] assumes that there remains a population on that territory to be governed.

(Portugal v India) [1960] ICJ Rep 6, 27. However, as a practical matter, there have been situations in the past where States have refused to recognize as States territories that are highly fragmented. For instance, the UK argued that 'the fragmentation of the territory of Bophuthatswana within South Africa, the pattern of the population and the economic dependence on South Africa more than justify our refusal to recognize Bophuthatswana': Minister of State, Foreign and Commonwealth Office, HC Deb 12 February 1977, vol 105, col 100 cited in Crawford (n 30) 47.

[70] *In re Duchy of Sealand*, Administrative Court of Cologne, 3 May 1978 (1989) 80 International Law Reports 683, 686; see also 685. That case concerned an entity established on a former anti-aircraft platform erected by the UK eight miles off its coast, attached by concrete pillars to the seabed. Article 60(8) of the UN Convention on the Law of the Sea (adopted 10 December 1982, entered into force 16 November 1994) 1833 UNTS 396 provides that '[a]rtificial islands, installations and structures do not possess the status of islands. They have no territorial sea of their own, and their presence does not affect the delimitation of the territorial sea, the exclusive economic zone or the continental shelf'.

[71] Crawford (n 30) 48. 'The only requirement is that the State must consist of a certain coherent territory effectively governed': 52. See section VII. B. below on leased land.

[72] ibid, 52.

[73] Vaughan Lowe, *International Law* (Oxford University Press, 2007) 138.

[74] US Ambassador Jessup, UNSCOR 383rd meeting (2 December 1948) 11 cited in Crawford (n 30) 48. Yet, in the context of belligerent occupation, Grant writes: 'Territory is not necessary to statehood, at least after statehood has been established... [it] appears to be the case that once an entity has established itself in international society as a state, it does not lose statehood by losing its territory or effective control over that territory': Tom D Grant, 'Defining Statehood: The Montevideo Convention and its Discontents' (1999) 37 Columbia Journal of Transnational Law 403, 435.

[75] Crawford (n 30) 52.

B. Permanent population

While 'defined territory' is one criterion of statehood, and though territory ultimately may disappear as a result of rising sea levels, it is more probable that the other indicia of statehood—a permanent population, an effective government, and the capacity to enter into relations with other States—will have been challenged long before this occurrence.

For low-lying islands such as Tuvalu and Kiribati, insufficient fresh water, as the water lens shrinks, has been cited as the most probable trigger for rendering these countries uninhabitable in the longer term.[76] If the area of the island diminishes, for example through coastal erosion and sea-level rise, then the size of the lens will shrink proportionately. The greatest threat to freshwater supplies is from overwash, which increases the salinity of the water and thus reduces its viability for drinking and agriculture.[77] This will increase with coastal erosion and more frequent storms. In Tuvalu, groundwater supplies have been contaminated by salt-water intrusion and waste and are no longer safe for human consumption.[78] Although each house on Funafuti has a rainwater tank, it is not always functional or attached to a tap. Furthermore, prolonged drought means that rainwater may become increasingly scarce. In October 2011, the government of Tuvalu declared a state of emergency on account of severe water shortages, necessitating the urgent assistance of Australia and New Zealand.[79] It is therefore likely that long before the land disappears, the bulk of the population will have moved on account of insufficient water supplies to sustain themselves and their livelihoods.

Just as international law does not require a State's territory to be a minimum size, nor is there a minimum population requirement.[80] Indeed, Tuvalu is the second smallest State by population (after the Vatican). The notion of a 'permanent' population simply means that it cannot be transitory. Thus, Shaw has argued that 'a nomadic population might thus not count for the purposes of territorial sovereignty'.[81] For present purposes, the relevant question is whether a State ceases to meet this criterion of statehood when a large proportion—or all—of its population lives outside the State's territory.[82]

There are already a number of Pacific countries with very large populations *outside* their territory and this does not affect their ability to continue to function as

[76] See, eg, Nishi Mimura and others, 'Small Islands' in Intergovernmental Panel on Climate Change (IPCC), *Climate Change 2007: Impacts, Adaptation and Vulnerability: Contribution of Working Group II to the Fourth Assessment Report of the Intergovernmental Panel on Climate Change* (Cambridge University Press, 2007) 689.

[77] Connell (n 18) 91.

[78] Tuvalu NAPA (n 33) 19.

[79] See n 30 above.

[80] On this point, see Thomas M Franck and Paul L Hoffman, 'The Right of Self-Determination in Very Small Places' (1976) 8 New York University Journal of International Law and Politics 331, 383: 'infinitesimal smallness has never been seen as a reason to deny self-determination to a population'. On the position of the Vatican, see Jorri C Duursma, *Fragmentation and the International Relations of Micro-States: Self-Determination and Statehood* (Cambridge University Press, 1996) 374, 411–12.

[81] Malcolm N Shaw, *International Law* (6th edn, Cambridge University Press, 2008) 199.

[82] See Ian Brownlie, *Principles of Public International Law* (6th edn, Oxford University Press, 2003) 70 where he states that it is linked to the territory; Peter Malanczuk, *Akehurst's Modern Introduction to International Law* (7th rev edn, Routledge, 1997) cited in Park (n 3) 5, who agrees with Brownlie.

States. For example, 56.9 per cent of Samoans and 46 per cent of Tongans live outside their own country.[83] Thus, the proportion of population living on the territory is not determinative of the population criterion for statehood. But if an exodus of population were accompanied by, or premised on, the imminent or eventual loss of territory, then would it assume a different significance? If no population remained on the territory, could the State continue to exist by retaining its own outpost on the territory (as is being contemplated in Kiribati) or elsewhere (as a government in exile or on territory that another State permits it to use)?[84] This is connected to the next section: at what point does a government cease to function?

C. Government

The existence of an effective government satisfies another requirement of statehood: independence. Crawford distinguishes between these two criteria as follows: 'government is treated as the exercise of authority with respect to persons and property within the territory of the State; whereas independence is treated as the exercise, or the right to exercise, such authority with respect to other States'.[85] He regards government as the most important criterion of statehood, 'since all the others depend upon it',[86] but notes that in practice its application may be much more complex (since it will be in borderline cases that its identification and scope will be tested).

States may nonetheless choose to recognize an entity as a State even where it is doubtful that the full signs of statehood exist. For example, in 1960 Congo was widely recognized as a State and was accepted as a UN member without dissent, even though it lacked an effective government.[87] Crawford concludes that this is because the requirement of government may be less stringent than thought, and, importantly, because it has two aspects: 'the actual exercise of authority, and the right or title to exercise that authority'.[88] In the Congo case, the conferral of independence by the former colonial power Belgium meant that there was no State against which the recognition of Congo could be unlawful. The assumption followed that where a former sovereign grants full independence, the new State has the right to govern its territory.

[83] See table in Stahl and Appleyard (n 36) 7, which draws on census information from Australia (2001), New Zealand (2001), and the US (2000) to show the population of Pacific island countries at home and the population abroad.

[84] Grote Stoutenburg refers to this as a 'population nucleus': Jenny Grote Stoutenburg, 'When Do States Disappear? Thresholds of Effective Statehood and the Continued Recognition of "Deterritorialized" Island States' (Threatened Island Nations: Legal Implications of Rising Seas and a Changing Climate Conference, New York, 23–25 May 2011) <http://www.law.columbia.edu/null/download?&exclusive=filemgr.download&file_id=5833> accessed 10 August 2011. In *In re Duchy of Sealand* (n 70) 683, the court held that for a group to constitute a 'population', they must form a cohesive, vibrant community. 'An association whose common purpose covered merely commercial and tax affairs was insufficient'.

[85] Crawford (n 30) 55. With respect to 'territory', it refers in this sense 'to the extent of governmental power exercised, or capable of being exercised, with respect to some territory and population. *Territorial sovereignty is not ownership of but governing power with respect to territory*': 56 (emphasis added).

[86] ibid, 56.

[87] ibid, 56ff.

[88] ibid, 57.

The case of secession is different because the seceding State has to establish its adverse claim, which includes demonstrating effective and stable exercise of governmental powers.[89] Indeed, this may explain why the presumption of continuity is so strong. First, premature recognition of another State could be seen as unlawful interference in the domestic affairs of the original State, which itself might undermine international stability.[90] Secondly, and related to the first point, there would otherwise be a void in international relations in which States would 'find it difficult or impossible to continue many mutually advantageous economic, administrative and technical relations with other nations'.[91]

Arguably, the case of 'disappearing islands' is more akin to the former: there would be no competing claim, and the presumption of continuity would apply until States no longer recognized the government (which could be in exile).[92] At the margins, the notion of continuity becomes quite subjective: '[i]n many instances the claim to continuity made by the State concerned will be determinative; other States will be content to defer to the position taken'.[93]

D. Capacity to enter into relations with other States

The capacity to enter into relations with other States is a conflation of the requirements of government and independence. It is, accordingly, a consequence, rather than a criterion, of statehood.[94] Crawford regards independence (sometimes also called 'sovereignty'[95]) as the central criterion for statehood, since it is the right to exercise 'in regard to a portion of the globe... to the exclusion of any other State, the functions of a State'.[96] It has two main elements: a separate existence within reasonably coherent borders, and not being subject to the authority of any other State.[97] The 'government in exile' concept reflects acceptance of the idea that a State can continue to exist even though its governmental functions cannot be performed from within its own territory (the assumption is, temporarily). Provided

[89] ibid, 58. See also at 59, drawing on Colin Warbrick, 'Recognition of States: Recent European Practice' in Malcolm D Evans (ed), *Aspects of Statehood and Institutionalism in Contemporary Europe* (Dartmouth Press, 1997) 14–16.

[90] On which, see Schachter (n 61) 259.

[91] International Law Association, 'Interim Report of the Committee on the Succession of New States to the Treaties and Certain Other Obligations of Their Successors' in *International Law Association Report of the Fifty-Second Conference* (1966) Annex E, 584 cited in Mushkat (n 61) 183. As Craven (n 53) 159 notes, 'states are not willing to jeopardize legal relations with an entity where there is clearly no successor state'.

[92] On the other hand, the absence of any other State staking a claim over the territory may make it easier for States to deny its continuity, since there is no adverse claim (and thus no risk of premature recognition which could constitute interference in the State's domestic affairs).

[93] Crawford (n 30) 668 referring also to Rein Mullerson, 'The Continuity and Succession of States, by reference to the Former USSR and Yugoslavia' (1993) 42 International and Comparative Law Quarterly 473, 477; Brigitte Stern, 'La succession d'Etats' (1996) 262 Recueil des Cours de l'Académie de Droit International 9, 82–6; Wladyslaw Czaplinski, 'La continuité, l'identité et la succession d'Etats—évaluation de cas récents' (1993) 26 Revue Belge de Droit International 374, 391–2.

[94] Crawford (n 30) 61.

[95] ibid, 89. Crawford says that 'it is better to use the term "independence" to denote the prerequisite for statehood and "sovereignty" the legal incident'.

[96] *Island of Palmas Case (United States of America v The Netherlands)* (1928) 2 RIAA 829, 838.

[97] Crawford (n 30) 66 (fn omitted).

that those functions are not interfered with or controlled by the host (or any other) State, its independence is preserved.

It is important to distinguish between independence as an 'initial qualification' for statehood, and its role for a State's 'continued existence'.[98] Crucially, for present purposes, the strong presumption of continuity of an existing State means that other States may continue to treat an entity as such, despite a lack of effectiveness[99] or even a 'very extensive loss of actual authority'.[100] It is likely that this presumption operates to avoid premature recognition of another State, which would be regarded as an unlawful interference in its domestic affairs. However, despite the lack of a competing State claim in the 'disappearing State' scenario, other contexts suggest that the continuity presumption would still operate.[101] Indeed, Crawford suggests that the concept of statehood is sufficiently flexible to accommodate entities that might otherwise be regarded as *sui generis*.[102]

As Thürer notes in the context of so-called 'failed States':

> Even when States have collapsed, their borders and legal personality have not been called in question. Such 'fictitious' States have not lost their membership of international organizations and, on the whole, their diplomatic relations have remained intact. Though they are unable to enter into new treaty obligations, the international law treaties they have concluded remain in force.[103]

Similarly, when a government operates in exile, the State continues to exist but its governmental functions are unable to be performed from within its own territory. Since the principle of territorial sovereignty means that a government may only act as a government in exile with the consent (express or implied) of the State in which it is hosted,[104] the powers of such a government are necessarily more circumscribed than when it operates from its own territory. For example, in cases concerning the scope of jurisdiction of the courts of governments in exile in Britain in the 1940s, it was observed that 'the sovereignty of any State is unrestricted on its own territory only, while on foreign territory it naturally yields to the sovereignty of the foreign State'.[105] Accordingly, 'this jurisdiction [of Dutch service courts] is only possible so far as it is authorized by the British legislature and can only be exercised in accordance with the statutory provisions referred to and subject to the conditions and safeguards specified by statute'.[106] As noted

[98] ibid, 63.

[99] ibid. The examples Crawford has in mind are unlawful invasion or annexation of a State.

[100] ibid, 89.

[101] See below on the Holy See and the Sovereign Order of Malta.

[102] See Crawford (n 30) 88: 'To suggest that entities such as "protected States" or "internationalized territories" are a priori excluded from statehood is unjustified and exaggerates the exclusivity of the international legal regime of statehood. The International Court of Justice has never made that assumption'.

[103] Thürer (n 64) 752.

[104] See further, Stefan Talmon, *Recognition of Governments in International Law: With Particular Reference to Governments in Exile* (Oxford University Press, 1998) 215ff. The host State can determine the extent to which the courts of the government in exile may exercise jurisdiction: 217.

[105] *Allied Forces (Czechoslovak) Case* (1941–42) 10 AD No 31, 123, 124 cited in Talmon (n 103) 217.

[106] *Amand v Home Secretary and Minister of Defence of Royal Netherlands Government* [1943] AC 147, 159 (Lord Wright) cited in Talmon (n 103) 217. However, the host State cannot prescribe how that jurisdiction is to be exercised: 218.

above, as long as the host State does not interfere with the functions of the government in exile, its independence is maintained.

V. Governments in Exile

There is a strong presumption in international law that States continue to exist even if there is a period without a government (or an effective one).[107] This shows the distinction between the 'State' and 'government', on which the legal position of the government in exile depends.[108] It perhaps also suggests that States are willing to tolerate a hiatus between the loss of indicia of statehood and acknowledgement that a State has ceased to exist.

The term 'government in exile' does not denote a special status or subject of international law, but rather reflects the domicile of a government (namely, 'the depository of a State's sovereignty and its representative organ in international relations').[109] History is replete with examples of governments operating in exile in the territory of other States.[110] The institution is most common in the case of belligerent occupation or illegal annexation, whereby the lawful government operates 'in exile', outside its own territory, until it once again becomes possible for that government to reassert its control in its own territory.[111] Traditionally, this has been understood as a temporary, time-bound mechanism.

The fact that governments can operate in exile suggests that the existence of territory, while essential to the original constitution of that entity as a State, is not integral to the exercise of certain governmental functions. As the French Foreign Minister wrote in 1814: 'A sovereign whose States are conquered... by the conquest only loses *de facto* possession and consequently retains the right to do everything that does not require that possession'.[112] Though a government's absence from its State does not automatically suspend or terminate existing treaties,[113] if it has to operate in exile then certain treaties may be terminated (or suspended) for reasons such as impossibility of performance or a fundamental change of circumstances.[114]

The scope of action of a government in exile depends on the rights that the host State is willing to grant it.[115] Functions that governments have continued to perform in exile include treaty-making, maintaining diplomatic relations, and

[107] Crawford (n 30) 34.

[108] ibid, 34 (fn omitted).

[109] Talmon (n 103) 16 (citation omitted).

[110] See ibid.

[111] Shain also describes the following as examples of governments in exile, although not all meet that criterion from an international law perspective: exiled groups aspiring to overthrow their home country's government (eg National Coalition Government of the Union of Burma, 1990); groups emerging as a result of political instability from the end of European colonialism (eg Gouvernement Provisoire de la République Algerienne, 1958); and groups seeking an independent political statute because of the occupation of their country by a neighbouring country (eg the Tibetan Government in Exile): Yossi Shain, *Governments-in-Exile in Contemporary World Politics* (Routledge, 1991).

[112] Charles Maurice de Talleyrand-Périgord, *Mémoires du prince de Talleyrand* (Calmann Levy, 1891) vol 2, 214–54 cited in ibid, 174.

[113] Talmon (n 103) 136; see also *Valk v Kokes* (1950) 17 ILR 114, 358.

[114] See Vienna Convention on the Law of Treaties (adopted 23 May 1969, entered into force 27 January 1980) 1155 UNTS 331, Arts 61(1), 62; Talmon (n 103) 136ff.

[115] Park (n 3) 7.

conferring immunities, privileges, and jurisdiction over nationals.[116] In particular, the exercise of diplomatic protection has included providing consular representation, lodging protests, arranging deportations of nationals, concluding amnesty agreements, and providing passports and identity documents to prevent nationals from being treated as stateless persons.[117] This last function is of particular relevance to the climate change-displacement context. In this respect, however, it is interesting to note that such documents have also been validly issued or extended by authorities in exile recognized in a lesser capacity than a government.[118]

As Park observes, presumably if a State extended an invitation to the government of an affected island State to operate within its territory, 'this would include a willingness to permit it to exercise its personal sovereignty over its nationals in the form of diplomatic and consular protection', and perhaps also 'an acceptance of its legislative jurisdiction'.[119] It is unlikely, however, to include jurisdiction to enforce its laws.[120] Its capacities would be restricted by its lack of territorial sovereignty.

The government in exile idea is premised on there still being an identifiable population over which the government has jurisdiction.[121] In the conventional case, the majority of those people will continue to reside in the State's territory, from which the government is temporarily severed. In the 'disappearing State' scenario, the need for the government to operate in exile is premised on the uninhabitability of the State's territory, at least for the majority of the population. Accordingly, given that the bulk of the population would reside in other sovereign States, they would be subject to the laws and jurisdiction of those States. The jurisdiction of the home State would therefore be the same as that which any State is able to exercise with respect to its nationals abroad (predominantly diplomatic protection). Once people begin to acquire dual nationality, the presumption of diplomatic protection may gradually favour the State in which the person resides (on the assumption

[116] Talmon (n 103) 15; see further, 146–9.

[117] ibid, 204–5 and citations there. The State's 'personal sovereignty over its nationals' can protect them and their interests abroad: see generally, 202–3. On the possible withdrawal of consular protection, see HC Deb, 8 November 1939, vol 353 cols 229–30, cited in ibid, 203.

[118] Talmon (n 103) 205. See further, JF Engers, 'The United Nations Travel and Identity Document for Namibians' (1971) 65 American Journal of International Law 571; (1968) 4 Revue Belge de Droit International 293–4 (discussing Belgium's acceptance of travel documents issued by unrecognized governments in exile). For example, during the Second World War the US recognized the French Committee of National Liberation (CFLN) not as the government of France, but 'as functioning within specific limitations during the war', namely 'as administering those French overseas territories which acknowledge its authority': see Talmon (n 103) 25 referring to a statement by the US on 26 August 1943.

[119] Park (n 3) 7 (fns omitted).

[120] ibid, referring to Talmon (n 103) 215–18, 238–43, who notes that only in very specific situations has some judicial competence been accorded (related to war-time events).

[121] Although Park suggests that if the whole population resided in another State, then 'the powers of the government would depend not only on the host country but also on the willingness of other States to accord or recognize the jurisdiction of the government, presumably without the possibility of reciprocity. The government's effectiveness would be questionable, and the criterion of "independence" would not appear to be met': Park (n 3) 7.

that this is where nationality is more 'effective').[122] Over time, the function of the government in exile will wane. In particular, if the government in exile were to merge with the organs of the host State, especially if done voluntarily,[123] then this would normally result in the first State's extinction (provided 'there is no other perceived international interest in asserting the continuity of the State').[124]

As Kälin notes, it is unlikely that small island States would readily relinquish their claims to statehood.[125] He suggests that they would likely try to retain 'at least a symbolic presence on their former lands, such as by building up a small island or surrounding it by dykes (even if that land would be too small or under-resourced to host any significant part of the population) and would continue to grant citizenship'.[126] If the link between government and citizen remains unbroken, then the issue is more about the extension of national protection outside the territory than about 'international protection' per se.

When asked about this, the President of Kiribati explained:

> I dream a lot. I dream that some of us would stay. If we had enough resources, we could build up one of these islands to a height a few metres above sea level to render it a place where we could survive. And those that wish to stay can stay. But I think these are the options. I'm going to Japan this week. I'm actually exploring a science fiction solution to the whole problem. A floating island—which will survive for the next thousand years; I'm intrigued, very very intrigued... it's a floating island.... But there is no doubt that we can build up these islands; we can build them up. But it's going to cost. We can put sea walls. We're doing it at the moment. So it can be done, but the question is it's very expensive, and for how long.[127]

State practice suggests that the international community would be willing to continue to accept maintenance of the status quo (recognition of on-going statehood) even when the facts no longer seem to support the State's existence. Since there is no self-executing mechanism for determining when a State no longer exists, the point at which a State such as Tuvalu or Kiribati could be said to have finally ceased to exist would depend not just on isolated acts of non-recognition by individual

[122] In dual nationality cases, the contemporary approach, based on the rule of real and effective nationality, is to 'search for "stronger factual ties between the person concerned and one of the States whose nationality is involved"', involving consideration of 'all relevant factors, including habitual residence, center of interests, family ties, participation in public life and other evidence of attachment': *Islamic Republic of Iran v United States of America, Case No A-18* (1984) 5 Iran–USCTR 251, 265.

[123] For example, if a self-governing colony reverts to imperial rule: see Crawford (n 30) 701.

[124] ibid, and citations there.

[125] Walter Kälin, 'Conceptualising Climate-Induced Displacement' in McAdam (n 4).

[126] ibid, 102. He says that, while overseas consulates remain, parents could still register newly born children as citizens: 91. The question would remain whether that nationality could be 'effective'. In the context of *de facto* statelessness, an 'effective' nationality means having 'the right of diplomatic protection exercised by a State of nationality in order to remedy an internationally wrongful act against one of its nationals, as well as diplomatic and consular protection and assistance generally, including in relation to return to the State of nationality': Summary Conclusions of UNHCR's Expert Meeting on 'The Concept of Stateless Persons under International Law' (Prato, 27–28 May 2010) para II.A.2.

[127] Interview with President Anote Tong (n 35). On Japan's 'Mega-Float' structures, see <http://www.srcj.or.jp/html/megafloat_en/whatmega/whatmega.html> accessed 2 February 2011.

States, but their cumulative effect.[128] According to Shaw, 'one has to consider the classical criteria of statehood together with assertions as to status made by the parties directly concerned and the attitudes adopted by third states and international organizations'.[129] In this regard, we are looking for 'a general acceptance by the international community as a whole that the situation has been resolved', rather than any particular length of time passing.[130]

Accordingly, '[a] State is not necessarily extinguished by substantial changes in territory, population or government, or even, in some cases, by a combination of all three'.[131] Indeed, its legal identity may be preserved to a degree even if it becomes a protectorate with some international legal personality.[132]

There remains the question whether States would be prepared to lose 'one of their own'. Kälin argues that it is 'difficult to imagine that any other UN member state would want to tarnish its own reputation by being seen as lacking any compassion for the dire fate of such island states by asking for their exclusion from that or other international organisations'.[133] There is precedent to suggest that, at least for an interim period, they would continue to recognize a State as such, even when the full indicia of statehood are lacking. Whether, and for how long, countries like Kiribati and Tuvalu could continue to remain members of the UN in such circumstances would therefore depend on the views of other States.[134]

Presumably, then, if recognition continues, then the 'deterritorialized' State could continue to interact as part of the community of nations.[135] What this would mean in real terms for the (former) inhabitants of that State is less clear, however. For whereas the continuing recognition of a non-existing State is to some degree academic, its citizens need somewhere to live, a legal status that is recognized, and the rights and duties that come with citizenship, including movement rights. How this would play out would depend on the way in which such movement is effected. Relocation of a population *en masse* may lead to a differently negotiated status from individual migration, where people may gradually be absorbed into their new community.

VI. Statelessness?

If a State does cease to exist, and in the absence of having acquired a new nationality, could its people be considered 'stateless' as a matter of international law?

[128] Crawford (n 30) 704–5. See also Talmon (n 103) 174ff.
[129] Shaw (n 81) 203–4, 960 cited in Park (n 3) 6. [130] Crawford (n 30) 704.
[131] ibid, 700. [132] ibid, 700–1. See section VIII. below on self-governing territories.
[133] Kälin (n 125) 102.
[134] See section IX. below on the Sovereign Order of Malta and the Vatican. There was a time when delegates from Tuvalu expressed concern that it would be the last State to join the United Nations and the first to leave, on account of disappearance: Enele S Sopoaga, 'Statement' (General Debate of the 56th Session of the United Nations General Assembly, New York, 16 November 2001) <http://www.tuvaluislands.com/news/archived/2001/2001-11-16UN.htm> accessed 11 August 2011.
[135] For a non-international law conceptualization of statehood, arguing that the State, sovereignty, and territory can be conceptually separated, see Fiona McConnell, 'De facto, Displaced, Tacit: The Sovereign Articulations of the Tibetan Government-in-Exile' (2009) 28 Political Geography 343.

A 'stateless person' is defined in the 1954 Convention relating to the Status of Stateless Persons as 'a person who is not considered as a national by any State under the operation of its law'.[136] As a preliminary issue, 'the permanent disappearance of habitable physical territory, in all likelihood preceded by loss of population and government, may mean the "State" will no longer exist for the purposes of this provision'.[137] As international experts at a UNHCR-hosted meeting on statelessness concluded, 'the situation is unprecedented and may necessitate progressive development of international law to deal with the preservation of the identity of the communities affected'.[138]

Accordingly, it is useful to consider the present operation of the law on statelessness to determine if, and to what extent, it would provide assistance in such circumstances. To begin with, it is important to note that even when a State becomes extinct according to conventional international law, the resultant legal status of the population on the territory is unclear.[139] There is no general right to nationality in customary international law, although there is certainly 'a strong presumption in favor of the prevention of statelessness in any change of nationality, including in a state succession'.[140] Although Article 15 of the Universal Declaration of Human Rights contains the right to a nationality, it lacks a correlative duty on the State to confer nationality.[141] Indeed, the absence of a right to nationality in the International Covenant on Civil and Political Rights has been ascribed to the complexity of the issue and States' inability to agree on its inclusion in the treaty.[142] The closest one comes to locating such a duty is the 'negative duty' arising under the statelessness treaties.[143]

[136] Convention relating to the Status of Stateless Persons (adopted 28 September 1954, entered into force 6 June 1960) 360 UNTS 117 ('1954 Convention').

[137] Summary Conclusions (n 126) para I.C.27. The author led the discussion on this topic.

[138] ibid.

[139] See generally, Paul Weis, *Nationality and Statelessness in International Law* (2nd rev edn, Sijthoff and Noordhoff International Publishers, 1979) 135ff.

[140] Jeffrey L Blackman, 'State Successions and Statelessness: The Emerging Right to an Effective Nationality under International Law' (1998) 19 Michigan Journal of International Law 1141, 1183. See also Park (n 3) 16–17.

[141] Universal Declaration of Human Rights (adopted 10 December 1948) UNGA res 217A (III) ('UDHR').

[142] Johannes MM Chan, 'The Right to a Nationality as a Human Right: The Current Trend Towards Recognition' (1991) 12 Human Rights Law Journal 1, 4–5. The only ICCPR provision relating to the right to a nationality is Art 24 on a child's right to acquire a nationality: International Covenant on Civil and Political Rights (adopted 16 December 1966, entered into force 23 March 1976) 999 UNTS 171 ('ICCPR'). See also the views of Hernández-Truyol and Hawk, who propose a 'deterritorialized, relational, and identity-based' citizenship based 'not on the nation's view of the individual, but rather the individual's view of themselves': Berta E Hernández-Truyol and Matthew Hawk, 'Traveling the Boundaries of Statelessness: Global Passports and Citizenship' (2005) 52 Cleveland State Law Review 97, 111. See also the notion of 'pragmatic citizenship' suggested in Elizabeth Mavroudi, 'Palestinians and Pragmatic Citizenship: Negotiating Relationships between Citizenship and National Identity in Diaspora' (2008) 39 Geoforum 307, 309: 'The assumption that there is a naturalised and potentially exclusive relationship between territory, national identity and citizenship, whereby national identity is neatly located in a clearly demarcated and bounded nation-state can be seen as problematic'.

[143] Blackman (n 140) 1176. See, in particular, the Convention on the Reduction of Statelessness (adopted 30 August 1961, entered into force 13 December 1975) 989 UNTS 175 ('1961 Convention') Arts 1(1), 1(3), 1(4), 4, 10.

While treaty law aims to prevent the inhabitants of an existing State from becoming stateless when a new State emerges on that territory,[144] there is divergent practice on whether nationality automatically changes or whether further provision has to be made by the new State for that to occur.[145] Crawford argues that the better view, in line with the decision of the Permanent Court of International Justice in the *Question concerning the Acquisition of Polish Nationality*, is that, subject to any stipulation to the contrary, people habitually resident in the territory of the new State acquire its nationality, for all international purposes, and lose their former nationality, although the new State may choose to delimit further who it will regard as its nationals.[146] While the issue of State succession does not apply to the Kiribati or Tuvalu context, the relevant point here is that existing international law lacks uniform practice in satisfactorily resolving the issue of nationality when one State ceases to exist. Though poorly ratified, the 1961 Convention on Reduction of Statelessness obliges States to ensure that any transfer of territory does not render people stateless.[147]

Perhaps unsurprisingly, the two international treaties on statelessness do not envisage the eventuality of literal, physical statelessness. In any event, the legal definition of 'statelessness' is carefully and deliberately circumscribed to apply only to *de jure* statelessness—premised on the *denial* of nationality through the operation of the law of a particular State.[148] This definition does not inquire into the effectiveness of that nationality, but only whether an individual has a nationality at all.[149] Thus, it does not even extend to the situation of *de facto* statelessness—where a person formally has a nationality which is ineffective in practice. Although the drafters of the two treaties showed a general sympathy towards their plight, they wanted to avoid granting benefits to people who simply renounced their nationality for personal convenience. Whereas the draft text of the 1954 Convention included a provision enabling States to extend protection to 'any person living outside his own country who, for reasons recognized as valid by the State in which he is a resident, has renounced the protection of the State, of which he is, or was, a national', this

[144] 1961 Convention, especially Arts 8–10; International Law Commission (ILC) Draft Articles on Nationality of Natural Persons in Relation to the Succession of States, annexed to UNGA res 55/153 (12 December 2000).

[145] Crawford (n 30) 52–3; Contrast the views of Ian Brownlie, 'The Relations of Nationality in Public International Law' (1963) 39 British Year Book of International Law 284, 320; Daniel P O'Connell, *State Succession in Municipal Law and International Law* (Cambridge University Press, 1967) vol 1, 497–528. See *AB v MB* (1951) 17 ILR 110, which refers to the 'absurd result of a State without nationals'.

[146] Crawford (n 30) 53 referring to *Questions concerning the Acquisition of Polish Nationality* (Advisory Opinion) PCIJ Series B No 7 (1923) 15; see also Nathaniel Berman, '"But the Alternative is Despair": European Nationalism and the Modernist Renewal of International Law' (1993) 106 Harvard Law Review 1792, 1834–42.

[147] 1961 Convention, Art 10. See also the ILC's Draft Articles on Nationality (n 144) Art 1, which contains a 'right to nationality'; Art 4 requires States to take measures to prevent statelessness as a consequence of succession.

[148] 1954 Convention, Art 1(1): 'For the purpose of this Convention, the term "stateless person" means a person who is not considered as a national by any State under the operation of its law'.

[149] Summary Conclusions (n 126) para I.A.3.

was ultimately reworded slightly and moved to the non-binding Final Act.[150] The 1961 Convention includes a resolution, appended to the Final Act of that treaty, stating that 'persons who are stateless *de facto* should as far as possible be treated as stateless *de jure* to enable them to acquire an effective nationality'.[151] However, the instruments' tight juridical focus leaves little scope for arguing for a broader interpretation that would encompass people whose State is at risk, or in the process, of disappearing (unless, of course, the State formally withdrew their nationality and through that act brought them within the legal concept of statelessness).

However, UNHCR's institutional mandate to prevent and reduce statelessness encompasses *de facto* statelessness as well.[152] In the 'sinking State' context, UNHCR has argued that even if the international community were to continue to acknowledge a State's on-going existence, despite signs that it no longer met the full indicia of statehood, its population could be regarded as *de facto* stateless. This view is based on the many practical constraints that the government would face in such a scenario, which would mean that 'their populations would be likely to find themselves largely in a situation that would be similar to if not the same as if statehood had ceased'.[153] From an institutional perspective, UNHCR has a mandate to engage with States about preventing statelessness and therefore to advocate on behalf of affected populations. In this regard, it has suggested that multilateral

[150] The Conference '*[r]ecommends* that each Contracting State, when it recognizes as valid the reasons for which a person has renounced the protection of the State of which he is a national, consider sympathetically the possibility of according to that person the treatment which the Convention accords to stateless persons'. See generally, Nehemiah Robinson, 'Convention relating to the Status of Stateless Persons: Its History and Interpretation' (Commentary, Institute of Jewish Affairs, World Jewish Congress, 1955, reprinted by UNHCR, 1997) commentary on Art 1. The British delegate also argued that the Refugee Convention already covered *de facto* stateless persons who were unable or unwilling to access the protection of th zeir government, and that opening the 1954 Convention to all *de facto* stateless persons would be tantamount to granting them benefits from which they were excluded as refugees: see Robinson, referring to SR.3, 2–3.

[151] 1961 Convention, res I. See also ibid, para II.A.2: the definition of *de facto* stateless persons is 'persons outside the country of their nationality who are unable or, for valid reasons, are unwilling to avail themselves of the protection of that country. Protection in this sense refers to the right of diplomatic protection exercised by a State of nationality in order to remedy an internationally wrongful act against one of its nationals, as well as diplomatic and consular protection and assistance generally, including in relation to return to the State of nationality'.

[152] See, eg, UNGA res 50/152 (9 February 1996) reiterated in UNGA res 61/137 (25 January 2007), UNGA res 62/124 (24 January 2008), UNGA res 63/148 (27 January 2009). The work of the United Nations High Commissioner for Refugees (UNHCR) extends in some cases to situations of *de facto* statelessness, such as in trying to get 'States to cooperate in the establishment of identity and nationality status of victims of trafficking, many of whom, especially women and children, are rendered effectively stateless due to an inability to establish such status, so as to facilitate appropriate solutions to their situations, respecting the internationally recognized human rights of the victims': Executive Committee of the High Commissioner's Programme, 'Statelessness: Prevention and Reduction of Statelessness and Protection of Stateless Persons', UN Doc EC/57/SC/CRP.6 (14 February 2006) para 7. See also UNGA res 50/152 (9 February 1996) paras 14–15; UNGA res 3274 (XXIX) (10 December 1974); UNGA res 31/36 (30 November 1976).

[153] UNHCR, supported by the International Organization for Migration and the Norwegian Refugee Council, 'Climate Change and Statelessness: An Overview' (Submission to the 6th Session of the Ad Hoc Working Group on Long-Term Cooperative Action (AWG-LCA 6) under the UN Framework Convention on Climate Change (UNFCCC), 1–12 June 2009) 2 <http://unfccc.int/resource/docs/2009/smsn/igo/048.pdf> accessed 3 June 2011.

comprehensive agreements could facilitate planned and orderly movement to other States, and that the early introduction of educational and other measures to prepare people for displacement could not only increase their resilience and adaptability once they move, but also while they remain on their islands.[154]

In the case of Tuvalu and Kiribati, at a certain point the objective evidence may indicate that continued habitation of those territories is imminently impossible. In keeping with the object and purpose of the treaty, and the resolution appended to the Final Act of the 1961 Convention that *de facto* stateless persons be treated in the same way as *de jure* stateless persons, one might argue that the benefits of the statelessness treaties should be extended to them. However, these instruments only bind States that have ratified them, and only in relation to stateless persons within their territory. Few States even have a status determination procedure to identify stateless persons, by contrast to refugees.[155] Accordingly, the statelessness treaties provide a very weak 'solution' in the present context, which is already highly contingent on other factors. As the preceding parts of the chapter have explained, it is far from clear that the international community would ever recognize that a State had ceased to exist. In light of the presumption of continuity of statehood, such recognition, if forthcoming at all, would likely occur long after the population had moved. The application of the law on statelessness may have little practical benefit such a long time after the fact.[156]

The international statelessness regime has a preventative as well as a remedial function, however. Attention would therefore be better focused on States' duty to prevent statelessness[157] and UNHCR's mandate to engage in actions to prevent and reduce statelessness.[158] As UNHCR has explained in the present context:

> To prevent temporary statelessness, acquisition of an effective nationality should be foreseen prior to the dissolution of the affected State. Dual nationality may therefore need to be permitted at least for a transitional period. As well, a waiver may be required of formal requirements for renunciation or acquisition of nationality which might be difficult

[154] ibid, 3.

[155] See The Equal Rights Trust (n 15).

[156] Finally, in terms of identity (rather than formal legal status), statelessness may be a problematic concept. In the case of exiled communities, such as Tibetans in India and Nepal, there is some evidence that the Tibetan Government in Exile is regarded as a pseudo 'home state', thus complicating conventional frames of reference about the geographical location of 'home': Fiona McConnell, 'Governments-in-Exile: Statehood, Statelessness and the Reconfiguration of Territory and Sovereignty' (2009) Geography Compass 1901, 1912. In the same way that many Pacific islanders reject the 'refugee' label, they may also reject the 'statelessness' one: see Ch 2.

[157] See UNGA res 61/137 (25 January 2007); UNGA res 62/124 (24 January 2008); UNGA res 63/148 (27 January 2009); UNGA res 64/127 (27 January 2010); UNGA res 65/194 (28 February 2011). UNHCR argues that the prevention of statelessness is a corollary of the right to a nationality: UNHCR (n 153) 2. That right is contained in its broadest form in the UDHR, Art 15; and in relation to children in the ICCPR, Art 24(3); Convention on the Rights of the Child (adopted 20 November 1989, entered into force 2 September 1990) 1577 UNTS 3, Art 7; International Convention on the Protection of the Rights of All Migrant Workers and Members of Their Families (adopted 18 December 1990, entered into force 1 July 2003) 2220 UNTS 3, Art 29. See also Convention on the Elimination of All Forms of Discrimination against Women (adopted 18 December 1979, entered into force 3 September 1981) 1249 UNTS 13, Art 9; International Convention on the Elimination of All Forms of Racial Discrimination (adopted 7 March 1966, entered into force 4 January 1969) 660 UNTS 195, Art 5(d)(iii); 1961 Convention.

[158] UNGA res 50/152 (9 February 1996) paras 14–15.

to fulfil for affected populations. Such arrangements would need to provide *inter alia* for the right of residence, military obligations, health care, pensions and other social security benefits. Citizens of affected States that might have been displaced earlier, possibly to third States not party to the agreement, may also need to be considered.[159]

VII. *En Masse* Relocation

A. Background

Historically, there have been at least 86 relocations of whole communities within the Pacific—37 resulting from environmental variability, and another 13 from human-induced environmental degradation.[160] For example, during the 1940s, the Fijian colonial administration admitted people from the Gilbert and Ellice Islands Colony (now Kiribati and Tuvalu) on account of pressures on their ecosystems. In the case of Kiribati, phosphate mining on Banaba prompted resettlement on the Fijian island of Rabi. In the case of Tuvalu, it seems that concerns about a growing population in a country with scarce resources were behind the partial relocation of some people from Vaitupu island (in Tuvalu) to the Fijian island of Kioa.[161] However, there appear to be no cases of *international* relocation in the post-colonial era (whether between Pacific islands, or from Pacific islands to other countries). This is not surprising. Indeed, as one scholar has explained:

> Movement of people among the colonies was orchestrated by the colonial services and did not require passports or contemporary protocols of residency, citizenship, visas and the like.... Colonial administrations could make decisions about land and community locations with fewer constraints than is currently possible, where land is enshrined in laws established to protect customary land rights in the newly independent nations. Secondly, colonial administrations could easily move people across what are now international boundaries, as long as the territories were colonised by the same metropolitan power.[162]

[159] UNHCR (n 153) 3. See also Park (n 3) 19; Executive Committee Conclusion No 95 (LIV) 'Internal Protection' (2003) para (v), which encourages 'States to cooperate with UNHCR on methods to resolve cases of statelessness and to consider the possibility of providing resettlement places where a stateless person's situation cannot be resolved in the present host country or other country of former habitual residence, and remains precarious'.

[160] John Campbell, Michael Goldsmith, and Kanyathu Koshy, 'Community Relocation as an Option for Adaptation to the Effects of Climate Change and Climate Variability in Pacific Island Countries (PICs)' (Asia-Pacific Network for Global Change Research, 2005) 21ff. Of these, some were evacuations (with the population eventually returning).

[161] See Richard Bedford and Charlotte Bedford, 'International Migration and Climate Change: A Post-Copenhagen Perspective on Options for Kiribati and Tuvalu' in Bruce Burson (ed), *Climate Change and Migration: South Pacific Perspectives* (Institute of Policy Studies, 2010) 90.

[162] Campbell (n 191) 77 referring to Martin G Silverman, 'Introduction: Locating Relocation in Oceania' in Lieber (n 224). Bedford argues that the relocation of people from Vaitupu to Kioa, within central Vanuatu, and from (now) Kiribati to the Solomon Islands was enabled by the British colonial system. 'Decisions could be made about land transfers with relatively little consultation and international boundaries were of little consequence. The Solomon Islands and the Gilbert and Ellice Islands fell under the control of the Western Pacific High Commission and moving people from one part to the other was relatively easy': summarized in Campbell, Goldsmith and, Kosny (n 160) 35.

Nevertheless, there has been a perennial discussion in the 'disappearing State' context of the *en masse* relocation of a State's population to another country. It was apparently first mooted in 1996 by Brian Fisher, then Executive Director of the Australian Bureau of Agricultural and Resource Economics, who promoted the economic 'appeal' of relocating small island States on account of its 'costs and benefits', compared to the cost of mitigation.[163] Both Kiribati and Tuvalu have raised the relocation option on occasion with Australia and New Zealand,[164] although it is no longer a priority, as examined below.

A dominant feeling among those who have been relocated is 'discontent', often over generations, deriving from 'the very strong relationship or bond that exists between most Pacific Island Communities and their land—in most cases they are inseparable'.[165] Field-based research in Pacific villages has demonstrated that 'relocation is a long-term process that requires considerable effort—from identifying suitable sites, through to negotiation and consultation both with relocating communities and those in the jurisdictions or land-owning communities in the destination. Hasty relocation, which may result if further research and negotiation is not conducted, is almost certainly bound to be problematic.'[166] This is examined in more detail below in relation to Nauru.

At the annual climate change conference in November 2000 (COP6), the Tuvaluan Prime Minister stated that the potential impacts of climate change on his country were 'so serious for our people, that the Cabinet, in which I am a member, has been exploring the possibility of buying land in a near-by country, in case we become refugees to the impacts of climate change'.[167] In 2001, Tuvalu reportedly held negotiations with Fiji about purchasing a large tract of land.[168] In 2008, it was reported that the Tuvaluan Prime Minister secretly met with Australian government officials to discuss a plan by which 'Tuvalu would be based in Australia but continue to function as a sovereign nation, in the hope of one day returning to its island home. The Tuvalu Government would continue to exercise its economic exclusion zone of about one million square kilometres and maintain its seat

[163] Colette Mortreux and Jon Barnett, 'Climate Change, Migration and Adaptation in Funafuti, Tuvalu' (2009) 19 Global Environmental Change 105, 105–6, referring to N Bita, 'Island Evacuation a Greenhouse Solution', *The Weekend Australian* (8 June 1996). Although see the remarks of Sir Kamuta Latasi below, who says he first raised the idea in 1995: interview with Sir Kamuta Latasi, Speaker of the Tuvaluan Parliament (and former Prime Minister) (Funafuti, Tuvalu, 27 May 2009). Connell notes that during the 1990s, Tuvalu consistently sought migration opportunities to relieve pressure from overcrowding, especially with the anticipated return of migrants working in the phosphate mining industry in Nauru: Connell (n 18) 95. See further, Ch 1.

[164] See, eg, Senate Foreign Affairs, Defence and Trade Committee, *A Pacific Engaged: Australia's Relations with Papua New Guinea and the Island States of the South-West Pacific* (Commonwealth of Australia, 2003) para 6.78; interview with President Anote Tong (n 35); interview with Sir Kamuta Latasi (n 163); Brad Crouch, 'Tiny Tuvalu in Save Us Plea over Rising Seas', *Sunday Mail* (London, 5 October 2008) <http://carbon-based-ghg.blogspot.com/2008/10/tiny-tuvalu-in-save-us-plea-over-rising.html> accessed 6 February 2011.

[165] Campbell, Goldsmith, and Koshy (n 160) 21.

[166] ibid, 6.

[167] Statement by the Hon Teleke P Lauti, COP6 (The Hague, November 2000), cited in UNFCCC, *Climate Change: Small Island Developing States* (Climate Change Secretariat, 2005) 13.

[168] Greg Bearup, 'Enterprise Island', *Sydney Morning Herald Good Weekend* magazine (Sydney, 5 May 2001) 44–7, cited in Connell (n 18) 95.

and vote at the United Nations'.[169] However, in my own interviews with members of the government of Tuvalu in May 2009, this was strongly downplayed. When directly questioned about relocation, I was told that Tuvalu planned to adapt so that people did not need to move.[170]

When I asked former Tuvaluan Prime Minister, Sir Kamuta Latasi, whether the government still had relocation plans, he stated:

> Well in actual fact, there's never been a new policy in this government. They have not yet formalized an immediate plan which should be presented to other countries for the future of the Tuvalu people. There is no such plan. But in the past, there have been plans and requests and believe you me, I was the one during the [Pacific Islands] Forum in 1995—in Brisbane, I was the Prime Minister. I specifically requested Paul Keating at that time: if Australia is willing to accept people from Tuvalu who wish to migrate to Australia, or even if Australia is prepared to give land for settlement; the response was a big 'no'. They said because of the policy Australia had, they would not make any exemptions for Tuvalu. And then I reset the question—what happens if I bring my people here as refugees; would that be accepted because of climate change refugees? They said, 'Oh well that is a remote area'. And then New Zealand responded by accepting our people to be fruit pickers ... The only country who helped us at the time was the Prime Minister of Niue who offered for some [Tuvaluan] families to travel to Niue and settle there and work there.... And about 8 or 10 families, they are all in Niue now. Some of them are citizens of New Zealand because Niue is under New Zealand and they have been very, very happy. But that's changed... it's out. They are trying to revive it now. So you see—the only country who responded to that need was a Pacific brother—Niue like us.[171]

Similarly, another official stated:

> They were proposals in—for Fiji—in the past. The government was thinking—the previous government—they were thinking of buying a piece of land in Fiji so that—it's not really to relocate the population but to build some infrastructure to accommodate travellers of Tuvalu and also some piece of land for Tuvaluans. If Tuvaluans want to stay there, they can buy a cheaper land from that. But it didn't happen.[172]

In interviews in Tuvalu, it was clear that few people want to contemplate relocation as a potential option. This resonates with the Tuvaluan Prime Minister's comments at COP14 in 2008, where he stated: 'We are not contemplating migration. We are a proud nation of people with a unique culture which cannot be relocated somewhere else. We want to survive as a people and as a nation'.[173]

[169] Crouch (n 164).

[170] Interview with Enele Sopoaga, Secretary for Foreign Affairs (Funafuti, Tuvalu, 25 May 2009).

[171] Interview with Sir Kamuta Latasi (n 163).

[172] Interview with Melton Tauetia, Climate Change Coordinator (Funafuti, Tuvalu, 21 May 2009).

[173] Prime Minister Apisai Ielemia, Statement delivered at the High-Level Segment of COP14 (Poznan, 11 December 2008) 4 <http://unfccc.int/files/meeting/cop_14/application/pdf/cop_14_his_statement_ielemia.pdf> accessed 3 December 2011. On problems of relocation, see UN Office of the High Commissioner for Human Rights, 'Discussion Paper: Protecting the Human Rights of Internally Displaced Persons in Natural Disasters: Challenges in the Pacific' (Davui Printery, 2011) 17–18. See also Ch 1.

Certainly, *en masse* relocation is not presently the preferred option of the governments of Kiribati or Tuvalu. Even though the President of Kiribati has not ruled out the prospect of buying land elsewhere, noting that Papua New Guinea and the Solomon Islands 'have a lot of resources; they've got a lot of land. In fact, we'd be happy to buy it.... I think with the reserve funds—it would be a good investment to buy it',[174] he regards this as politically untenable and would instead prefer to see gradual migration. As noted above, the government of Tuvalu now regards relocation as a matter of last resort because people should not have to leave their homes, especially since they are 'amongst the lowest contributors to factors causing climate change'.[175] Indeed, the Niue Declaration on Climate Change, adopted by the Pacific Islands Forum in 2008, recognizes 'the importance of retaining the Pacific's social and cultural identity, and the desire of Pacific peoples to continue to live in their own countries, where possible'.[176]

Most recently and most vocally, on coming to office in 2008 the President of the Maldives boldly stated that he was seeking to purchase land in India or Australia to which to relocate his nation.[177] It appears that this was more of a profile-raising exercise than a serious intention, although subsequently the Indonesian Maritime Minister announced that Indonesia was considering renting out some of its 17,500 islands to 'climate change refugees'.[178] It is unclear whether this was in direct response to the Maldives' expression of interest.

Finally, as a postscript to this section, it is interesting to note that the small Alaskan community of Newtok, with 321 inhabitants,[179] has voted three times to relocate internally on account of increased climate-related ecological hazards and their detrimental impact on the local infrastructure. In the mid-1990s, the Newtok Native Corporation identified and successfully obtained a relocation site on Nelson Island.[180] However, a lack of cross-agency coordination and the absence of funding earmarked for relocation has hindered the move to date. Accordingly, the Newtok Planning Group 'has had to compartmentalize their efforts instead of

[174] Interview with President Anote Tong (n 35).

[175] The Niue Declaration on Climate Change, 39th Pacific Islands Forum, Forum Communiqué, Annex B (19–20 August 2008) 23.

[176] ibid.

[177] Randeep Ramesh, 'Paradise Almost Lost: Maldives Seek to Buy a New Homeland', *The Guardian* (London, 10 November 2008) <http://www.guardian.co.uk/environment/2008/nov/10/maldives-climate-change> accessed 21 November 2008. This is not formal government policy, however: <http://www.abc.net.au/radionational/programs/lawreport/climate-change-indian-ocean/3695390> accessed 2 December 2011.

[178] 'Indonesian Islands for Rent', *The Straits Times* (Singapore, 6 May 2009) <http://singapuranews.multiply.com/journal/item/6357> accessed 14 October 2011; 'Indonesia Offers Pacific Climate Refugees Island Rental', *Pacific Beat* (Radio Australia, 3 June 2009) <http://www.radioaustralia.net.au/pacbeat/stories/200906/s2588395.htm> accessed 21 June 2009 citing Dr Syamsul Maarif, Secretary-General of the Maritime Affairs Ministry; see also Steve Holland, 'Indonesia's Rent-an-Island Answer to Climate Change' (*ABC News*, 3 June 2009) <http://www.abc.net.au/news/stories/2009/06/03/2588165.htm> accessed 21 June 2009. In an article entitled 'Threatened Islands Take the Lead in Fighting Climate Change', 30 March 2010, *Deutsche Welle* wrote that the Indonesian Fisheries Ministry was (still) considering renting out some of its islands to victims of climate change: <http://www.dw-world.de/dw/article/0,,5410956,00.html> accessed 7 July 2011.

[179] Bronen (n 52) 373.

[180] The land was exchanged by the federal government on 17 November 2003: ibid, 384.

executing one streamlined relocation plan'.[181] Even though the Newtok relocation would occur within the same country, this example highlights the practical and legal obstacles to achieving group resettlement. It also underscores the importance of community consultation and participation in finding appropriate solutions in particular contexts. It demonstrates that a policy that may be rejected in one situation may be considered appropriate in another.

B. Legal issues

As a matter of principle, there is nothing in international law that would prevent the reconstitution of a State such as Kiribati or Tuvalu within an existing State, such as Australia. In reality, though, the political likelihood of this happening is remote. Principles of international law relating to the cession of territory would need to be adhered to in order for any such transfer to be legal, in the sense of transferring sovereignty (as opposed to a mere private property transaction).[182] As Crawford observes,

> international law defines 'territory' not by adopting private law analogies of real property but by reference to the extent of governmental power exercised, or capable of being exercised, with respect to some territory and population. Territorial sovereignty is not ownership of but governing power with respect to territory. There is thus a good case for regarding government as the most important single criterion of statehood, since all the others depend upon it.[183]

At the core of the principle of self-determination lies 'the right of a community which has a distinct character to have this character reflected in the institutions of government under which it lives'.[184] However, this does not give a community a right to claim the land of an existing State if their own is threatened. Ultimately, the principle of self-determination must give way to territorial integrity and State sovereignty, such that no State has a claim on another to cede territory in order to permit its continued existence.

Theoretically, it would be possible for one State to 'lease' territory from another, although one might query the extent to which power could then be freely exercised in a manner sufficient to meet the other requirements of statehood.[185] While a State might be afforded jurisdiction over that territory, it would not be unencumbered

[181] ibid, 387.

[182] See Soons (n 17) 230. This is the way the Fijian islands of Rabi and Kioa were purchased by Kiribati and Tuvalu respectively.

[183] Crawford (n 30) 56.

[184] Ian Brownlie, 'The Rights of Peoples in Modern International Law' in James Crawford (ed), *The Rights of Peoples* (Clarendon Press, 1988) 5.

[185] This will depend on the terms of agreement under which a territory is leased: see, eg, Cuba–US Agreement for the leasing of Guantanamo Bay and discussion in 'Territorial Leasing and Boundary Conflicts: Michael Strauss' (*Exploring Geopolitics*, August 2009) <http://www.exploringgeopolitics.org/Interview_Strauss_Michael_Territorial_Leasing_Boundary_Conflicts_Disputes_Borders_Guantanamo_Bay.html> accessed 11 August 2011; Treaty between Great Britain and China (Treaty of Nanking) (signed 29 August 1842) 30 British and Foreign State Papers 389; Convention of Friendship between Great Britain and China (Convention of Peking) (signed 24 October 1860) 50 British and Foreign State Papers 10.

by the 'landlord' State's territorial jurisdiction unless expressly obtained from the previous sovereign.[186] For example, bilateral agreements may permit one State or entity (such as the UN) to exercise some jurisdiction on the territory of another State (eg territorial concessions granted for diplomatic premises, military bases, and so on), but this does not involve cession per se: the territory on which a degree of jurisdiction is exercised by another entity remains the sovereign territory of the host State. To function as a State, an entity must have a government 'in general control of its territory, to the exclusion of other entities not claiming through or under it'.[187]

There is also much more to relocation than simply securing territory. Those who move need to know that they can remain and re-enter the new country, enjoy work rights and access to health care there, have access to social security if necessary, be able to maintain their culture and traditions,[188] and that the legal status of children born there is clear. The acquisition of land alone does not secure immigration or citizenship rights, but is simply a private property transaction.[189] Unless individuals personally acquire such rights (and in some cases, even if they do but retain dual

[186] Brownlie (n 184) 5; see also Rosalyn Higgins, *The Development of International Law through the Political Organs of the United Nations* (Oxford University Press, 1963) 24. Note that the case of New Iceland in Canada was not a State within a State, as has sometimes been suggested: see, eg, Steina Sommerville (nee Stefansson) in *The Interlake News* in 1946, who dubbed New Iceland 'The Twelve Year Republic'; Rosemary Rayfuse (n 17). It was created as an Icelandic 'reserve' by Canadian Order-in-Council No 875 of 8 October 1875 to provide land to Icelandic immigrants who had left considerable environmental degradation and poverty at home, and was ultimately dissolved by Order-in-Council No 2306 of 9 October 1897. This process of creation and dissolution alone indicates that it was not a sovereign equal of Canada, and historians have consistently emphasized that the settlers had no intention of creating an Icelandic colony: Walter J Lindal, *The Icelanders in Canada* (National Publishers, 1967) 135. This is reflected in an address given on the occasion of the Canadian Governor-General's visit to New Iceland on 14 September 1877: 'We have gathered under the flag of our new land, and as British subjects... We accept gladly our new way of life as British subjects with the opportunity to acquire all the freedom and rights which pertain thereto. As British subjects, we desire that these rights be granted to us, and we are firmly resolved to preserve them. We are prepared to do our share in the maintenance of public order, and in the defense of our country, to perform the duties which England expects of every citizen': Address by Friðjón Friðriksson cited in Nelson Gerrard, '"A Matter of Honour": The Constitution of New Iceland: Then and Now' (Building a New Relationship Conference, University of Manitoba, 27 October 2000) <http://www.sagapublications.com/articles.html> accessed 10 November 2009.

[187] Crawford (n 30) 59 (fn omitted).

[188] See, eg, Human Rights Council, 'Report of the Office of the United Nations High Commissioner for Human Rights on the Relationship between Climate Change and Human Rights' UN Doc A/HCR/10/61 (15 January 2009) 17–18; Sub-Commission on the Promotion and Protection of Human Rights, 'Prevention of Discrimination: Prevention of Discrimination and Protection of Indigenous Peoples: Expanded Working Paper by Françoise Hampson on the Human Rights Situation of Indigenous Peoples in States and Other Territories Threatened with Extinction for Environmental Reasons', UN Doc E/CN.4/Sub.2/2005/28 (16 June 2005); Stephen Humphreys (ed), *Human Rights and Climate Change* (Cambridge University Press, 2010).

[189] Examples include the purchase of Rabi island in Fiji by the Banabans (from Kiribati) and the purchase of Kioa island in Fiji by the Vaitupu people of Tuvalu. These islands are too small to be considered relocation sites for the whole population of Kiribati and Tuvalu, and there are cultural reasons why they would not appeal. One reason given in interviews was that there is great rivalry between the nine island groups of Tuvalu, and relocation to Kioa would mean that the people of Vaitupu island would gain superiority. Crawford notes that 'the persistent analogy of territorial sovereignty to ownership of real property is misguided', indicating the vastly different functions that State links to territory serve: Crawford (n 30) 717.

nationality[190]), there is little in international law that would prevent a host country from expelling them should it wish to do so, provided there is another country obliged to admit them. This poses an on-going risk as long as the home State continues to exist. Even if the latter does 'disappear', its relocated citizens would not automatically have the same rights as the nationals of their new host country. It is only with formal cession of land at the State-to-State level that one State acquires the lawful international title to it and nationals can move to that area as part of their own national territory. The likelihood of this happening today is remote.[191] A human rights approach to relocation not only focuses attention on these questions, but makes their resolution a central aspect of proposed 'solutions'. If *en masse* relocation to another country is to be a viable option, then issues other than land alone need to be considered in order to provide security for the future.[192]

Finally, there remains the fundamental question of how to balance the human rights of relocating groups with those of the communities into which they move. 'Any relocation that involves moving away from a group's traditional territory and into that of another is likely to be highly fraught and will require considerable consultation and negotiation'.[193] The effects of dislocation from home can last for generations, and can have significant ramifications for the maintenance and enjoyment of cultural and social rights by resettled communities.[194]

C. Proposed relocation from Nauru

Even when such legal issues are resolved, relocation may still not be a popular option. As the following example from the Pacific region illustrates, concerns about the maintenance of identity, culture, social practices, and land tenure are very real to those whose movement is proposed, and these may not be readily understood by

[190] For example, Britain can revoke citizenship from nationals (albeit in limited circumstances) if doing so would not render them stateless: British Nationality Act 1981 (as amended in 2002 and 2006), s 40.

[191] Although following the 2010 Haitian earthquake, the African Union was reported to be considering a proposal to create a new State for them in Africa, citing 'a sense of duty and memory and solidarity' given that Haitians are descendants of African slaves: Barry Malone and Duncan Miriri, 'African Union to Consider "Land for Haitians" Plan' (*Reuters*, 31 January 2010) <http://www.reuters.com/assets/print?aid=USTRE60U0IV20100131> accessed 11 August 2011. Arguably, this is a special case based on historical links.

[192] Furthermore, as Campbell discusses, the ability of States to give away land itself may raise serious human rights considerations for those already inhabiting (or with claims to) that land: Campbell, (n 162). See also Bellagio Deliberations (n 8) para 32 (fn omitted): 'Any relocation plans need to ensure the enjoyment of the full range of relevant rights and a secure status for those relocated. Particular attention would need to be given to rights to enjoy and practice one's own culture and traditions and to continue to exercise economic rights in their areas or countries of origin. In particular, individuals ought to have access to information about the reasons and procedures for their movement and, where applicable, on compensation and relocation. They have a right to participate in the planning and management of any planned movement and to enjoy their rights to life, dignity, liberty, security and self-determination. Additionally, the needs and interests of host communities need to be respected and carefully balanced in this process'.

[193] Campbell, Goldsmith, and Koshy (n 160) 5.

[194] Campbell (n 162).

outsiders. This, in turn, may lead to misunderstandings and misguided policies, which can have negative long-term, inter-generational effects.

In the 1960s, as a result of the immense environmental destruction caused by phosphate mining, it was proposed that the population of Nauru be resettled in Australia. Nauru had been a British mandate territory administered on behalf of the League of Nations. In 1919, Australia, the United Kingdom (UK), and New Zealand entered into an Agreement to jointly control the administration of Nauru, predominantly to facilitate phosphate mining. When the UN's international trusteeship system succeeded the League's mandate system, Nauru became a trust territory of Australia, New Zealand, and the UK (the 'partner governments').[195]

Increasing concern by the partner governments, the British Phosphate Commissioner, the UN, and the people of Nauru about the impact of phosphate mining on the country's continued habitability led to a series of negotiations about resettlement elsewhere.[196] As early as 1953, the UN Visiting Mission stated that 'without wanting to be dogmatic, ... resettlement in some other location, as expressed by the Nauruans themselves, may be the only permanent and definite solution',[197] and that it saw 'no other alternative'.[198]

Sites were originally investigated in and around Papua New Guinea but did not meet the three necessary requirements: 'employment opportunities enabling Nauruans to maintain their standard of living; a community which would accept the Nauruans; and willingness and readiness on the part of the Nauruans to mix with the existing people'.[199] On 12 October 1960, the partner governments agreed to offer permanent residence and citizenship in those countries to any Nauruans willing 'to transfer to those countries and ... likely to be able to adapt themselves to life there'.[200] The partner governments proposed resettlement on the following terms: (a) citizenship; (b) equal opportunity and freedom of social contact; (c) education of young people to the fullest extent of their capabilities, plus an annual allowance of £600 for five years, after which they would be assisted to look for suitable employment; and (d) for adults able to undertake employment, and for whom it could be found, passage, a house, maintenance for six weeks, further training or the requisite tools for self-employment, and eligibility for all social welfare benefits.[201] As Tabucanon and Opeskin observe, however, the partner governments'

[195] UNGA res 180 (II), 'Draft Convention on Genocide' (21 November 1947).

[196] *Case concerning Phosphate Lands in Nauru (Nauru v Australia)* (Preliminary Objections submitted by Australia) in *ICJ Pleadings, Certain Phosphate Lands in Nauru (Nauru v Australia)* vol 2, para 59.

[197] Nauruan Memorial, para 13, Annex 8, vol 4 cited in Preliminary Objections submitted by Australia (n 196) para 60.

[198] Nauruan Memorial, paras 32–5, Annex 8, vol 4 cited in Preliminary Objections submitted by Australia (n 196) para 60.

[199] Preliminary Objections submitted by Australia (n 196) para 60.

[200] ibid, para 61 citing Annex 4.

[201] Gil Marvel Tabucanon and Brian Opeskin, 'The Resettlement of Nauruans in Australia: An Early Case of Failed Environmental Migration' (2011) 46 The Journal of Pacific History 337, 344, referring to Trusteeship Council, 'Examination of Conditions in Trust Territories' (1961) 15 International Organization 671, 685–6.

concern for Nauru may have been less about altruism and more an attempt to protect on-going control over phosphate resources.[202]

While Australian government documents state that '[i]t was envisaged that the transfer should take place gradually over a period of 30 or more years and that some material assistance to that end would be given',[203] the Nauruan view was that '[i]t was never envisaged that all Nauruans would take up the offer. Many would stay, and it was understood that Nauru would always remain a spiritual home for those resettled'.[204]

The resettlement offer was rejected by the Nauru Local Government Council (the only institution of self-government in Nauru, comprised of the 14 district chiefs).[205] It argued that the very nature of the scheme 'would lead to the assimilation of the Nauruans into the metropolitan communities where they settled'.[206] The Nauruans instead requested an island of their own in a temperate zone. Following the consideration and rejection of a variety of sites,[207] in 1963 Australia offered them Curtis Island (near Gladstone, Queensland). Australia planned to acquire the land on Curtis Island from a small group of private owners, and grant the Nauruans freehold title over 'New Nauru'.[208] Pastoral, agricultural, fishing, and commercial activities were to be established, 'and the entire costs of resettlement including housing and community services such as electricity, water and sewerage etc would be met out of funds provided by the Governments of Australia, New Zealand and the United Kingdom. It was estimated that the cost would be in the region of 10 million pounds'.[209] While Australia made clear that 'Australian sovereignty would not be surrendered over any mainland or island location',[210] those resettled would 'be enabled to manage their own local administration and to make domestic laws or regulations applicable to their own community', subject to their acceptance of 'the privileges and responsibilities of Australian citizenship'.[211]

Nauru again rejected the offer, deeming these political arrangements to be unsatisfactory. The Nauruan representatives feared that they would not be able to maintain their distinct identity and would be 'assimilated without trace into the Australian landscape'.[212]

[202] Tabucanon and Opeskin (n 201) 7.

[203] Preliminary Objections submitted by Australia (n 196) para 61.

[204] *Case concerning Phosphate Lands in Nauru (Nauru v Australia)* (Written Statement of Nauru) in *ICJ Pleadings, Certain Phosphate Lands in Nauru (Nauru v Australia)* vol 2, para 19

[205] Tabucanon and Opeskin (n 201) 8.

[206] Preliminary Objections submitted by Australia (n 196) para 61.

[207] For a detailed overview of possible resettlement sites and the reasons why they were rejected, see Tabucanon and Opeskin (n 201).

[208] ibid, 11.

[209] Preliminary Objections submitted by Australia (n 196) para 63 citing statement by the Australian Minister for Territories (20 August 1964).

[210] ibid, para 62.

[211] Trusteeship Council Official Records, 13th Session (May/June 1963) UN Doc T/SR.1203–1224, 6 cited in *Case concerning Phosphate Lands in Nauru (Nauru v Australia)* (Memorial of the Republic of Nauru) in *ICJ Pleadings, Certain Phosphate Lands in Nauru (Nauru v Australia)* vol 1, para 169.

[212] Memorial of the Republic of Nauru (n 211) para 171.

Your terms insisted on our becoming Australians with all that citizenship entails, whereas we wish to remain as a Nauruan people in the fullest sense of the term even if we were resettled on Curtis Island. To owe allegiance to ourselves does not mean that we are coming to your shores to do you harm or become the means whereby harm will be done to you through us. We have tried to assure you of this from the beginning. Your reply has been to the effect that we cannot give such an assurance as future Nauruan leaders and people may not think the same as we do.[213]

Nauruan and Australian perspectives on the issue of relocation reveal quite different views as to why it failed. Nauru claimed that resettlement in Australia was offered as a quick-fix solution that would cost the Australians far less than rehabilitating the land.[214] It saw it as 'an attempt to break up the Nauruan identity and their strong personal and spiritual relationship with the island',[215] ignoring Nauruan land tenure laws and 'the right of the Nauruan people at international law to permanent sovereignty over their natural wealth and resources'.[216] The Nauruans maintained that they were never 'seeking full sovereign independence' over Curtis Island, but that 'anything which did not preserve and maintain [their] separate identity was quite unacceptable'.[217]

By contrast, Australia pinned the failure of the resettlement negotiations precisely on the issue of sovereignty. Seemingly frustrated by what it envisaged as 'a genuine and generous attempt to meet the wishes of the Nauruan people',[218] the Australian representative informed the Trusteeship Council that '[i]f an area was chosen which was now Australian territory and which could be made available, the basis of the administrative arrangements would be that, subject to the resettled Nauruans accepting the privileges and responsibilities of Australian citizenship, they should be enabled to manage their own local administration and to make domestic laws or regulations applicable to their own community'.[219] However,

213 Nauru Talks 1964, 1–2, Annexes, vol 3, Annex 1 cited in ibid.

214 Written Statement of Nauru (n 204) para 22. See also para 18: 'Resettlement was simply a *quid pro quo* for depriving the Nauruan community of suitable and productive living space as a consequence of the devastation of their land (cf, Nauru *Memorial*, para 177). It was also, perhaps, a way of avoiding the issue of rehabilitation'.

215 ibid, para 20.

216 ibid, para 74.

217 'Statement by Hammer DeRoburt, OBE, GCMG, MP, Head Chief, Nauru Local Government Council', Appendix 1 to Memorial of the Republic of Nauru (n 211) para 21. However, as Tabucanon and Opeskin (n 201) 12 note, *The Age* newspaper at the time stated that Nauru wanted to establish Curtis Island as a sovereign State, tied to Australia by a treaty of friendship, and controlled by Australia only in matters of defence, quarantine, and possibly external affairs and civil aviation: 'Island Offer Rejected by Nauru', *The Age* (Melbourne, 21 August 1964). This may have been based on the 1962 Treaty of Friendship between New Zealand and Western Samoa. Similarly, Tate argues that the three fundamental conditions of resettlement on Curtis Island were that the Nauruans be granted full independence, enjoy territorial sovereignty over their new homeland, and retain sovereignty over Nauru: Merze Tate, 'Nauru, Phosphate, and the Nauruans' (1968) 14 Australian Journal of Politics & History 177, 181, cited in Tabucanon and Opeskin (n 201) 12.

218 UNGAOR, 18th Session, 4th Committee, 1513th meeting (12 December 1963) UN Doc A/C.4/SR.1513, 565, para 4 cited in Memorial of the Republic of Nauru (n 211) para 170.

219 Trusteeship Council Official Records (n 210) 6 cited in Memorial of the Republic of Nauru (n 211) para 169.

the government could not 'see its way clear to transferring sovereignty of territory which is at present part of Australia'.[220]

The issue resurfaced in 2003, when the Australian Foreign Minister, Alexander Downer, was reported as saying that he was considering the resettlement of all Nauruans in Australia, or giving them a vacant island to move to. This was dismissed by the President of Nauru, who again said it would undermine Nauru's identity and culture.[221] Mr Downer later downplayed the idea, observing that other Pacific nations might expect similar treatment.

Cultural misunderstandings about the importance of land and cultural identity remain at the heart of discussions today about relocating entire Pacific communities in response to climate change impacts. While most suggestions to relocate communities are no doubt well intentioned, there are significant implications of doing so from a top-down approach. As Campbell notes, the effects of dislocation from home can last for generations, and can have significant ramifications for the maintenance and enjoyment of cultural and social rights by resettled communities.[222]

When asked whether he would like to be able to retain some form of self-governance for Kiribati if the whole population ultimately had to leave, the President of Kiribati stated:

> Quite frankly that's an issue that I've never really focused on. Because I think, retaining national integrity—in my speeches what I've said is that I leave that for the rest to sort it out. I focus on getting our people to survive. But these issues—I think, at some point in time they will have to be addressed. But if you're scattering your people in different parts of the globe, how do you retain national unity?[223]

VIII. Self-Governing Alternative

Relocation does not, in and of itself, necessarily preclude claims that the State continues to exist, especially if some of the original population remains in the home State.[224] Indeed, one of the ideas proposed by the President of Kiribati is the establishment of a small government outpost on the State's only high ground,

[220] UNGAOR (n 218) para 4 cited in Memorial of the Republic of Nauru (n 211), para 170. Tabucanon and Opeskin (n 201) 14 regard issues of national identity, Australian racism, and control of economic resources as the three key reasons why resettlement failed.

[221] Kathy Marks, 'Australia Moots Radical Future for Bankrupt Nauru', *The Independent* (London, 20 December 2003) <http://www.independent.co.uk/news/world/australasia/australia-moots-radical-future-for-bankrupt-nauru-577190.html> accessed 11 August 2011: 'Mr Downer said Canberra was "very concerned" about the situation in the tiny island state, which is bankrupt and widely regarded as having no viable future'.

[222] Campbell (n 162). On the difficulties of voluntary internal relocation, which are likely to be magnified in the context of international relocation (particularly if involuntary), see Kai Schmidt-Soltau, 'Protected Areas and Resettlement: What Scope for Voluntary Relocation?' (2007) 35 World Development 2182.

[223] Interview with President Anote Tong (n 35).

[224] Indeed, as Nauru observed about its own negotiations with Australia: 'But Nauru would, at that point, still have remained under Trusteeship. Resettlement would not have granted to Australia or the British Phosphate Commissioners any further title to the land than that which they could claim under the Trusteeship. By the act of resettlement, Nauru was not to be annexed to Australia. As a self-determination unit, the Nauruan community could still seek control in Nauru both politically,

the island of Banaba, so as to retain the State and its control over resources, such as those generated by its extensive exclusive economic zone (EEZ). Ironically, most of the Banaban population was relocated in the mid-1940s to Rabi island in Fiji as a result of phosphate mining.[225] According to the former Solicitor-General of Kiribati, Banaba could only sustain a couple of hundred people in any event, with fresh water supplies a particular problem.[226]

A more radical alternative, however, would see the deliberate, earlier dissolution of the independent, sovereign State, but with the aim of preserving the 'nation'—as an identifiable national, linguistic, and cultural community—for longer. For many Tuvaluans and I-Kiribati, the issues of key importance to them are the retention of 'home'—land, community, and identity—rather than preserving the political entity of the State itself.[227] Indeed, a claim to self-determination does not necessarily involve a claim to statehood and secession.[228]

There are a number of ways in which a move away from fully fledged statehood to a self-governing alternative could be undertaken. For present purposes, the option considered is one based on a well-established model within the Pacific: self-governance in free association with another State. The rationale behind this model is to respect 'the individuality and the cultural characteristics of the territory and its peoples' and give the associated territory 'the right to determine its internal constitution without outside interference',[229] while certain functions (such as defence) are carried out by another State. Crawford describes association as 'one of the more significant possibilities of self-government communities (especially island communities) that are too small to be economically and politically viable standing alone'.[230] It is also a familiar model in the Pacific, describing the relationship of the Cook Islands and Niue vis-à-vis New Zealand.[231]

That there is no single concept of self-governance is borne out in the different approaches of Niue and the Cook Islands.[232] The Cook Islands has continually stressed its independence. For example, the 2001 Joint Centenary Declaration of the Principles of the Relationship between the Cook Islands and New Zealand (reiterating the views expressed in an exchange of letters in May 1973 between those two governments) emphasized the 'relationship of partnership and free asso-

through independence, and economically, in respect of the phosphate industry': Written Statement of Nauru (n 204), para 18.

225 See, eg, Martin G Silverman, 'Making Sense: A Study of a Banaban Meeting' in Lieber (n 162); Elfriede Hermann, 'Emotions, Agency and the Dis/placed Self of the Banabans in Fiji' in Toon van Meijl and Jelle Miedema (eds), *Shifting Images of Identity in the Pacific* (KITLV Press, 2004).

226 Interview with David Lambourne, Solicitor-General of Kiribati (Tarawa, Kiribati, 8 May 2009).

227 Interview with Tebao Awerika, Ministry of Foreign Affairs in Kiribati (Tarawa, Kiribati, 12 May 2009).

228 Brownlie (n 184) 6.

229 'Principles Which Should Guide Members in Determining Whether or Not an Obligation Exists to Transmit the Information Called for in Article 73 e of the Charter of the United Nations', UNGA res 1541 (XV) (15 December 1960) Principle VII.

230 Crawford (n 30) 626.

231 Crawford writes that the Cook Islands and Niue 'are not States but have some separate international status by virtue of the relevant association agreements': ibid, 492.

232 The status of associated territories depends on the specific arrangements made, and their implementation: ibid, 632.

ciation between the Cook Islands and New Zealand as equal States independent in the conduct of their own affairs'.[233] It noted that '[r]esponsibility at international law rests with the Cook Islands in terms of its actions and the exercise of its international rights and fulfilment of its international obligations',[234] and that '[t]he Government of the Cook Islands possesses the capacity to enter into treaties and other international agreements in its own right with governments and regional and international organisations'.[235]

By contrast, Niue has resisted being treated like an independent State.[236] Indeed, its constitution commits New Zealand to providing it with 'necessary economic and administrative assistance'. Nonetheless, both Niue and the Cook Islands are separate administrative entities within the Realm of New Zealand—their governments have full executive powers and their parliaments can make their own laws. New Zealand is not permitted to make laws for them unless authorized or invited by the legislature to do so. By agreement, Niueans and Cook Islanders hold New Zealand citizenship (and do not have additional Niuean or Cook Islands citizenship)[237] and can freely enter, live, and work in New Zealand (and thus also Australia).[238]

There are historical reasons for these relationships. The Cook Islands and Niue are former British protectorates which were annexed as dependent territories by New Zealand at the turn of the 20th century.[239] Through acts of self-determination overseen by the UN,[240] in 1965 and 1974 respectively, their populations chose to become self-governing territories in free association with New Zealand, which is 'a status distinct from that of full independence'.[241] While they remain separate administrative and cultural entities, they have the security of New Zealand

[233] Joint Centenary Declaration of the Principles of the Relationship between the Cook Islands and New Zealand (agreed 6 April 2001) Preamble.

[234] ibid, para 4.

[235] ibid, para 5.

[236] See Alison Quentin-Baxter, 'Niue's Relationship of Free Association with New Zealand' (1999) 30 Victoria University of Wellington Law Review 589, 593.

[237] On the 'extension of [this] privilege', which is 'constitutionally a matter totally within the control of the state of New Zealand', see Yves-Louis Sage, 'Self-Determination: "Shared Sovereignty": A Comparative Analysis' in Kennedy Graham (ed), *Models of Regional Governance for the Pacific: Sovereignty and the Future Architecture of Regionalism* (Canterbury University Press, 2008) 128.

[238] See Cook Islands Constitution Act 1964 (NZ); Niue Constitution Act 1974 (NZ); Constitution of the Cook Islands (Schedule to the Cook Islands Constitution Act); Constitution of Niue (Schedule 2 of the Niue Constitution Act). The 1973 Trans-Tasman Travel Arrangement permits New Zealand citizens to visit, live, and work in Australia and vice versa. See also the Special Category Visa (SCV) for New Zealand citizens since 1994; see generally Australian Government, Department of Immigration and Citizenship, 'Fact Sheet 17: New Zealanders in Australia' (2010) available at <http://www.immi.gov.au/media/fact-sheets/17nz.htm> accessed 31 August 2010.

[239] Niue was annexed in 1901; the Cook Islands' annexation was approved by the New Zealand Parliament on 27 September 1900, and New Zealand's boundaries were formally extended from 11 June 1901.

[240] See respectively 'Question of the Cook Islands', UNGA res 2064 (XX) (16 December 1965); 'Question of Niue', UNGA res 3285 (XXIX) (13 December 1974).

[241] 'Because the way of life of their peoples was so different from that of people in New Zealand, no New Zealand Government put forward the option of integration with New Zealand for serious consideration. Self-government in free association with New Zealand seemed to offer the best prospect': Alison Quentin-Baxter, 'The Problems of Islands' (2000) 31 Victoria University of Wellington Law Review 427, 430. See also Crawford (n 30) 629–32.

citizenship, economic assistance, and political support if requested. Both countries have devolved responsibility for external affairs and defence to New Zealand, however these confer no rights of control to New Zealand and can only be acted upon at the request of and on behalf of the governments of the Cook Islands and Niue.

Given the absence of such strong historical ties with Kiribati and Tuvalu, it is questionable whether New Zealand or Australia would be willing to enter into similar free association relationships with them. Alternatives such as federation[242] or incorporation[243] might be perceived as more attractive, given the economic benefits that could be gained by merger, such as control over the extensive EEZs of Kiribati and Tuvalu.[244] In any event, the political likelihood of the Tuvaluan and I-Kiribati populations determining by referendum to move to a self-governance model, let alone to dissolve the State altogether through merger, seems remote in light of how recently independence was obtained.[245]

Even so, in interviews with government officials and community members in Kiribati and Tuvalu, it became clear that what mattered most to people was the maintenance of community and culture, rather than 'the State' per se. National boundaries established during the colonial period were designed to secure and promote European interests, rather than local culture and identity.[246] As such, '[t]he idea of sovereign statehood or the state as being an instrument through which they align themselves is anathema to their notions of sovereignty. Sovereignty to them is limited to their own religious, tribal, landholding and language groups'.[247] Furthermore, people do not have a sense of the State's role in their daily life. Most daily activities revolve around the church and traditional community structures, and the 'government's influence and impact on their lives is minimal to almost non-existent'.[248] Similarly, since 80–90 per cent of resources and land come under

[242] A federal State is 'a sole person in the eyes of international law': Montevideo Convention on the Rights and Duties of States, Art 2. As a system of political organization in which a State is comprised of different national groups, Brownlie regards federalism as 'probably better able than any other system to provide a regime of stable autonomy which provides group freedoms within a wider political cosmos and keeps the principle of nationality in line with ideas of mutuality and genuine coexistence of peoples': Brownlie (n 184) 6. See further, Crawford (n 30) 483–9.

[243] This is the basis on which the Cocos (Keeling) Islands joined Australia. Principle IX of the Annex to UNGA res 1541 (XV) (15 December 1960) assumes that the people of the State that integrates into another should be treated as equal citizens of the integrating State, accorded full citizenship rights and freedom of movement: see Crawford (n 30) 624.

[244] See Soons (n 17). Rayfuse (n 17) 11 has suggested that conditions of merger could include a requirement that any revenue generated from these territorial acquisitions be placed into a trust fund to pay for the resettlement of the merging State's population (including on-going costs that might normally be borne by the State, such as pensions, although it should be noted that there is very little social security in Kiribati or Tuvalu). Interestingly, there were plans to finance the long-term future of relocated Nauruans through the establishment of an investment fund from phosphate royalties. Its purpose would have been to fund future expenses for housing and economic development once phosphate mining had ceased: Tabucanon and Opeskin (n 201) 7.

[245] Interview with David Lambourne (n 226); interview with Tebao Awerika (n 227).

[246] John Henderson, 'Governance and Constitutional Issues: Reflections on Current Problems and Future Options' in Michael Powles (ed), *Pacific Futures* (Pandanus Books, 2006) 13.

[247] Transform Aqorau, 'Rethinking Sovereignty in the Pacific' in Powles (n 246) 219.

[248] ibid, 220.

customary ownership, there is little sense of State ownership.[249] 'In a real sense therefore, the idea that there is some supranational authority exercising power and responsibilities in a spatial jurisdictional zone is not widely accepted'.[250] Interestingly, though, 'generally Pacific Islanders do not identify themselves as being from their country until they are outside it'.[251] This is significant in the context of relocation.

If whole communities were to relocate, then it is fundamental that the principle of internal self-determination be respected to ensure that the relocating groups could retain control over their political, economic, social, and cultural development. Falk's proposed regime for the self-determination of indigenous peoples may provide a useful schema in the present context:[252]

1. Recognition of internal self-determination through an appropriate form of autonomy enabling for protection and self-control over life in all aspects;
2. Some dimension of external self-determination to form solidarity networks with other affected groups and organizations;
3. Promotion of human rights in relation to the specific circumstances of those displaced;
4. Acceptance 'of some international personality for [affected] peoples, so that they can present their claims and grievances in areas outside the national legal system';[253]
5. Special attention to the protection of sacred sites; and
6. Notions of restitution in the form of land rights and financial assistance to overcome distress and, in particular to those who find themselves at the brink of extinction.

As Kälin has observed, ensuring that 'populations of affected small island States can continue to retain their identities as communities, and exist as viable communities even after the loss of most or all of their territory', remains the 'biggest challenge'.[254] A considerable amount of thought, negotiation, and participatory planning must occur if this issue is to be addressed—let alone resolved—in a human rights-sensitive manner. As Chapter 9 reiterates, approaching the issue of relocation through a human rights lens means that in all phases of displacement—the pre-displacement and planning phase, the period of actual displacement, and the resettlement or relocation phase—people's needs and rights must be recognized and respected. Communities must have access to information about the reasons and procedures for their movement, participate in the planning and management of their movement, and have the rights to life, dignity, liberty, and security respected at all stages. As Campbell and others have observed, without careful

[249] ibid. [250] ibid. [251] ibid, 219.

[252] Richard Falk, 'The Rights of Peoples (in Particular Indigenous Peoples)' in Crawford (n 184) 35. See also Ch 9. [253] ibid.

[254] Kälin (n 125) 102.

planning, negotiation, and consultation with both relocating and host communities, relocation is likely to be fraught.[255]

IX. Conclusion

State practice suggests that there is likely to be a presumption of a State's continuity for some time, even as the legal indicia of statehood begin to wane. However, at some future point this may cease as the objective characteristics of statehood start to recede,[256] and States, unilaterally or collectively, may gradually withdraw their recognition of an entity as a State.[257] As noted above, though, this is not imminent, and the utility of the law on statelessness for addressing the needs of displaced people is slight. It may, however, provide a hook for institutional engagement, such as through UNHCR's mandate for the prevention of statelessness.[258]

International legal personality is not confined to States, and other entities, such as international organizations, groupings of States, Taiwan, and the Sovereign Order of Malta, operate to differing degrees at the international level. While they do not have the same extensive 'full' powers of States, they have certain functional powers that enable them to operate at the international level.[259] Furthermore, the Vatican City is a State, even though its position as such is 'peculiar' and 'the criteria for statehood in its case are only marginally (if at all) complied with'.[260] Its rec-

[255] Campbell, Goldsmith, and Koshy (n 160). See also Graeme Hugo, 'Lessons from Past Forced Resettlement for Climate Change Migration' in Etienne Piguet, Antoine Pécoud, and Paul de Guchteneire (eds), *Migration and Climate Change* (Cambridge University Press and UNESCO Publishing, 2011); Yntiso Gebre, 'Resettlement and the Unnoticed Losers: Impoverishment Disasters among the Gumz Hosts in Ethiopia' (2003) 62 Human Organisation 50; Anthony Oliver-Smith, 'Communities after Catastrophe: Reconstructing the Material, Reconstituting the Social' in Stanley E Hyland (ed), *Community Building in the Twenty-First Century* (School of American Research Press, 2005). For useful elements in a strategic relocation plan, see Bronen (n 52) 402.

[256] Rayfuse (n 17) 13 writes: 'in an international community still based on the Westphalian notion of states, it may not be appropriate or realistic to envisage the permanent establishment and continuing existence of deterritorialised states ad infinitum. Rather, it may be useful to view this status as transitional, lasting perhaps one generation (30 yrs) or one human lifetime (100 yrs), by which time it is likely that much else in the international legal regime, including the existing law of the sea regime, will have to be reconsidered and reconfigured, in any event'.

[257] Although the better view is that recognition is declaratory, rather than constitutive, of statehood, it is acknowledged that 'the present state of the law makes it possible that different states should act on different views of the application of the law to the same state of facts': David J Harris, *Cases and Materials on International Law* (6th edn, Sweet and Maxwell, 2004) 145.

[258] See Ch 8; Park (n 3).

[259] It is not certain that small island States such as Kiribati and Tuvalu would ever fall into this category, only because States sometimes continue to recognize statehood even when its criteria 'are only marginally (if at all) complied with': Crawford (n 30) 223. This is the case with the State of the Vatican City. The strength and influence of its government, the Holy See, compensates for its very small territory and lack of a permanent population, in the same way that in certain 'failed States', the existence of territory and people compensate for the virtual absence of a government: Crawford (n 30) 223. The question is whether, in the absence of a permanent population within a diminishing territory, other States would be prepared to continue to recognize Tuvalu and Kiribati as on-going States or not.

[260] Crawford (n 30) 223.

ognition by other States is therefore of 'considerable importance'.[261] The strength and influence of its government, the Holy See, compensates for its very small territory and lack of a permanent population, in the same way that in certain 'failed States', the existence of territory and people compensate for the virtual absence of a government.[262] The question is whether, in the absence of a permanent population within a diminishing territory, other States would be prepared to treat Tuvalu and Kiribati as continuing States or not.

If Tuvalu and Kiribati were at some point regarded as having acquired a different kind of international legal personality, other than as a State, then in the absence of acquisition of a new nationality, their former nationals could be said to meet the definition of a 'stateless person' in Article 1 of the 1954 Statelessness Convention: someone 'not considered as a national by any State under the operation of its law'. This is because in international law, when a State ceases to exist, so does nationality of that State.[263] States parties to the 1954 Convention would thus be obliged to afford former nationals the rights contained within it, including 'as far as possible facilitat[ing] the[ir] assimilation and naturalization'.[264] While this would finally bring those displaced within an existing legal category, it is far from adequate as a means of addressing potential displacement from small island States. It is reactive, rather than proactive; it requires people to leave their homes and be present in the territory of a State party to the Convention in order to claim its benefits; and, in the absence of any status determination procedure for stateless persons, there is no clear means by which those benefits could be accessed.

While there is no simple legal 'solution' to the 'disappearing States' phenomenon and the status of those displaced, it is important to be aware of the human rights implications of certain mooted alternatives, in particular with respect to (and for) individual and community decision-making and choices. Historical examples from the Pacific show that relocation *en masse*, while theoretically a means of maintaining cultural integrity, has been fraught with difficulties in practice and risks being imposed as a top-down 'solution' that strips individuals and communities of agency. By contrast, self-governance in free association with another State is an option that would preserve a degree of autonomy and sense of 'nation' and culture for some time, but it is questionable whether this move away from full statehood would presently appeal to recently independent States such as Kiribati and Tuvalu, and, moreover, to potential partner States like Australia and New Zealand.

Paradoxically, planned and staggered migration over time—the solution favoured by Kiribati if *in situ* adaptation to climate change is not possible—may ultimately start to erode longer-term claims to continued sovereignty and statehood. This is because the State's 'disappearance' may begin once the bulk of the permanent population has moved abroad and obtained a legal status in a

[261] ibid.

[262] ibid.

[263] As Weis (n 139) 136 notes: 'In the case of universal succession, the predecessor State is extinguished and its nationality ceases to exist. All persons who were nationals of the predecessor State cease to be such.'

[264] 1954 Convention, Art 32.

new country, either through naturalization or by being born there (especially if overseas-born children acquire the nationality of their new country and do not retain their parents' nationality). Additionally, though the 'population' criterion of statehood does not require that a majority of nationals lives within the State's territory, a substantial loss of population would start to erode the effectiveness of the State's government as its economic base declined.[265]

On the other hand, relieving population pressure on environmentally fragile atoll States such as Kiribati and Tuvalu may in fact enable a smaller community to remain there for much longer, and in turn maintain the 'population' criterion of statehood. The negative impacts of overcrowding on the adequacy of infrastructure, access to resources, employment opportunities, and environmental sustainability mean that relieving population pressure through migration could help to enhance the quality of life for those who remain. It might also generate additional income via remittances to assist with in-country adaptation and service provision. By comparison to relocation or humanitarian assistance schemes, planned migration offers individuals and households the greatest choice about when to move, and affords them the opportunity to establish 'pockets' of their communities abroad which others can join over time. It also enables potential host States to better plan for immigration and develop culturally sensitive policies towards migrants, rather than trying to spontaneously accommodate displaced people who do not easily fit existing legal categories.

[265] But see the Tibetan government in exile's links to its various 'constituencies'—the diaspora in exile and the population remaining in the homeland. McConnell argues that it has 'attempted to construct a Tibetan "population" in exile and has striven to institutionalise state–citizen relationships including the introduction of "voluntary" taxation, exile parliamentary elections and the provision of welfare services to Tibetans in India and Nepal': McConnell (n 156) 1907.

6

Moving with Dignity: Responding to Climate Change-Related Mobility in Bangladesh

I. Introduction

Through a case study of Bangladesh,[1] this chapter examines legal and policy responses to displacement that is predominantly internal and rapid-onset in nature. As a low-lying, densely-populated delta nation, with a significant proportion of its population living in coastal or flood-prone areas,[2] Bangladesh is one of the countries most vulnerable to climate change. Over the next century, it is predicted that climate change will cause Bangladesh to experience increased temperatures; monsoon rain and riverbank erosion; more intense cyclones; and more severe droughts.[3] This, in turn, will affect food production,[4] water supplies, health, and people's ability to remain in their homes and earn a living. Sea-level rise may alter the sediment balance and salinity of the water and soil in coastal

[1] This chapter is based on a previously published article: Jane McAdam and Ben Saul, 'Displacement with Dignity: International Law and Policy Responses to Climate Change Migration and Security in Bangladesh' (2010) 53 German Yearbook of International Law 233. I am grateful to Ben Saul for permitting me to draw heavily on that piece in this chapter.

[2] Global risk analysts, Maplecroft, use GIS data to identify countries according to their vulnerability to manage risks posed by future climate change. Of 166 countries, Bangladesh is rated 12th and is defined as extremely vulnerable: see <http://maplecroft.com/portfolio/doc/climate_change/Climate_Change_Poster_A3_2010_Web_V01.pdf> accessed 3 May 2010. 60 per cent of Bangladesh's population lives in flood-prone areas: World Bank, 'Data: Climate Change and South Asia' <http://web.worldbank.org/WBSITE/EXTERNAL/COUNTRIES/SOUTHASIAEXT/0,,contentMDK:22408900~pagePK:2865106~piPK:2865128~theSitePK:223547,00.html> accessed 25 February 2010. Some 40 million people live in coastal areas, with 5 million residing in highly vulnerable areas within 100 km of the coast and less than 12 metres above sea level: Koko Warner and others, *In Search of Shelter: Mapping the Effects of Climate Change on Human Migration and Displacement* (CARE International and others, 2009) 13.

[3] See generally, Government of the People's Republic of Bangladesh, *Climate Change and Bangladesh* (September 2007) <http://www.bdresearch.org.bd/home/climate_knowledge/cd1/pdf/Bangladesh%20and%20climate%20change/Climate%20change%20impacts%20,vulnerability,%20risk/Climate%20Change%20And%20Bangladesh.pdf> accessed 14 July 2011; Intergovernmental Panel on Climate Change (IPCC), *Climate Change 2007: Impacts, Adaptation and Vulnerability: Contribution of Working Group II to the Fourth Assessment Report of the Intergovernmental Panel on Climate Change* (Cambridge University Press, 2008) Ch 10.

[4] Disruptions to agriculture lead to higher food prices. The Economist Intelligence Unit's Bangladesh Country Report of February 2008 stated that the price of the staple food, rice, had nearly doubled since January 2007 following floods and a cyclone: 'Country of Origin Information Report: Bangladesh' (UK Border Agency, 11 August 2009) para 2.03.

regions, threatening freshwater supplies, fish stocks, water drainage, and arable land. The rapid pace of change means that people's traditional coping strategies will be greatly challenged.

Of Bangladesh's 160 million people, 81 per cent live on less than $2 a day and 50 per cent live on less than $1.25 a day.[5] Three quarters of the population live in rural areas which are environmentally vulnerable, and half of rural households are landless. When an already marginalized population loses its means of production, rebuilding homes and recommencing economic activity is particularly difficult, and vulnerability to poverty is increased.[6]

Drawing on fieldwork undertaken in Bangladesh and eastern India in mid-2010, this chapter examines the extent to which climate change is likely to impact on displacement and migration in and from Bangladesh, and the legal and policy frameworks which might respond to this. It challenges alarmist predictions that large numbers of Bangladeshi 'climate refugees'[7] will be displaced across international borders by the middle of the 21st century, and, in turn, threaten international or regional security.[8] Based on current patterns of movement (which are the most likely indicators of future movement),[9] most displacement in Bangladesh that can be linked to climate change is likely to be internal rather than cross-border in character. Importantly, reliance on alarmist estimates may impact negatively on the creation of principled and appropriate legal and policy responses by diverting attention away from the areas of most need.

The chapter is underscored by four premises. First, it is inherently fraught to speak of 'climate change' as the 'cause' of human movement, even though its impacts may exacerbate existing vulnerabilities. Secondly, while there remains a need for further empirical work to better understand present and past displacement and migration in Bangladesh—to provide an evidentiary basis from which rational and appropriate responses to future movement may be devised—'the best way to prepare for the consequences of climate change in 2050 or 2100 is to improve the ability to deal effectively with Bangladesh's existing vulnerabilities

[5] Oxford Poverty and Human Development Initiative, 'Country Briefing: Bangladesh: Multidimensional Poverty Index (MPI) at a Glance' (July 2010) <http://www.ophi.org.uk/> accessed 25 August 2010.

[6] Alice Poncelet, 'Bangladesh Case Study Report: "The Land of Mad Rivers"' (EACH-FOR, 2009) 10.

[7] See, eg, 'PM Warns of Climate Refugee Crisis', *The Daily Star* (Dhaka, 22 September 2010) <http://www.thedailystar.net/newDesign/latest_news.php?nid=26005> accessed 9 July 2011; 'Hasina Highlights Unfortunate Plight of Climate Migrants', *The New Nation* (Dhaka, 25 September 2010) <http://www.highbeam.com/doc/1P3-2145800121.html> accessed 18 October 2011. The Bangladeshi non-governmental organization (NGO) network, Equity BD, which has from time to time used the 'climate refugee' terminology (see, eg, Equity BD, 'Climate Change Induced Forced Migrants: In Need of Dignified Recognition under a New Protocol' (December 2009)) says it now rejects this: interview with Md Shamsuddoha and Rezaul Karim Chowdhury from Equity BD (Dhaka, Bangladesh, 19 June 2010).

[8] See McAdam and Saul (n 1) 253–65; Ch 8.

[9] See, eg, Graeme Hugo, 'Climate Change-Induced Mobility and the Existing Migration Regime in Asia and the Pacific' in Jane McAdam (ed), *Climate Change and Displacement: Multidisciplinary Perspectives* (Hart Publishing, 2010) 9; Jon Barnett and Michael Webber, *Accommodating Migration to Promote Adaptation to Climate Change* (Commission on Climate Change and Development, 2009) 17.

now'.[10] Thirdly, while movement can be a sign of vulnerability, it can also be a means to achieve security and attain human rights, especially when it is able to be planned.[11] Finally, no single legal or policy response will adequately address the complexity of climate change-related displacement. Rather, a combination of strategies is needed. For example, adaptation needs to be financed and developed, but this does not mean that migration options should be overlooked. Indeed, as Chapter 7 examines in more detail, migration is 'a potentially positive adaptive strategy of individuals, households and communities' which may bring significant development,[12] economic, and social benefits for migrants, their families (via remittances), and the areas to which they move.[13] Whether internal or cross-border in nature, migration should be 'mainstreamed' into policy as a rational adaptation strategy, rather than considered as a sign that adaptation has failed.[14]

The first part of the chapter describes the impacts of climate change on displacement and migration in Bangladesh. The second part examines the nature of such movement, and the final section sets out a number of options for law and policy reform with respect to climate change-related movement within and from Bangladesh.

II. The Impacts of Climate Change on Movement in Bangladesh

There is scientific consensus that the effects of climate change are aggravating many of the 'natural' environmental hazards already faced by Bangladesh, including sudden-onset events such as flooding, cyclones, storm surges, water-logging, salinity intrusion, and riverbank erosion, and slow-onset processes like coastal erosion (predominantly through rising seas, but also hydrological dynamics) and land loss.[15] Sea-level rise from climate change is anticipated to worsen many of these processes and to subsume up to 30 per cent of Bangladesh's coastal land by 2080.[16]

[10] International Organization for Migration (IOM), *Assessing the Evidence: Environment, Climate Change and Migration in Bangladesh* (IOM, 2010) 9.

[11] ibid, 26.

[12] See World Bank, *World Development Report: Development and Climate Change* (International Bank for Reconstruction and Development/World Bank, 2010).

[13] IOM (n 10) x.

[14] See, eg, Roger Zetter, 'Legal and Normative Frameworks' (2008) 31 Forced Migration Review 62. For example, there is evidence that people who move into areas of new land along rivers or in coastal zones in Bangladesh help to accelerate the conversion of new sediment deposits into viable agricultural land: IOM (n 10) xii.

[15] See generally, *Climate Change and Bangladesh* (n 3); IPCC (n 3) Ch 10.

[16] James Pender, 'Community-Led Adaptation in Bangladesh' (2008) 31 Forced Migration Review 54, citing research by the UK Institute of Development Studies. The IPCC's Second Assessment Report suggested that 17.5 per cent of land could be lost with a one metre sea-level rise: IPCC, *IPCC Second Assessment: Climate Change 1995: A Report of the Intergovernmental Panel on Climate Change* (IPCC, 1995) 34. See also Cleo Paskal, 'How Climate Change is Pushing the Boundaries of Security and Foreign Policy' (2007) Chatham House Briefing Paper 4, citing modelling published by UNEP, University of Dhaka, World Bank, and others.

Whereas once predictable floods were beneficial to agriculture, more recent severe and irregular flooding has adversely impacted on livelihoods. Flooding currently displaces between 500,000 and one million Bangladeshis per year,[17] with water drainage capacity decreasing.[18] Historically, the single biggest reason for displacement in Bangladesh has been riverbank erosion, with some 160,000 hectares eroded since 1973. Some experts place the figure at one million displaced people a year, and around 70 per cent of those living in risk areas are landless (that is, do not own their own land).[19] Erosion is likely to worsen as monsoon rains increase.[20]

The effects of climate change will interact with existing underlying causes of political, economic, and social instability in Bangladesh, exacerbating the risks of displacement and insecurity. Poverty, underdevelopment, limited land availability, and conflict over land are already potent drivers of migration. Climate impacts will further limit economic opportunities, including by impacting on river resources (such as fishing and agriculture) and straining government capacity to assist displaced people.

A. Causation

While the impacts of climate change are likely to contribute to displacement in Bangladesh, amplifying existing susceptibilities, it is difficult to isolate 'climate change' as a cause of movement. First, in a country heavily affected by frequent 'natural' disasters, it can be difficult to disentangle climate change-related events from 'ordinary' environmental processes. Since 1990, for instance, over 100,000 people have moved permanently from the coastal island of Kutubdia in south-eastern Bangladesh to cities such as Chittagong, Cox's Bazar, and Dhaka, due to coastal erosion and salinity.[21]

Cyclone Aila, which hit coastal areas of Bangladesh in May 2009, is another case in point. Cyclones are a natural occurrence in Bangladesh. On average, Bangladesh is hit by a tropical cyclone every three years,[22] but there is a sense within Bangladesh that the frequency of major cyclones is already increasing.[23]

[17] Warner and others (n 2) 13; Global IDP Project, Background Information on the IDP Situation in Chittagong Hill Tracts (CHT) in Bangladesh (13 June 2001) <http://reliefweb.int/node/82278> accessed 14 October 2010.

[18] Interview with Abu M Kamal Uddin, CDMP (Dhaka, Bangladesh, 16 June 2010).

[19] See K Maudood Elahi, 'Impacts of Riverbank Erosion and Flood in Bangladesh: An Introduction' in K Maudood Elahi, K Saleh Ahmed, and M Mafizuddin (eds), *Riverbank Erosion, Flood and Population Displacement in Bangladesh* (Riverbank Impact Erosion Study, Jahangirnagar University, 1991); Waliul Islam and M Ataur Rahman, 'Riverbank Erosion in Bangladesh: An Overview of Field Observation' (REIS Newsletter (3), Riverbank Impact Erosion Study, Jahangirnagar University, 1987).

[20] Interview with Abu M Kamal Uddin, CDMP (n 18); Government of the People's Republic of Bangladesh, *Bangladesh Climate Change Strategy and Action Plan* (September 2009) 14.

[21] Interview with Abul Kalam Azad, Research Director, Bangladesh Institute of International and Strategic Studies (Dhaka, Bangladesh, 15 June 2010).

[22] *Bangladesh Climate Change Strategy and Action Plan* (n 20) 11.

[23] Interview with Rizwana Hasan, Chief Executive of the Bangladesh Environmental Lawyers Association (BELA) (Dhaka, Bangladesh, 16 June 2010): major cyclones have increased on average from once in every four years to annually.

Though Cyclone Aila was particularly intense and displaced over 100,000 people, experts are divided on how to quantify the extent to which climate change contributed to the destruction and displacement it caused. Even if the increased intensity of cyclones due to climate change could be accurately measured, it hardly makes sense for policymakers to seek to differentiate between those 'naturally' displaced by Aila and those displaced by its climate-related intensification.

Secondly, climate change tends to multiply pre-existing stressors, such as poverty, underdevelopment, unsustainable agricultural practices, and population pressure, rather than cause movement on its own. For example, it is clear that commercial activities such as shrimp cultivation contribute to the weakening of river embankments. Pipes pumping water are run through the middle of the banks, making them structurally fragile, and local agricultural production is destroyed as rice paddy fields are deliberately flooded with saline water from shrimp cultivation.[24] Some NGOs therefore argue that much of the responsibility for ensuing displacement rests with the government, which has welcomed commercial investment and industry (selling off land to developers despite insisting no land was available for the landless poor), while failing to assist those whose livelihoods are compromised by large-scale farming activities.[25] The head of the Bangladesh Environmental Lawyers Association explained that 'when [Cyclone] Aila came, all the embankments collapsed.... Now you can link the cyclone to climate change—you may, you may not—but the fact remains that the people are displaced not because the cyclone hit, but because the embankment collapsed.'[26] Similarly, while increased rainfall in areas such as Cox's Bazar may be a cause of destructive landslides, the role of human activities such as illegal logging, vegetation clearing, cultivation, and mud excavation for construction are also significant contributing factors.[27]

As explained in Chapter 1, some scholars argue that it is arbitrary to identify 'climate change' as a driver of forced migration while omitting other causes such as those mentioned above.[28] From a policy perspective, it would seem both practically impossible and conceptually arbitrary to attempt to differentiate between those displaced persons who deserve 'protection' on account of climate change, and those who are victims of 'mere' economic or environmental hardship. As a Bangladeshi government official explained:

[24] See, eg, ibid; 'Farmers Resist Saline Water Based Shrimp Farming', *The Daily Star* (Dhaka, 11 May 2009) <http://www.thedailystar.net/newDesign/news-details.php?nid=87724> accessed 15 July 2011; Quazi Amanullah, '"No" to Saline-Water Based Shrimp Farming', *The Daily Star* (Dhaka, 9 April 2009) <http://www.thedailystar.net/newDesign/news-details.php?nid=83329> accessed 15 July 2011, cited in IOM (n 10) 19.

[25] See, eg, interview with Rizwana Hasan, BELA (n 23); interview with Dr Hameeda Hossain, Founder of human rights NGO Ain O Salish Kendra (ASK) (Dhaka, Bangladesh, 14 June 2010).

[26] Interview with Rizwana Hasan, BELA (n 23).

[27] Interview with Abu M Kamal Uddin, CDMP (n 18).

[28] See Alexander Betts and Esra Kaytaz, 'National and International Responses to the Zimbabwean Exodus: Implications for the Refugee Protection Regime' (2009) UNHCR New Issues in Refugee Research, Research Paper No 175.

Let's say for example, one person is able to carry only 40 kg on his shoulders. That's his limit, and he's a poor man. Now on the top of that, I come, and I give him one kilogram. So now the question is: who is responsible for killing him? Is this the 40 kilograms he was already carrying on his head, or the one kilogram I have now put on the top of that?[29]

Thirdly, the way climate change impacts are felt varies from society to society, since people's ability to cope with them is affected by broader political, economic, and social conditions. This extends to people's mobility decisions as well. Since it is difficult to anticipate the likely scale and success of adaptation measures, it is difficult to quantify the extent to which adaptation will ameliorate the need for people to move in Bangladesh over time.

III. The Likely Nature of Movement

A. Introduction

The factors discussed above make it impossible to accurately predict how many people will move as a result of climate change in Bangladesh. Even so, past and present patterns of movement in Bangladesh relating to natural disasters and environmental degradation suggest that the vast majority of movement will be internal.[30]

Alarmist predictions that some 30 million people[31] will be displaced from Bangladesh by 2050 as a result of climate change therefore must be treated with caution. As the International Organization for Migration (IOM) observes, they are based on sea-level rises which fall outside the 'harshest' scenarios of the Intergovernmental Panel on Climate Change (IPCC); count land loss but not accretion; assume the absence of any adaptation measures to reduce vulnerability;[32] and overlook 'localized, fine-grained' contexts.[33] The common assumption that displacement will involve large-scale *cross-border* movements to countries such as India, or further afield in South-East Asia or Europe, simply does not reflect existing patterns of movement from natural disasters in Bangladesh, which provide the best indicator of future movement. While there remains a need for further, targeted research on mobility in Bangladesh—especially across a larger sample size than has been possible in some of the published studies—existing studies do enable conclusions to be drawn about the shortcomings of existing national and

[29] Interview with Abu M Kamal Uddin, CDMP (n 18).

[30] Interview with SM Munjurul Hannan Khan, Deputy Secretary, Ministry of Environment and Forests and National Focal Point for the UNFCCC and IPCC, Bangladesh (Dhaka, Bangladesh, 15 June 2010).

[31] Interview with Mihir Kanti Majumder, Secretary, Ministry of Environment and Forests, Bangladesh (Dhaka, Bangladesh, 15 June 2010); Kurt M Campbell and others, *The Age of Consequences: The Foreign Policy and National Security Implications of Global Climate Change* (Center for Strategic and International Studies and Center for a New American Security, 2007) 5. Biermann and Boas calculate that there will be 26 million 'climate refugees' by that time solely on account of rising seas: Frank Biermann and Ingrid Boas, 'Preparing for a Warmer World: Towards a Global Governance System to Protect Climate Refugees' (2010) 10 Global Environmental Politics 60, 70.

[32] IOM (n 10) 17 (fn omitted).

[33] ibid, 6.

international responses, and for recommendations to be made about future policy development.

The next sections set out a taxonomy of movement in Bangladesh. This taxonomy is necessarily generalized, and there is considerable fluidity between the categories. People may not go through all of the phases listed below, or they may move between them as their circumstances change. It is rare for migration to be a single occurrence: rather, there is often initial flight from a disaster, subsequent movement to find a safe place to stay, and then migration further afield to seek work. With successive disasters, this pattern may be repeated, such that it is best described as cyclical rather than linear in nature. For example, of 200 households surveyed in one study on riverbank dwellers, the cumulative number of displacements was 892—an average of 4.46 times per household.[34] One woman said that: 'Mentally we get prepared for the next move as soon as we get settled in a new area.'[35]

B. Internal displacement from natural disasters

Displacement due to sudden-onset events is predominantly localized within Bangladesh. People tend to move very short distances and seek to return to their homes as soon as they can, although this is sometimes impossible when areas are inundated repeatedly.[36] Since people living in vulnerable areas are often very poor, they typically lack the resources to move long distances and do not have support networks in other countries to assist them on arrival. For example, in a comprehensive study on riverbank erosion, households on average moved only one kilometre away from their homes, with 10 kilometres the furthest distance. Indeed, some people said that even if they could, they would not move far 'because of attachment to the land where their forefathers had lived' and out of a deep respect for their ancestral homes.[37] Such studies therefore suggest that mass cross-border migration in the aftermath of sudden events is unlikely. Additionally, the study found that people typically moved just four days prior to their homes being washed away.[38] Only one family had the resources to enable them to relocate 90 days before the riverbank started to collapse.

The data on secondary movements after initial displacement is less extensive. Studies in India suggest that floodplain-dwellers tend to move temporarily to

[34] Chowdhury R Abrar and Syed Nurullah Azad, *Coping with Displacement: Riverbank Erosion in North-West Bangladesh* (RDRS Bangladesh, North Bengal Institute, and Refugee and Migratory Movements Research Unit, Dhaka, 2004) 32.

[35] ibid, 88, citing a woman from Aditmari, Lalmonirhat.

[36] See IOM (n 10) 12.

[37] Abrar and Azad (n 34) 46. See also Janet Seeley and others, '"The Family is Suffering": Challenges Faced by Migrants' Families Who Stay Behind in a Village in Rural Northwest Bangladesh' in Chowdhury R Abrar and Janet Seeley (eds), *Social Protection and Livelihoods: Marginalised Migrant Workers of India and Bangladesh* (The University Press, 2009) 41; Allan Findlay and Alistair Geddes, 'Critical Views on the Relationship between Climate Change and Migration: Some Insights from the Experience of Bangladesh' in Etienne Piguet, Antoine Pécoud, and Paul de Guchteneire (eds), *Migration and Climate Change* (Cambridge University Press and UNESCO Publishing, 2011) 149–50.

[38] Abrar and Azad (n 34) 33–4.

high ground for work and shelter, and permanently where livelihoods are more severely affected.[39] A recent study on Vietnam suggested that repeated floods may induce people to migrate on a more permanent basis because of the disruption of multiple temporary moves.[40] The evidence that has been compiled in Bangladesh suggests that although long-distance and more permanent migration is an uncommon response in the immediate post-displacement phase—since basic survival is the prime concern during this period—it may become a longer-term survival strategy.[41] Temporary and circular migration provide a 'safety net' by opening up alternative livelihood opportunities and allowing remittances to be sent back to remaining family members. Much of it is rural–rural,[42] especially during periods when agricultural labour is in high demand.[43] Rural–urban migration is also common, and is likely to be increasingly so because of 'the lack of available land, high population density and shortage of year-round work across rural Bangladesh, allied to the "pull" factor of employment opportunities in urban areas'.[44] One study says it accounts for two thirds of internal movement, compared to only 10 per cent of intra-rural migration.[45] For example, post-flood rural to urban movement is used as a means of replenishing assets damaged by the flood to assist households to avoid a cycle of debt.[46] Family ties in the cities are also a pull factor.[47]

Cyclical and seasonal labour movement has long been a feature of rural life in Bangladesh, so it would be erroneous to suggest that this is a new, climate change-related phenomenon. However, climate change is likely to exacerbate existing movement patterns to the extent that extreme weather events become more frequent and severe, some areas of land become less resilient to these events, and some households or communities relocate because they can no longer adapt to repeated displacement.[48] For example, following Cyclone Aila there was a significant

[39] See Frank Laczko and Christine Aghazarm, *Migration, Environment and Climate Change: Assessing the Evidence* (IOM, 2009) 272 referring to SL Kayastha and RP Nadava, 'Flood Induced Population Migration in India: A Case Study of Ghaghara Zone' in Leszek A Kosinski, and K Maudood Elahi (eds), *Population Redistribution and Development in South Asia* (Dordrecht Reidel, 1985).

[40] See Warner and others (n 2) 15, citing EACH-FOR, 'Synthesis Report' (2009) 33.

[41] Abrar and Azad (n 34) 113.

[42] Especially for women who marry elsewhere: IOM (n 10) 23.

[43] ibid.

[44] ibid, referring to Rita Afsar, 'Internal Migration and the Development Nexus: The Case of Bangladesh' (Refugee and Migratory Movements Research Unit, Dhaka and DFID, 2003) <http://www.migrationdrc.org/publications/working_papers/WP-C2.pdf> accessed 1 September 2010.

[45] HZ Rahman, M Hossain, and B Sen, *Dynamics of Rural Poverty in Bangladesh* (Bangladesh Institute of Development Studies, 1996), cited in Richard Black, Dominic Kniveton, and Kerstin Schmidt-Verkerk, 'Migration and Climate Change: Towards an Integrated Assessment of Sensitivity' (2011) 43 Environment and Planning A 431, 442.

[46] See MZ Mizanur Rahman, 'Emigration and Development: The Case of a Bangladeshi Village' (2000) 38 International Migration 119, cited in IOM (n 10) 10.

[47] Interview with Abul Kalam Azad, Bangladesh Institute of International and Strategic Studies (n 21).

[48] As Black and others have explained, one way of estimating the impact that climate change will have on future movement is by examining existing migration patterns and determining their susceptibility to climate change: Black, Kniveton, and Schmidt-Verkerk (n 45) 437 and table at 440.

increase in seasonal movement.[49] It remains difficult to ascertain how much of this movement was permanent, and how much remains cyclical.

Many displaced people who move to the cities end up in urban slums. Dhaka is already home to 13 million people,[50] and some estimates place the slum population as high as 85 per cent of the city's population.[51] Though existing data suggests that many slum dwellers have come from disaster-prone rural areas in search of work,[52] further research on this is needed.[53] Dhaka slum residents come from 28 out of 64 districts in Bangladesh, and other cities also host large slum populations.[54] Dhaka's population is estimated to grow to 20 million by 2020, with around 400,000 new and mostly poor migrants arriving there from rural areas each year.[55] Many of the new arrivals appear driven by the inter-related problems of poverty, environmental factors, and climate change impacts. Some slums in Dhaka (such as Bhola) are even named after coastal areas which have been adversely affected by climate change impacts,[56] yet it remains difficult to distinguish those who have moved on account of poverty from those who have been affected by climate change.[57] The underlying factors are interlinked.

C. Cross-border migration

The available evidence suggests that the likelihood of mass cross-border displacement from Bangladesh on account of climate change is significantly exaggerated. In the riverbank erosion study mentioned earlier, for instance, only five per cent of 595 flood-affected rural households could afford to send people

[49] IOM (n 10) 12. See also 'Bangladesh: Cyclone Aila Survivors Take Another Hit' (IRIN, 7 April 2010) <http://www.irinnews.org/Report.aspx?ReportId=88716> accessed 2 September 2010; ActionAid and others, *Bangladesh: In-Depth Recovery Needs of Cyclone Aila Affected Areas* (25–31 October 2009) <http://reliefweb.int/sites/reliefweb.int/files/resources/F6603B7EF22A16B4C125768D004B1190-Full_Report.pdf> accessed 15 July 2011.

[50] UN-Habitat, *State of the World's Cities 2008/2009* (Earthscan, 2008).

[51] Statistical overview of Bangladesh, based on 2001 estimates: <http://www.unhabitat.org/categories.asp?catid=248> accessed 11 July 2011. Others cite figures of 40 per cent: interview with Rizwana Hasan, BELA (n 23).

[52] For example, a 1998 survey among 230 households in Serajganj found that 5,500 of 30,000 slum dwellers had moved on account of riverbank erosion: David Hutton and C Emdad Haque, 'Human Vulnerability, Dislocation and Resettlement: Adaptation Processes of River-bank Erosion-induced Displacees in Bangladesh' (2004) 28 Disasters 41. 75 per cent of them were living next to people from their original district, suggesting that they may have moved *en masse*.

[53] IOM (n 10) 25. In my fieldwork, conversations with slum dwellers showed that some had moved on account of environmental degradation, but this was a very small sample and so cannot necessarily be taken as representative.

[54] Centre for Urban Studies and others, 'Slums of Urban Bangladesh: Mapping and Census 2005' (May 2006), cited in IOM (n 10) 24.

[55] World Bank Office Dhaka, 'Dhaka: Improving Living Conditions for the Urban Poor', Bangladesh Development Series Paper No 17 (World Bank, 2007) xiii.

[56] Interview with Ahmed Swapan Mahmud, Executive Director of VOICE (Dhaka, Bangladesh, 14 June 2010).

[57] Interview with SM Munjurul Hannan Khan, Ministry of Environment and Forests (n 30). See McAdam and Saul (n 1) 250–3.

abroad,[58] with temporary and permanent internal migration the main responses. This also accords with data from fieldwork in Bangladesh, in which a common thread was that poverty, in addition to a strong sense of attachment to land, family, and culture, inhibits movement abroad.[59]

Certainly, there is no evidence of mass movement across international borders. As IOM notes, 'irregular migrants would be likely to face considerable difficulty in accessing post-disaster humanitarian support' if they did cross a border,[60] given their unlawful status. However, the close ethnic, linguistic, religious, and cultural links between Bangladesh and West Bengal in India, not to mention family networks and economic opportunities, mean that some movement is inevitable—in both directions.[61] Anecdotally, it is very easy for Bangladeshis (the majority of whom are Bengali) to assimilate in West Bengal, since they share the same language and are physically similar.[62] This explains why in some areas affected by Cyclone Aila, for instance, 'the Hindu families will tell you that their family head has moved to India' because of existing family networks there and the close cultural ties between Bangladesh and Indian West Bengal.[63] Similarly, some Indians sought relief from Cyclone Aila in Bangladesh.

Interestingly, though, in contrast to Bangladesh, there is a near-complete inattention to climate change-related displacement in India.[64] This is despite the fact that areas such as the Sundarbans have been heavily affected by climate impacts.[65] One explanation is that the affected areas of India comprise a relatively small part of territory and population, whereas affected areas in Bangladesh are proportionately much larger and its displacement from them represents a more critical 'survival' issue.[66] The focus of displacement discourse in India has also tended to be around development-induced displacement rather than climate change.[67]

[58] See Israt Rayhan and Ulrike Grote, 'Coping with Floods: Does Rural–Urban Migration Play Any Role for Survival in Rural Bangladesh?' (2007) 1 Journal of Identity and Migration Studies 82, 89, cited in IOM (n 10) 10.

[59] Interview with Ahmed Swapan Mahmud, VOICE (n 58); interview with Senior Official, Department of Environment, Ministry of Environment and Forests, Bangladesh (Dhaka, 14 June 2010); interview with Mihir Kanti Majumder, Ministry of Environment and Forests (n 31); interview with Abul Kalam Azad, Bangladesh Institute of International and Strategic Studies (n 21).

[60] IOM (n 10) 28.

[61] ibid, 29.

[62] Interview with SM Munjurul Hannan Khan, Ministry of Environment and Forests (n 30); interview with Devjyot Ghoshal, Indian journalist (Kolkata, India, 24 June 2010); interview with Sabyasachi Basu Ray Chaudhury, Calcutta Research Group (Kolkata, India, 25 June 2010). It is apparently relatively simple to obtain documents on the black market, such as an Indian driver's licence, which provides 'evidence' of Indian nationality and thus enables people to access Indian assistance.

[63] Interview with Rizwana Hasan, BELA (n 23). However, Devjyot Ghoshal noted that this is not a regular phenomenon and the border between Bangladesh and India, deep in the Sundarbans, is hard to identify: email from Devjyot Ghoshal to author (6 October 2010).

[64] Interview with Devjyot Ghoshal, Indian journalist (n 62); interview with Sabyasachi Basu Ray Chaudhury, Calcutta Research Group (n 62).

[65] Interview with Tushar Kanjilal, Tagore Society for Rural Development (Kolkata, India, 24 June 2010), who suggested that there is indifference to people's plight: 'who cares if 4.5 million people are wiped out?'; see also Tushar Kanjilal, *Who Killed the Sundarbans?* (Tagore Society for Rural Development, 2000); interview with Devjyot Ghoshal, Indian journalist (n 62).

[66] Interview with Devjyot Ghoshal, Indian journalist (n 62).

[67] ibid.

In many places, the Indo-Bangladeshi border remains porous, despite the fact that India has constructed a fence along parts of it in recent decades.[68] Seasonal migration into India has long occurred, while water scarcity in some areas (as a result of Indian development on river systems) has propelled cross-border movement.[69] Yet, for poor Bangladeshis who do move, India is not necessarily the promised land. For example, in Delhi, a significant proportion of one group of urban poor—rag-pickers—are said to be Bangladeshis; some poor Bangladeshi migrants are working in low-paid jobs previously filled by tribal Biharis;[70] and yet others are among the absolute poor with 'no identity', who work as fruit and vegetable vendors, umbrella sellers, and so on.[71] As one Indian analyst notes, the closest Indian destination, West Bengal, is not necessarily attractive given its relatively poor economic performance under a communist government for many decades.[72] The entry of poor Bangladeshi migrants into India (for example, as domestic help or for construction) is often facilitated by Indian sponsors who bring them into shanties in Indian cities and bribe police, and whereas Indian NGOs have campaigned for the rights of poor Indian workers, poor Bangladeshis may be left without such protection.[73]

It is very difficult to obtain accurate figures about the number of Bangladeshi migrants in India. On the one hand, India has claimed that there are up to 20 million illegal Bangladeshi immigrants there.[74] On the other hand, the Bangladeshi government denies that there is any irregular cross-border migration to India, on the basis that the Bangladeshi economy is as good as India's and there is, accordingly, no incentive to move.[75] The Indian census of 2001 indicates that only about 280,000 Bangladeshis migrated to India between 1991 and 2001,[76] but this would not capture a lot of the irregular movement that undoubtedly occurs.[77]

International movement may occur via a 'domino' effect, however. Highly skilled, professional, and business migration from Bangladesh is likely to increase as internal rural–urban movement places acute pressure on the infrastructure of

[68] This was the general view of those interviewed, with the exception of Chowdhury R Abrar: interview with Professor Chowdhury R Abrar, Refugee and Migratory Movements Research Unit, University of Dhaka (Dhaka, Bangladesh, 14 June 2010).

[69] Bangladesh Institute of International and Strategic Studies, Management of Water Resources and Water Security: The Case of the Ganges-Brahmaputra-Meghna (GBM) River Basins Conference (Dhaka, 15 June 2010).

[70] Interview with Professor Chowdhury R Abrar, University of Dhaka (n 68).

[71] Interview with Sabyasachi Basu Ray Chaudhury, Calcutta Research Group (n 62).

[72] ibid. [73] ibid.

[74] Reaz Rahman, 'Bangladesh's Fight against Terror: Lessons Learnt and Future Perspectives' in Mufley R Osmany and Mohammed Humayun Kabir (eds), *Global War on Terror: Bangladesh Perspective* (Bangladesh Institute of International and Strategic Studies, 2007) 19; Lailufar Yasmin, 'Bangladesh–India Relations: Neighbourliness Beyond Terrorism' in Imtiaz Ahmed (ed), *Terrorism in the 21st Century: Perspectives from Bangladesh* (The University Press, 2009) 94; Deepa M Ollapally, *The Politics of Extremism in South Asia* (Cambridge University Press, 2009) 194.

[75] Discussed in interview with Chowdhury R Abrar, University of Dhaka (n 68).

[76] Cited in Rizwana Shamshad, 'Politics and Origin of the India-Bangladesh Border Fence' (17th Biennial Conference of the Asian Studies Association of Australia, Melbourne, July 2008) 11.

[77] See, eg, interview with Chowdhury R Abrar, University of Dhaka (n 68).

cities like Dhaka and 'pushes' the relatively wealthy to move abroad.[78] The incentive for less skilled migrant workers (who are usually *not* among the poorest of the poor) to go abroad, such as those millions who work in construction or as domestic help in the Middle East or Malaysia, is also likely to increase as a result of the socio-economic stresses exacerbated by climate change in Bangladesh. However, since both groups of people can avail themselves of existing migration opportunities, it is unlikely that their movement will ever be cast as climate change-related, even though it might be indirectly driven by it. As discussed in Chapter 1, it will be rendered an invisible phenomenon, even if large numbers of people are forced to move over time.

Some regard this 'domino' effect as entirely appropriate. As one prominent human rights advocate wryly observed, 'not every poor farmer wants to go to Australia'.[79] Indeed, relocating someone from a remote rural area to a major metropolis in another country could have negative consequences on both sides. As a Bangladeshi Parliamentarian astutely observed:

> I think you also have to be careful about whether you can just have an open-ended migration protocol or whether it is managed migration. I think the Bangladesh position is it has to be managed and should ideally be a win-win, so an individual who's displaced [from rural Bangladesh] isn't actually the individual who is migrating to another country, let's say the UK or USA.... You can't look at it as a simple one-for-one where someone is displaced and that very person goes out. I don't think it's going to work that way.[80]

The discussion above illustrates that most people displaced by the impacts of climate change in Bangladesh will not move very far from their homes; many will move intra-rurally; many others will move to urban areas within Bangladesh; and only very few of the poor will move *irregularly* across an international border, and typically only when they have family links there. Those who move in a *regular* or lawful manner across borders will tend to be wealthier professionals, or less skilled workers who are nonetheless financially able to migrate for work abroad. In most cases, international migration will manifest as a pragmatic livelihood strategy rather than as a plea for formal international legal 'protection' from State harm or State failure in Bangladesh itself.

[78] As a Bangladeshi parliamentarian noted, with Dhaka 'bursting at the seams' from rural–urban migration, and 'virtually not functioning' due to its inadequate infrastructure, there may come a point where movement abroad becomes more likely: interview with Saber Chowdhury MP, All Parliamentary Committee on Climate Change (Dhaka, Bangladesh, 21 June 2010). See also Shammunul Islam and M Mizanur Rahman, 'Climate Change: Pressure on Urbanisation', *The Daily Star* (Dhaka, 10 July 2011) <http://www.thedailystar.net/newDesign/news-details.php?nid=193545> accessed 15 July 2011.

[79] Interview with Sultana Kamal, Executive Director of ASK (Dhaka, Bangladesh, 21 June 2010).

[80] Interview with Saber Chowdhury MP, All Parliamentary Committee on Climate Change (n 78). The President of Kiribati takes a very similar approach: interview with President Anote Tong (Tarawa, Kiribati, 12 May 2009).

IV. Options for Law and Policy Reform

This section examines a number of options for law and policy reform that would strengthen protection against climate change-related displacement within and from Bangladesh, and which would enhance opportunities for dignified livelihoods within Bangladesh or planned migration abroad. Those displaced within Bangladesh remain citizens of that country and entitled to the protections that flow from that status. It is therefore likely that they will be regarded as a domestic concern and not within the purview of international attention. Most people internationally displaced by climate change will be regarded by other States as mere 'environmental' or 'economic' migrants, categories which currently are not recognized (let alone protected) under international law. Such people accordingly have limited rights under international law and are not entitled to any particular legal status.

A. Strengthen in-country adaptation

Improving international financial and technical assistance to Bangladesh in adapting to climate change could play a critical role in preventing further displacement. Indeed, there seems to be a direct correlation between relief assistance and migration. In a study of Bangladesh's severe 1998 floods, it was found that people who felt adequately assisted and compensated were less likely to move.[81] Thus, the prompt provision of adequate assistance can reduce longer-term migration.

Such assistance can help to build community resilience by providing alternative livelihoods, supplying technical solutions, and encouraging disaster risk reduction. The success of such an approach depends not only on securing international agreement on financing adaptation,[82] but also on addressing corruption within Bangladesh and ensuring the effective delivery of assistance to those who need it most.[83] As Irene Khan has observed, 'insecurity of livelihood [in Bangladesh] is not simply a consequence of nature. To a greater extent it is a result of bad laws and policies, unscrupulous employers and corrupt officials.'[84] Some in Bangladesh also express concern that foreign aid funding might be diverted into financing climate

[81] Bimal Kanti Paul, 'Relief Assistance to 1998 Flood Victims: A Comparison of the Performance of the Government and NGOs' (2003) 169 The Geographical Journal 75, cited in IOM (n 10) 11. This was very clear in other context as well: the Queensland floods and Cyclone Yasi in early 2011. Extensive planning and warnings prevented considerable loss of life. See further, Koko Warner, 'Assessing Institutional and Governance Needs Related to Environmental Change and Human Migration' (Study Team on Climate-Induced Migration, German Marshall Fund of the United States, June 2010).

[82] See, eg, Atiq Rahman, 'Vulnerable Communities Demand Climate Justice', *The Daily Star* (Dhaka, 7 December 2009) <http://www.thedailystar.net/newDesign/news-details.php?nid=116655> accessed 15 July 2011.

[83] Interview with Sultana Kamal, ASK (n 79).

[84] Irene Khan, *The Unheard Truth: Poverty and Human Rights* (WW Norton and Co, 2009) 74.

change adaptation,[85] resulting in one group of the poor 'losing out' in favour of another.[86]

Adaptation measures are already developing in Bangladesh which may help to prevent displacement. Many people now raise their houses on plinths and adjust their farming techniques, including by using flood-resistant strains of rice or by developing 'floating gardens' to deal with water-logging.[87] Seasonal migration has developed as an important coping strategy.[88] Efforts are also being made to raise embankments and to protect them with forest cover, and to establish coastal green belts.[89] A Bangladeshi parliamentarian highlighted the need to invest in new technological solutions, such as harvesting silt for land reclamation.[90]

As one official explained:

> adaptation doesn't happen automatically. Adaptation happens in the mindset of the people. Those bamboo protections—those are not adaptation; those are the *means* for adaptation.... Tree plantation itself is not adaptation; it's a *means* for adaptation. Now the people, over time, have to accept or discard [these different options].

In other words, 'the people will decide' what are effective, sustainable adaptation measures.[91]

However, Bangladesh 'cannot adapt indefinitely'.[92] As the head of a prominent Bangladeshi NGO explained, the problem with this kind of analysis is that it assumes any adaptation is good. Instead, the pertinent question should be 'is this adaptation with dignity?' While it is possible for people to grow vegetables in floodwaters, is wading through stagnant water to collect them evidence of a dignified existence?

> If this is your house, and because of the saline water inundation, or because of the submerging of land areas, you are today surrounded by water and because you are surrounded by water and you don't have any other place you go, you start doing floating gardens in the flood water. That is adaptation, but that denies the right to a dignified life, because I have

[85] Interview with Ahmed Swapan Mahmud, VOICE (n 56).

[86] By contrast, in Tuvalu the view was expressed that if socio-economic problems were *not* presented as climate change-related, then funding might dry up. See Ch 1.

[87] IOM (n 10) 18.

[88] Afsar (n 44) 2.

[89] Interview with Senior Official, Department of Environment, Ministry of Environment and Forests (n 59); interview with Mihir Kanti Majumder, Ministry of Environment and Forests (n 31).

[90] Interview with Saber Chowdhury MP, All Parliamentary Committee on Climate Change (n 78).

[91] Interview with Abu M Kamal Uddin, CDMP (n 18).

[92] Interview with Mihir Kanti Majumder, Ministry of Environment and Forests (n 31); see also interview with Atiqul Islam Chowdhury, Executive Director, Coastal Association for Social Transformation Trust (COAST) (Dhaka, Bangladesh, 18 June 2010): 'adaptation has limits'. As Kälin and Schrepfer have observed: 'Migration as adaptation, however, has its clear limits as in particular the most vulnerable communities do not have the means to adapt in this manner. Furthermore, where migration turns into forcible movement, e.g. because a long-lasting and expanding drought no longer allows for the survival of pastoralists or nomads and forced them to move to the big cities, one should no longer speak of adaptation': Walter Kälin and Nina Schrepfer, 'Protecting People Crossing Borders in the Context of Climate Change: Normative Gaps and Possible Approaches' (Study for the Swiss Ministry of Foreign Affairs, April 2011) 17, forthcoming as a paper in the UNHCR Legal and Protection Policy Research Series.

a right to live in a normal situation. I don't accept living surrounded by, you know, this stagnant, dirty, filthy water that gives me diseases.[93]

In her view, 'nobody talks about human rights in the adaptation process'.[94] Because of Bangladesh's pre-existing environmental vulnerability, people are used to coping with natural disasters and have become, by necessity, very resilient. Yet, this resilience is mistaken as evidence that adaptation is working. Adaptation with dignity requires not only a focus on defending sustainable livelihoods, but doing so in a way that enables people to live with their human rights respected. This is reminiscent of the President of Kiribati's idea of migration with dignity.[95] Framing responses to climate change in human rights terms provides an empowering framework for rights holders, and an accountability framework for duty holders to respect and protect rights.[96] There is a risk that focusing too intently on physical adaptation may obscure this kind of approach.

For these reasons, migration and planned resettlement options should be regarded as normal forms of adaptation—as paths to a more dignified life—rather than as signs that adaptation has failed.[97] While movement can indicate vulnerability, it can also be a way of achieving security and protection, especially when it can be planned.[98] Seasonal migration has long been an important strategy for coping with Bangladesh's natural environmental conditions,[99] and more structured opportunities for migration should be 'mainstreamed' into adaptation strategies.[100]

B. Implement international standards on internal displacement

In its 2009 Climate Change Strategy and Action Plan, the government of Bangladesh set out three long-term action points to address internal and cross-border migration:

A1. Development of a monitoring mechanism of internal and external migration
A2. Development of a protocol to provide adequate support for their re-settlement and rehabilitation
A3. Building of capacity through education and training to facilitate their re-settlement in new environment.[101]

[93] Interview with Rizwana Hasan, BELA (n 23).
[94] ibid.
[95] Interview with President Anote Tong (n 80). See also the comments of Kiribati's Foreign Secretary, Tessie Lambourne: 'We are proud people. We would like to relocate on merit and with dignity': cited in Laurie Goering, 'Kiribati Officials Plan for "Practical and Rational" Exodus from Atolls' (*Reuters AlertNet*, 9 December 2009) <http://www.alertnet.org/db/an_art/60714/2009/11/9-181804-1.htm> accessed 13 December 2009.
[96] This is a variation on Irene Khan's description of poverty and human rights: Khan (n 84) 13.
[97] See, eg, Zetter (n 14) 62. [98] IOM (n 10) 26. [99] Afsar (n 44) 2.
[100] On the need to mainstream migration into National Adaptation Programmes of Action (NAPAs), see IOM, *World Migration Report 2010: The Future of Migration: Building Capacities for Change* (IOM, 2010) 77–9.
[101] Government of the People's Republic of Bangladesh (n 20) 17: 'The settlement of these environmental refugees will pose a serious problem for the densely populated Bangladesh and migration

Since most climate change-related displacement in Bangladesh is likely to be local, one of the most appropriate policy responses to assist the government to meets its goals would be to strengthen the legal frameworks governing internal displacement.

The Guiding Principles on Internal Displacement are a set of international soft law principles designed to apply *inter alia* to people fleeing their homes due to natural or human-made disasters (which would encompass climate change).[102] Although the Guiding Principles are not legally binding, they draw on binding rules from international humanitarian law, refugee law, and human rights law to elucidate 'best practice' national standards at all stages of displacement—from preventing displacement, to addressing needs during displacement, and in the return and recovery phases.

Fieldwork revealed little awareness among Bangladeshi officials or policymakers of the existence or content of the Guiding Principles, a finding confirmed by IOM in Bangladesh.[103] Some interviewees doubted whether a more formal legal framework could be easily adopted any time soon.[104] Others observed that Bangladesh faces difficulties in securing basic rights for the very poor non-displaced and displaced alike,[105] and so implicitly queried the feasibility of special treatment for internally displaced persons (IDPs). Laws in Bangladesh are often flouted and new laws may not be well implemented and enforced.

The Guiding Principles could provide Bangladesh with a blueprint for assisting and protecting people displaced internally by climate impacts, within a rule of law and human rights-based framework. In other contexts, countries such as Colombia have incorporated substantial parts of the Guiding Principles into domestic law, while in Africa there is now a regional treaty for the protection of IDPs.[106] By encouraging Bangladesh to implement domestically the Guiding Principles, the international community could help build the capacity of Bangladesh to deal rationally and responsibly with the plight of IDPs in the following ways.

(1) Pre-displacement phase

Where displacement is anticipated outside an emergency context (eg planned relocation), the Guiding Principles call for individuals to have access to information about the reasons and procedures for their movement, and, where applicable, on compensation and relocation. Those affected should be able to participate in the planning and management of movement, and have the rights to life, dignity, liberty, and security respected.[107] This is important to observe in any context where

must be considered as a valid option for the country. Preparations in the meantime will be made to convert this population into trained and useful citizens for any country.'

[102] Guiding Principles on Internal Displacement, UN Doc E/CN.4/1998/53/Add.2 (11 February 1998) ('Guiding Principles'). However, UNHCR's institutional mandate as the lead agency responsible for IDPs only applies to IDPs displaced *by conflict*. See further, Ch 8.

[103] Interview with Rabab Fatima and Anita Wadud, IOM (n 10).

[104] ibid.

[105] Interview with anonymous UN official (Dhaka, Bangladesh, 17 June 2010).

[106] African Union Convention for the Protection and Assistance of Internally Displaced Persons in Africa, 22 October 2009 (Kampala Convention, not yet in force), Art 5(4).

[107] Guiding Principles (n 102) Principles 7–9.

movement can be planned. It has a particular resonance, for example, for countries like Kiribati and Tuvalu.[108]

In Bangladesh, there is considerable scope to operationalize these principles, such as through wider implementation of early-warning systems (which already successfully exist for cyclones); awareness strategies to assist people to prepare for displacement (such as by shifting livestock, building materials, and grain to safe areas ahead of time, and saving for emergencies); and consideration of the potentially adverse impacts and consequences of building structures to mitigate riverbank erosion (such as inserting piping for shrimp cultivation).[109] The authorities could also do more to identify vulnerable areas, provide information to local people, and invest in monitoring mechanisms to anticipate displacement.[110] At present, there is relatively poor data available in Bangladesh concerning displacement risks.[111]

Importantly, it would be helpful to decentralize processes by 'increasing reliance on local level understandings of problems' through participatory decision-making and the allocation of government resources to match local needs.[112] This approach recognizes the value of indigenous knowledge and coping mechanisms, while developing further strategies to improve their lives.

(2) During displacement

During the displacement phase, the Guiding Principles set out a human rights framework for ensuring that a wide range of civil, political, economic, social, and cultural rights are respected, including through the provision of humanitarian assistance.[113]

To this end, Bangladesh could start by enacting comprehensive disaster management legislation, which remains absent despite Bangladesh 'ranking number one in terms of stress from cyclone due to climate change, and number six in terms of stress from flood from climate change'.[114] Two Bangladeshi scholars, Abrar and Azad, have recommended the development of an institutional rapid response mechanism to provide logistical support to those displaced. Since people tend to remain at home until the last moment, which increases threats to life and property, they identify the need for teams of volunteers to be mobilized to assist with evacuation to pre-determined places where there is food, shelter, healthcare, sanitation, and water available. Healthcare workers and adequate medicine must be immediately dispensed to the site, and mobile medical units should respond to emergencies (including by being equipped to deal with the outbreak of diseases).[115] Similarly, mobile schools could be established to reach displaced children.[116]

[108] See Ch 5, section VII.
[109] Abrar and Azad (n 34) 122–3.
[110] ibid, 130.
[111] Interview with Senior Official, Department of Environment, Ministry of Environment and Forests (n 59).
[112] Abrar and Azad (n 34) 107.
[113] Guiding Principles (n 102) Principle 25.
[114] Interview with Rizwana Hasan, BELA (n 23).
[115] Abrar and Azad (n 34) 123–4.
[116] ibid, 126–7.

Though Bangladesh has 'standing orders' on disaster management,[117] they are discretionary and unenforceable. They also lack specificity and human rights protections. For example, while they mention the rehabilitation of affected people, they provide little detail on the rights which are guaranteed, or the timeframes or processes by which protection or assistance will occur. Thus, as one Bangladeshi NGO stated, 'in terms of natural disasters, we don't have right-based law. We don't have duty-oriented laws also; it just says you will be rehabilitated, but exactly what should be done to prevent disasters is not there.'[118] There is thus a need to strengthen the 'portability' of rights for those internally displaced in Bangladesh, such as by reference to more specialized 'soft law' instruments which provide for rights-based approaches.

Food insecurity is a major concern in the immediate aftermath of displacement, especially since most households do not have any food in reserve.[119] One NGO survey showed that *monga* (famine) had forced people in certain border areas in the Thakurgaon district to cross over to India to work as labourers.[120] Yet, the experience of *monga* in northern Bangladesh illustrates that it is often not absolute scarcity of food that causes *monga*, but rather ineffective distribution, local inflation, uncoordinated stocking, poor coordination in transportation, and a lack of coordination between government agencies.[121] NGOs providing micro-credit have also been criticized for being too inflexible in refusing to write off the debts of those affected by storms or environmental problems.[122]

Some existing responses have been counter-productive. For example, the replacement of rural 'Food for Work' programmes with international aid via the World Food Progamme has been criticized for exacerbating livelihood insecurity and migration to urban centres.[123] Improved management of responses to displacement is thus an important element in subsequent patterns of movement. As noted above, a study conducted after the severe floods in 1998 showed that those who felt adequately compensated were less likely to move.[124] Thus, the prompt provision of adequate assistance can reduce longer-term migration and avert the need for more disruptive resettlement elsewhere.

(3) Resettlement or relocation

The Guiding Principles also deal with return, resettlement, and integration. They note that the competent authorities should ensure that IDPs are able to return home voluntarily—or resettle elsewhere in the country—in safety and with dignity, and be able to participate fully in the planning and management of their return or

[117] Bangladesh Ministry of Disaster Management and Relief (Disaster Management Bureau), Standing Orders on Disaster (1999).

[118] Interview with Rizwana Hasan, BELA (n 23).

[119] Abrar and Azad (n 34) 36.

[120] ibid, 71, citing Manab Kallyan Parishad (Human Welfare Association) survey, reported in *Bangladesh Observer* (7 October 2003).

[121] Abrar and Azad (n 34) 71.

[122] Interview with Chowdhury R Abrar, University of Dhaka (n 68).

[123] Abrar and Azad (n 34) 103.

[124] Paul (n 81) cited in IOM (n 10) 12.

resettlement and subsequent integration.[125] The authorities should facilitate access for international humanitarian organizations to help IDPs in this process.[126] The authorities are also obliged to assist IDPs to recover property and possessions, or to be compensated for losses.[127]

According to Abrar and Azad, the rehabilitation phase is presently the least well supported in Bangladesh. Displaced people are left largely to their own devices and lack effective State assistance. People often move into even more vulnerable locations than the ones from which they fled.[128] Some have no choice but to resettle on *char* lands[129]—areas of fertile, alluvial silt deposit—knowing that flood or erosion is likely to displace them again.[130] While it might be assumed that newly accreted areas could provide free land to groups of landless people,[131] this is not the case: in most instances, the land is already subject to prior claims stemming from earlier periods of river erosion.

There is very little coordinated national resettlement, relocation, or rehabilitation planning in Bangladesh for those displaced from their lands,[132] and no government department has a specific responsibility for IDPs.[133] One official acknowledged that the government should provide opportunities to move, and gave a vague indication that resettlement might be considered in the future:

> We are going to prepare . . . a nationalistic plan for relocation. We will have also from here [a] contingency plan for [a] few districts which are erosion-prone. And we will also go for, to disaster and climate-resilient habitat. . . . I will be doing a study, I will be talking to the people, I will be trying to find out what is the best option on the table, relocation . . . Land, how much is there, whether we could offer this land to our landless people.[134]

The only laws for the rehabilitation of IDPs in Bangladesh relate to those affected by the 1971 war of independence.[135] Government programmes for redistributing vacant land to the landless have been described as 'dysfunctional' and tarnished by corrupt middlemen.[136] The Standing Orders on disaster management, mentioned above, only provide for emergency responses and do not encompass more permanent solutions such as planned relocation or rehabilitation.[137] One Bangladeshi

[125] Guiding Principles (n 102) Principle 28. [126] ibid, Principle 30.

[127] ibid, Principle 29. [128] Abrar and Azad (n 34) 124.

[129] There are around two million, mostly landless, people living in *char* lands: see citations in IOM (n 10) 13, fn 42. [130] Abrar and Azad (n 34) 38–9.

[131] See Haakon Lein, 'The Poorest and the Most Vulnerable? On Hazards, Livelihoods and Labelling of Riverine Communities in Bangladesh' (2009) 30 Singapore Journal of Tropical Geography 98, 107, noting that: 'Char lands are not, as has been implied, open frontier zones capable of absorbing large groups of poor, landless people migrating from more densely populated rural areas. Although land appears to be available, there is always someone who owns or claims ownership to emerging char land.' Cited in IOM (n 10) 13.

[132] Interview with Rizwana Hasan, BELA (n 23); interview with SM Munjurul Hannan Khan, Ministry of Environment and Forests (n 30); interview with Rabab Fatima and Anita Wadud, IOM (n 10).

[133] Interview with Mihir Kanti Majumder, Ministry of Environment and Forests (n 31).

[134] Interview with Abu M Kamal Uddin, CDMP (n 18).

[135] Interview with Rizwana Hasan, BELA (n 23).

[136] Interview with Ahmed Swapan Mahmud, VOICE (n 56).

[137] Interview with Rabab Fatima and Anita Wadud, IOM (n 10).

parliamentarian emphasized that political action on resettlement is needed in order to overcome bureaucratic inertia.[138]

The reluctance to plan for the resettlement of IDPs is influenced by a number of factors. First, the government's focus in climate change policy is on *in situ* adaptation to climate change—harnessing funds to implement in-country programmes to assist people (particularly in rural areas) to maintain their lands and livelihoods. That focus is also partly based on a belief that there is little vacant land available for resettlement:

> Relocation within the country will not happen, because we do not have any space . . . people also do not want to leave their parents and sisters [at] home . . . . So we have to try, as much as possible, to fit them in their own space, you know, protect, reduce the vulnerability of their areas, make the embankments stronger, put some more efforts on tree plantation and those sort of natural protection, things like that. And we will need some investment to do so . . . .[139]

The lack of available land for resettlement is typically not due to land shortages, but rather chronic maladministration of public lands (known as *khas* land). Up to 88 per cent of *khas* land and 95 per cent of *khas* water bodies—that might otherwise be placed at the disposal of displaced people—'are under illegal possession of the powerful elites and other vested interest groups'.[140] Eviction from *khas* land is common, and lost access to common properties means that people are cut off from their major food source.[141] The failure to provide solutions for displaced people also places stress on Bangladesh's biodiversity, such as where people increasingly encroach on protected forest and marine areas like the Sundarbans.[142]

Secondly, there is a strong government narrative that movement is a matter of individual choice, which should not be dictated by State policy.[143] Underlying this approach is undoubtedly Bangladesh's unfortunate experience of State-driven resettlement in the Chittagong Hill Tracts, described by one official as a 'political invasion' which caused 'irreversible damage'.[144] The relocation of up to 600,000 Bengali settlers into Chittagong Hill Tracts[145] during the period of military rule in Bangladesh incited violence which was only settled by a Peace Accord in 1997 (although unrest continues).[146] The conflict was driven by competition over land and political power, and resulted in the dispossession of tribal peoples. The authorities are wary of repeating past mistakes.

[138] Interview with Saber Chowdhury MP, All Parliamentary Committee on Climate Change (n 78).

[139] ibid; see also interview with SM Munjurul Hannan Khan, Ministry of Environment and Forests (n 30).

[140] ASK, *Human Rights in Bangladesh 2008* (ASK, 2009) 127; see also interview with Sultana Kamal, ASK (n 79).

[141] ASK (n 140) 127.

[142] Interview with Mihir Kanti Majumder, Ministry of Environment and Forests (n 31).

[143] Interview with SM Munjurul Hannan Khan, Ministry of Environment and Forests (n 30).

[144] Interview with Abu M Kamal Uddin, CDMP (n 18).

[145] Dan Smith and Janani Vivekananda, *A Climate of Conflict: The Links between Climate Change, Peace and War* (International Alert, 2007) 16.

[146] ATM Amin, 'Regional Security Environment and the Menace of Terrorism' in Ahmed (n 74) 122.

However, that was a very different context and should not prevent current democratic authorities in Bangladesh from finding suitable public or private lands for resettlement. In accordance with the Guiding Principles on Internal Displacement, resettlement would be voluntary, sensitive to people's human rights, and based on the provision of full information about options.[147] Resettlement sites need to be selected in full consultation with existing communities in those areas, as well as with the potential new settlers.[148] For relocation to have the best chance at working, it has to be owned by the affected communities, not imposed from above.[149] An effective resettlement policy would require a comprehensive audit of vacant public lands (perhaps in the context of the wider need for national land zoning in Bangladesh),[150] a strategy for the prioritization of competing needs in such land, and the setting of timeframes for its utilization.

One of the most important factors in the success of any planned resettlement is the availability of new livelihoods, including opportunities to earn income and build capital.[151] Some previous European-supported programmes, completed in 2008, to create clustered 'ideal villages' for resettling up to 100 displaced people in rural areas failed largely on account of the lack of employment opportunities, resulting in people selling their new land and moving elsewhere.[152]

Skills development programmes can enable people to acquire new trades to generate income, and information about marketing agricultural produce would also assist.[153] For example, in one survey a woman explained how she used skills acquired from an NGO about planting trees and vegetables to help her cope with the adverse conditions in the rainy season, when she could sell the produce at a reasonable profit.[154]

However, few government or international programmes in Bangladesh specifically target the displaced. Local and international NGOs in Bangladesh tend to focus on coping with disasters per se, rather than preparing for displacement and rehabilitation.[155] While government schemes focusing on destitute women, widows, very poor women, divorced women, and the landless may assist some displaced people, this is not because of their displacement but rather their underlying

[147] Evidence shows that adaptation strategies will be most successful where they are based on meaningful community participation, such as in constructing embankments: see Habibullah and others, 'Participatory Water Management: A Strategy for Climate Change Adaptation in Coastal Bangladesh' (2009), cited in IOM (n 10) 17.

[148] As the Inter-American Court of Human Rights observed, it is necessary to 'obtain their free, prior, and informed consent, according to their customs and traditions': *Saramaka People v Suriname* (IACtHR, 28 November 2007) Series C No 172, para 134.

[149] See Ch 5.

[150] Interview with Saber Chowdhury MP, All Parliamentary Committee on Climate Change (n 78).

[151] Abrar and Azad (n 34) 107.

[152] Interview with Md Shamsuddoha and Rezaul Karim Chowdhury, Equity BD (n 7).

[153] Abrar and Azad (n 34) 124–5.

[154] ibid, 71.

[155] See also Oxfam's River Basin Programme, CARE's Flood Proofing Project, and RDRS's (a local Bangladeshi NGO) Char Development Project: ibid, 104–5. There is a rise in Islamic NGOs in some areas which promote religious values and run programmes mainly for men. Abrar and Azad suggest that these well-funded NGOs could better assist communities by widening the focus of their programmes: (n 34) 104–5.

circumstances.[156] There have been some resettlement programmes for destitute and abandoned women and widows, but they do not include any support for or facilitation of livelihoods. This means people often leave settlements relatively quickly because of the lack of income-generating activities.[157]

Indeed, this final phase—the livelihood management phase—was identified by people displaced by riverbank erosion as the most fundamental, and the one in which support would be most welcomed.[158] They indicated a desire for government insurance schemes with 'soft' premiums to assist them to rebuild their lives when property is lost. Since pre-existing loans can become a major burden for the displaced, the government could develop a policy whereby loans could be written off for people who lose a certain proportion of assets. It has also been suggested that NGOs reconsider their credit policies so as to provide assistance to the displaced, both by offering loans to them as well as easing the terms and conditions of payment (as a form of corporate social responsibility). NGOs can play an important role in creating market opportunities for marginalized groups.[159]

C. Encourage international labour mobility and lawful migration pathways

As explored in greater detail in the next chapter, international migration is one of a number of strategies being pursued by the Bangladeshi government.[160] While it will not provide a mobility pathway for everyone, particularly the very poor, for others it may provide an important strategy for livelihood diversification and risk management.

At present, adaptation remains the predominant focus in Bangladesh, and there is an unfortunate tendency not to view migration itself as a legitimate adaptation strategy. Bangladesh's 2009 Climate Change Strategy and Action Plan includes plans to develop a migration monitoring mechanism, and support for resettlement, rehabilitation, and capacity-building through education and training to facilitate resettlement in a new environment.[161] The plan states that 'migration must be considered as a valid option for the country. Preparations in the meantime will be made to convert this population into trained and useful citizens for any country.'[162]

A United Nations (UN) official in Dhaka said that while climate change displacement and migrant labour are not presently conceptually linked in Bangladesh, 'the more we advance in years we're going to see a greater linkage between the two'.[163] IOM believes that 'migration management' should be 'one element of a

[156] Abrar and Azad (n 34) 101. For example, destitute women may qualify for Vulnerable Group Feeding and Vulnerable Group Development cards (respectively receiving 30 kg and 10 kg of rice a month), but the number of applications for these far exceeds the number that can be allocated.

[157] Abrar and Azad (n 34) 102, referring to the Guchcha Gram, Adarshya Gram, Asrayan, and Abashon programmes.

[158] Abrar and Azad (n 34) 125.

[159] Abrar and Azad (n 34) 127–8.

[160] See Ch 7 for further analysis.

[161] Bangladesh Climate Change Strategy and Action Plan (n 20) 59.

[162] ibid, 17. This is reminiscent of the President of Kiribati's idea of 'migration with dignity'.

[163] Interview with anonymous UN official (n 105).

holistic approach to addressing the human security implications of environmental events and processes, including the consequences of climate change'.[164] One Bangladeshi official believed that bilateral migration agreements with countries such as Australia, New Zealand, and Canada may be a way forward,[165] even if a global 'umbrella' agreement is also pursued.

As discussed in section III.C. above, in Bangladesh—and probably many other contexts—climate change-related movement is likely to have a domino effect. The people who do move abroad may not be directly affected by the impacts of climate change, but indirectly, as cities become overpopulated, resources become increasingly strained, and life becomes increasingly intolerable. As one person explained,[166] this is not inappropriate: to relocate a poor farmer to a capital city in an industrialized country would not serve either well, yet to enhance migration options for the educated and well-resourced may in turn open up greater opportunities for those moving from rural to urban areas within Bangladesh. Despite the risks of a 'brain drain', many of those interviewed cited the very positive contribution that expatriates have had on Bangladesh's economy, society, and political life.

V. Conclusion

This study of the links between climate change and displacement in Bangladesh illustrates the complexity of human movement, and the corresponding challenges of designing appropriate legal and policy responses to it. The causes of displacement are multi-dimensional and difficult to disentangle. Climate change impacts affect movement in different ways depending on whether they are slow or rapid-onset processes, and involve time projections which are long in terms of policymaking.

First, slow-onset climate change impacts do not attract the same level of attention by policymakers and the media as sudden disasters such as cyclones, tidal surges, or floods. For example, riverbank erosion, which can occur very quickly[167] and may affect people's lives and livelihoods more profoundly (since erosion leads to land loss in a way that flooding generally does not),[168] has not attracted the same level of media or government attention as the more dramatic disasters. Furthermore, given the difficulties experienced in raising money to assist those affected by severe flooding in Pakistan in 2010, compared to the amounts raised in response to the Asian Tsunami in 2004 and the Haitian earthquake in January 2010, it would seem that obtaining funding will be difficult unless a permanent adaptation or insurance fund of some kind is established.[169]

164 IOM (n 10) 26. See also interview with Chowdhury R Abrar, University of Dhaka (n 68), calling for more, regularized seasonal migration into India.

165 Interview with SM Munjurul Hannan Khan, Ministry of Environment and Forests (n 30).

166 Interview with Sultana Kamal, ASK (n 79).

167 People in Dimla in North-West Bangladesh explained that river water can rise very suddenly such that within two to three hours the whole area is flooded: Abrar and Azad (n 34) 31.

168 ibid, 32.

169 See further, Ch 1.

Secondly, in terms of developing policy, the time projections are long: 2050 is a common marker. As IOM notes, considerable adaptation measures are feasible over such an extended timeframe. While 'narratives of mass displacement are understandable in highlighting the potential long-term risks of failing to curb CO2 emissions globally, they should be approached with considerable caution as they risk undermining the case for investment and adaptation measures in vulnerable coastal regions to deal with very real existing vulnerabilities'.[170] At the same time, migration has not typically been seen as an adaptation strategy in itself. As at July 2011, of the 45 National Adaptation Programmes of Action (NAPAs) submitted worldwide by the Least Developed Countries (LDCs) (including by Bangladesh), only 10 mentioned migration or resettlement in their priority projects.[171] A further 11 raised it as a possible adaptation or policy strategy.[172] Even though the NAPAs are designed to address urgent and immediate needs, such needs include 'those for which further delay could increase vulnerability or lead to increased costs at a later stage',[173] and it could certainly be argued that forward-planning to forestall population movements meets these criteria.

Thirdly, the Bangladeshi government may have inadvertently placed itself in a tenuous position when it comes to lobbying for financial and practical assistance for climate change-related movement. Suggestions that climate change will overwhelm and incapacitate the country sit uneasily with the government's perception of itself as a world leader in disaster early-warning responses and management. Further, for historical and political reasons, the official line of the Bangladeshi government is that there is no unauthorized migration between Bangladesh and India.

[170] IOM (n 10) 18.

[171] Bhutan, *Bhutan National Adaptation Programme of Action* (no date, submitted May 2006), Project Profile 1 'Disaster Management Strategy: Planning for Food Security and Emergency Medicine to Vulnerable Communities' 30, Project Profile 4 'Landslide Management and Flood Prevention (Pilot Schemes in Critical Areas)' 44, Project Profile 5 'Flood Protection of Downstream Industrial and Agricultural Area' 48; Gambia, *National Adaptation Programme of Action (NAPA) on Climate Change* (November 2007), Project 9 'Restoration/Protection of Coastal Environments', 82, Project 10 'Protection of Human Settlements by Coastal Protect Measures on Safer Islands', 72–3; Government of the Maldives, *National Adaptation Programme of Action* (2007), Project 1 'Integration of Future Climate Change Scenarios in the Safer Island Strategy to Adapt Sea Level Rise and Extreme Weather Risks Associated with Climate Change', 47; Mozambique, *National Adaptation Programme of Action (NAPA)* (December 2007), Project 1 'Strengthening of an Early Warning System' 29; Samoa, *National Adaptation Programme of Action* (2005), Project Profile 7 'Implement Coastal Infrastructure Management Plans for Highly Vulnerable Districts', 48; Sao Tomé and Príncipe, *National Adaptation Programmes of Action on Climate Change* (December 2006), Project Title 'Displacement of Local Communities', 48; Solomon Islands, *National Adaptation Programme of Action* (November 2008), 'Human Settlement', 80, 'Climate Change Adaptation on Low-Lying and Artificially Built-Up Islands in Malaita and Temotu Provinces', 86; United Republic of Tanzania, *National Adaptation Programme of Action (NAPA)* (January 2007) ix; Tuvalu, *Tuvalu's National Adaptation Programme of Action* (May 2007), NAPA Project Profile 6, 'Strengthening Community Disaster Preparedness and Response Potential', 51–2; Uganda, *Climate Change: Uganda National Adaptation Programmes of Action* (2007), Project No 4 'Community Water and Sanitation Project', 56. See Appendix for a detailed overview. See also Susan F Martin, Managing Environmentally Induced Migration' in Laczko and Aghazarm (n 39) 361ff.

[172] See, eg, NAPAs of Burundi, Eritrea, Ethiopia, Guinea-Bissau, Kiribati, Malawi, Mauritania, Nepal, Senegal, Togo, and Vanuatu. See Appendix for a detailed overview.

[173] See <http://unfccc.int/national_reports/napa/items/2719.php> accessed July 2011.

Talking up the numbers of people who might be displaced by climate change does not sit well with the government's position that no-one needs to leave Bangladesh. Finally, as a country which has not ratified the Refugee Convention,[174] Bangladesh may be in a relatively weak position politically should it wish to suggest that other countries have an obligation to protect its people who are displaced on account of climate-related impacts.

Fourthly, Bangladesh's present approach to climate change policy (including adaptation) tends not to be situated within a human rights-based framework. There is a need to develop a suite of policies that focus on preventing displacement (through adaptation, education, and support for livelihoods); assisting those who are displaced (particularly through the rights-based framework of the Guiding Principles on Internal Displacement); and viewing migration itself as an acceptable form of adaptation (including through planned, rights-respecting resettlement schemes involving the participation of affected communities).

Finally, Bangladesh, its neighbours, and the international community should establish more bilateral and regional 'economic' migration opportunities for Bangladeshis—without expending unnecessary energy pursuing a global 'protection'-oriented treaty that would be ill-suited to the complexity of the movements involved.[175] Such policies would help to address the genuine human security problems arising from climate change-related displacement in Bangladesh, in particular social tensions over scarce resources, problems of urbanization, and cross-border sensitivities about irregular migration.

[174] Convention relating to the Status of Refugees (adopted 28 July 1951, entered into force 22 April 1954) 189 UNTS 137, read in conjunction with the Protocol relating to the Status of Refugees (adopted 31 January 1967, entered into force 4 October 1967) 606 UNTS 267.

[175] See Ch 7.

7

'Protection' or 'Migration'? The 'Climate Refugee' Treaty Debate

I. Introduction

As explained in Chapter 1, the way a phenomenon is conceptualized is central to the way its regulation is approached. Questions of definition have clear governance implications, informing the appropriate location of environmental migration both procedurally—as an international, regional, or local, developed and/or developing country concern/responsibility—and thematically—for instance, within the existing refugee protection framework or under the UN Framework Convention on Climate Change (UNFCCC).[1]

One of the biggest shortcomings of much of the scholarship being generated on 'climate migration' is a tendency to treat climate change-related movement as a single phenomenon that can be discussed in a general way. As Walter Kälin's climate displacement typology highlights,[2] a number of very different scenarios are captured within this rubric, and it is only through examining them separately, with attention to their distinctive and common features, that any meaningful policy or normative frameworks can be developed.[3] Certainly, a one-size-fits-all approach will not adequately respond to the variety of movement encapsulated within the 'climate migration' framework.[4]

The next three chapters explore possible policy responses to climate change-related movement. The present chapter responds to proposals for a new refugee-like instrument for people displaced by climate change. First, it analyses the conceptual and pragmatic challenges of constructing such an instrument, not least given the

[1] UN Framework Convention on Climate Change (adopted 9 May 1992, entered into force 21 March 1993) 1771 UNTS 107 ('UNFCCC').

[2] See Ch 1.

[3] Walter Kälin, 'Conceptualising Climate-Induced Displacement' in Jane McAdam (ed), *Climate Change and Displacement: Multidisciplinary Perspectives* (Hart Publishing, 2010). Part of Kälin's chapter is an attempt to disaggregate these scenarios in order to identify appropriate institutional, legal, and political responses to them. See discussion of this typology in Ch 1.

[4] Bedford and Bedford, for example, see a mix of strategies as 'essential': Richard Bedford and Charlotte Bedford, 'International Migration and Climate Change: A Post-Copenhagen Perspective on Options for Kiribati and Tuvalu' in Bruce Burson (ed), *Climate Change and Migration: South Pacific Perspectives* (Institute of Policy Studies, Victoria University of Wellington, 2010) 93. They should include options for voluntary individual/family movement, and for planned resettlement of whole communities.

very disparate types of movement that can be encompassed within that concept. Secondly, it examines an alternative paradigm for responding to movement: migration. This builds on calls by some affected States to expand existing migration opportunities for their nationals, and their resistance to the 'climate refugee' concept. The final chapter analyses institutional responses to climate change-related movement. The following chapter examines the guiding legal frameworks and principles that should underpin State and institutional responses to cross-border displacement in the context of climate change and disasters, including any soft-law framework that is developed.

At the outset, it should be noted that the policy options canvassed in these and other chapters[5] are not mutually exclusive. For example, migration options should be explored for pre-emptive movement, but this should not rule out a parallel humanitarian response for rapid-onset disasters or for people facing slow-onset change who are unable or unwilling to migrate. Guiding principles may be a preliminary step towards a binding legal instrument, or simply a helpful tool to assist governments in responding to potential and/or actual displacement. A range of options should be utilized which are country/region-specific and attuned to their particular needs, including the locus and timing of movement.

II. A 'Climate Refugee' Treaty

A. Introduction

The absence of a clear international legal framework to respond to people displaced by climate change has resulted in calls from a variety of sectors for a new international instrument to protect so-called 'climate refugees'. Proposals vary from creating a protocol to the Refugee Convention,[6] a protocol to the UNFCCC, or a stand-alone treaty, to provide so-called 'climate refugees' with international protection, including a legal status and resettlement/integration solutions.

Drawing on the fieldwork examples already discussed in previous chapters (Tuvalu, Kiribati, and Bangladesh), this chapter argues that advocacy for a new treaty to address climate change-related movement is presently misplaced for a number of reasons. First, as emphasized throughout this book, it is difficult to isolate climate change from other factors as a primary cause of movement. This may create problems in defining the legal scope and application of the instrument, and ensuring that those intended to be covered by it actually are. Secondly, a 'climate refugee' treaty would privilege those displaced by climate change over other forced migrants (such as those escaping poverty), perhaps without an adequate (legal and/

[5] For example Ch 5 examined relocation; Ch 6 examined internal movement and adaptation.

[6] Convention relating to the Status of Refugees (adopted 28 July 1951, entered into force 22 April 1954) 189 UNTS 137, Art 1A(2), read in conjunction with Protocol relating to the Status of Refugees (adopted 31 January 1967, entered into force 4 October 1967) 606 UNTS 267 (together 'Refugee Convention'). See discussion in Ch 2 about the limitations of the Refugee Convention regime, including the requirement that an international border is crossed.

or moral) rationale as to why.[7] Thirdly, it may premise protection on individual status determination, which is unsuited to mass displacement scenarios. Fourthly, defining 'climate refugees' may harden the category and exclude some people from much-needed assistance. Finally—and perhaps most significantly—there seems to be little political appetite for a new international agreement on protection. As one official in Bangladesh pessimistically observed, 'this is a globe for a rich man'.[8]

B. Background

The purpose of this chapter is not to deny the real impacts that climate change is already having on communities, or that migration is a normal adaptive response to such change. Rather, it queries the utility—and, importantly, the policy consequences—of pinning 'solutions' to climate change-related displacement on a multilateral instrument, in light of the likely nature of movement, the desires of affected communities, and the fact that a treaty will not, without wide ratification and implementation, 'solve' the humanitarian issue.[9] The argument is developed by examining some conceptual and pragmatic difficulties in attempting to construct a refugee-like instrument for people fleeing the effects of climate change, and by critiquing whether there are *legal* benefits, as opposed to *political* benefits, to be gained by advocating for such an instrument.

As the fieldwork in particular has highlighted, a *universal* treaty may be inappropriate in addressing the concerns of particular communities. The role of 'international movement' as a response to climate change is conceived of differently in Tuvalu, Kiribati, and Bangladesh because of their particular geographical, demographic, cultural, and political circumstances, and it may be that localized or regional responses are better able to respond to their needs. Such approaches can take into account the particular features of the affected population in determining who should move, when, in what fashion, and with what outcome. Staggered migration, circular migration, or the promise of a place to migrate to, should it become necessary, might be welcome measures that could appeal both to host and affected communities alike.[10] Furthermore, by contrast to many other triggers of

[7] For criticism of this approach, see Alexander Betts and Esra Kaytaz, 'National and International Responses to the Zimbabwean Exodus: Implications for the Refugee Protection Regime' (2009) UNHCR New Issues in Refugee Research, Research Paper No 175; see also Alexander Betts, 'Towards a Soft Law Framework for the Protection of Vulnerable Irregular Migrants' (2010) 22 International Journal of Refugee Law 209. This is discussed in Ch 1.

[8] Interview with Abu M Kamal Uddin, Comprehensive Disaster Management Programme (CDMP) (Dhaka, Bangladesh, 16 June 2010).

[9] See also Walter Kälin and Nina Schrepfer, 'Protecting People Crossing Borders in the Context of Climate Change: Normative Gaps and Possible Approaches' (Study for the Swiss Ministry of Foreign Affairs, April 2011) 58, forthcoming as a paper in the UNHCR Legal and Protection Policy Research Series, referring to my work: 'We share these concerns. They lead us to conclude, that—at least for the time being—a soft law approach is overall more appropriate.'

[10] This is the preferred approach of the government of Kiribati, for example. See, eg, interview with President Anote Tong (Tarawa, Kiribati, 12 May 2009); comments of Kiribati's Foreign Secretary, Tessie Lambourne cited in Laurie Goering, 'Kiribati Officials Plan for "Practical and Rational" Exodus from Atolls' (*Reuters AlertNet*, 9 December 2009) <http://www.alertnet.org/db/an_art/60714/2009/11/9-181804-1.htm> accessed 12 January 2010.

displacement, the slow onset of some climate change impacts, such as rising sea levels, provides a rare opportunity to *plan* for responses, rather than relying on a remedial instrument in the case of spontaneous (and desperate) flight.

The context of the chapter also needs to be understood lest the argument be misconstrued. There is a widespread, non-critical assumption, predominantly among those not versed in international law, that a treaty will provide *the* answer to climate change-related displacement. This argument is flawed for a number of reasons, explored below. Certainly, the chapter should not be construed as an outright objection to a possible future treaty regime by which States might accept a duty to assist people displaced in part by climate change, and agree to responsibility-sharing mechanisms.[11] Indeed, as fieldwork in the Pacific and Bangladesh has shown, people are already moving in response to environmental changes,[12] and States will ultimately need to develop coordinated responses that acknowledge the need for cross-border movement in certain circumstances and which regularize the status of those who move, either through humanitarian or migration schemes. Recognition of a duty to assist could help to encourage international cooperation on sharing the responsibility for displaced people and facilitate the establishment of institutional mandates (such as by creating a lead United Nations (UN) agency or focal point).[13]

Rather, my concern is that if a treaty becomes the *main* focus of international policy development, attention may shift from the more immediate, alternative, and additional responses that may enable people to remain in their homes for as long as possible (which is the predominant wish among affected communities), or to move safely within their own countries, or to migrate in a planned manner over time. As Mortreux and Barnett have pointed out,

[11] For example, the Agreement on the Cooperation for the Sustainable Development of the Mekong River Basin (entered into force by signature, 5 April 1995) 2069 UNTS 3 is a regional treaty between four States that establishes a framework for cooperation 'in all fields of sustainable development, utilization, management and conservation of the water and related resources of the Mekong River Basin...in a manner to optimize the multiple-use and mutual benefits of all riparians and to minimize the harmful effects that might result from natural occurrences and man-made activities' (Art 1). It also establishes an institutional framework, the Mekong River Commission, 'to provide an adequate, efficient and functional joint organizational structure to implement this Agreement and the projects, programs and activities taken thereunder in cooperation and coordination with each member and the international community, and to address and resolve issues and problems that may arise from the use and development of the Mekong River Basin water and related resources in an amicable, timely and good neighbourly manner' (Preamble). Thus, at a minimum, it commits States to negotiate on the issues, through 'consultation and evaluation' and 'a dynamic and practical consensus' (Ch II definitions).

[12] See also 'Carteret Islanders Become First Climate Refugees: PNG Relocates Families as Island Home Disappears' (*Pacific Islands Report*, 4 May 2009) <http://archives.pireport.org/archive/2009/may/05-04-09.htm> accessed 10 June 2009, reporting a story from the *Papua New Guinea Post-Courier* (1 May 2009); UN Office of the High Commissioner for Human Rights, 'Discussion Paper: Protecting the Human Rights of Internally Displaced Persons in Natural Disasters: Challenges in the Pacific' (Davui Printery, 2011). More generally, see the special issue on climate change and displacement: (2008) 31 Forced Migration Review.

[13] See Ch 8. The President of Kiribati has argued that 'the international community should pay for the relocation costs', perhaps by being 'made to pay something into some kind of rehabilitation fund for the victims': interview with President Anote Tong (n 10).

Existing discussions about displacement have tended to obstruct the space needed for meaningful analysis and careful debate about the magnitude and timing of risks, and the best ways to avoid and respond to them. In particular, the discourses of displacement effectively circumvent careful consideration of adaptive measures that could be supported by the international community to prevent forced migration, and neglect the role of individuals in negotiating climate change and determining their own responses based on their own needs and values.[14]

Drafting a treaty necessitates an intent focus on defining who is 'within' or 'outside' its scope of application, and while such deliberations are occurring, other opportunities may be missed. As one commentator has observed of the current international climate change negotiations, concentrating on a treaty can ironically encourage *inaction* on climate change, since discussions tend to get bogged down in linguistic detail rather than substance, and the very process of negotiation provides an excuse to do nothing until an outcome is achieved—an outcome that is likely to be a considerable compromise given that consensus is required.[15]

The arguments in this chapter do not foreclose international action to address cross-border displacement that may be linked to climate change impacts. Indeed, as Chapter 9 examines, shifting the focus from the 'cause' to the 'effect' may be one way of reconceptualizing the class of persons in need of protection. My principal concern with some present treaty proposals is the lack of empirical evidence underpinning key assumptions about movement, which extends to assumptions about the wishes of those who may need to move. This chapter builds on field-based research to identify how best to serve the needs and protect the human rights of affected communities.

C. The treaty proposals

A variety of actors have called for a new international treaty on climate change displacement, or a protocol to the Refugee Convention or the UNFCCC, to create a new class of refugee-like protected persons. At the State level, for example, the Maldives in 2006 proposed amending the Refugee Convention to extend the definition of a 'refugee' in Article 1A(2) to include 'climate refugees'.[16] In December 2009, in the lead-up to the Copenhagen climate change conference, the Bangladeshi Finance Minister similarly stated: 'The convention on refugees could be revised to protect people. It's been through other revisions, so this should

[14] Colette Mortreux and Jon Barnett, 'Climate Change, Migration and Adaptation in Funafuti, Tuvalu' (2009) 19 Global Environmental Change 105, 106.

[15] See comments by Dr Jon Barnett, Panel Member (Deakin Lectures 2010: Innovation, Energy and Climate Change in the Developing World, Melbourne, 10 June 2010) <http://abc.com.au/rn/backgroundbriefing/stories/2010/2947337.htm> accessed 11 July 2010.

[16] Republic of the Maldives Ministry of Environment, Energy and Water, *Report on the First Meeting on Protocol on Environmental Refugees: Recognition of Environmental Refugees in the 1951 Convention and 1967 Protocol Relating to the Status of Refugees* (Male, 14–15 August 2006) cited in Frank Biermann and Ingrid Boas, 'Protecting Climate Refugees: The Case for a Global Protocol' (2008) 50(6) Environment 8.

be possible.'[17] A Bangladeshi non-governmental organization (NGO) network, Equity and Justice Working Group Bangladesh (EquityBD), called for a new protocol to the UNFCCC 'to ensure social, cultural and economic rehabilitation of the "climate refugees" through recognizing them as "Universal Natural Persons"'.[18] In September 2010, the Bangladeshi Prime Minister, Sheikh Hasina, proposed a joint South Asian initiative to mobilize international support under the UNFCCC to ensure the social, cultural, and economic rehabilitation of climate change-induced displaced people.[19] In April 2011, she was reported as calling for the 'establishment of an international regime under the UN to tackle the situation'.[20]

Some scholars have also proposed new legal instruments to address climate change-related movement. Biermann and Boas suggested a UNFCCC Protocol on the Recognition, Protection, and Resettlement of Climate Refugees.[21] A group of legal scholars from the University of Limoges published a Draft Convention on the International Status of Environmentally-Displaced Persons.[22] Docherty

[17] See Harriet Grant, James Randerson and John Vidal, 'UK Should Open Borders to Climate Refugees, Says Bangladeshi Minister' *The Guardian* (London, 4 December 2009) <http://www.guardian.co.uk/environment/2009/nov/30/rich-west-climate-change/print> accessed 8 December 2009.

[18] EquityBD, 'Climate Change Induced Forced Migrants: In Need of Dignified Recognition under a New Protocol' (December 2009). An earlier version of this is Md Shamsuddoha and Rezaul Karim Chowdhury, 'Climate Refugee: Requires Dignified Recognition under a New Protocol' (EquityBD, April 2009) <http://www.equitybd.org/English/Press%20040409/English%20Position%20paper.pdf> accessed 10 November 2009. EquityBD no longer uses the 'refugee' terminology: interview with Md Shamsuddoha and Rezaul Karim Chowdhury (Dhaka, 19 June 2010).

[19] 'PM Warns of Climate Refugee Crisis', *The Daily Star* (Dhaka, 22 September 2010) <http://www.thedailystar.net/newDesign/latest_news.php?nid=26005> accessed 10 June 2011.

[20] 'PM for International Regime to Address Cross Border Migration', *The Daily Star* (Dhaka, 19 April 2011) <http://www.thedailystar.net/newDesign/latest_news.php?nid=29412> accessed 27 April 2011.

[21] Frank Biermann and Ingrid Boas, 'Preparing for a Warmer World: Towards a Global Governance System to Protect Climate Refugees' (November 2007) Global Governance Working Paper No 33, which subsequently became Frank Biermann and Ingrid Boas, 'Preparing for a Warmer World: Towards a Global Governance System to Protect Climate Refugees' (2010) 10 Global Environmental Politics 60. See also Biermann and Boas (n 16); for criticism of their approach, see Mike Hulme, 'Commentary: Climate Refugees: Cause for a New Agreement?' (2008) 50(6) Environment 50. For another UNFCCC-based proposal, see Angela Williams, 'Turning the Tide: Recognizing Climate Change Refugees in International Law' (2008) 30 Law & Policy 502; CRIDEAU and CRDP, Faculty of Law and Economic Science, University of Limoges, 'Draft Convention on the International Status of Environmentally-Displaced Persons' (2008) 4 Revue Européene de Droit de l'Environnement 375 and updated version May 2010 <http://www.cidce.org/> accessed 8 August 2011.

[22] 'Draft Convention on the International Status of Environmentally-Displaced Persons' (n 21). Art 2(2) of the original draft defined 'environmentally-displaced persons' as 'individuals, families and populations confronted with a sudden or gradual environmental disaster that inexorably impacts their living conditions and results in their forced displacement, at the outset or throughout, from their habitual residence and requires their relocation and resettlement'. A 'right to resettlement' was elaborated in Art 9: States parties are to establish 'transparent and open legal procedures for the demand and grant or refusal of the status of environmentally-displace [sic] person based on the rights set forth in the present chapter'. The May 2010 version reads: '"Environmentally-displaced persons" are individuals, families and populations confronted with a sudden or gradual environmental disaster that inexorably impacts their living conditions, resulting in their forced displacement, at the outset or throughout, from their habitual residence.' The new version contains no express references to relocation or resettlement, although the new Art 9 reads: 'All persons confronted by sudden or

and Giannini proposed an 'independent' or 'stand-alone' convention defining the term 'climate change refugee' and containing 'guarantees of assistance, shared responsibility, and administration'.[23] Byravan and Chella Rajan raised the prospect of a treaty creating 'climate exile' status, giving people in 'physically unviable' States the 'right to migrate to a particular or previously agreed upon country'.[24] An Australian-based project also seeks to elaborate 'a draft convention for persons displaced by climate change', which would 'establish an international regime for the status and treatment of such persons'.[25] The Council of Europe Parliamentary Assembly's Committee on Migration, Refugees and Population has suggested 'adding an additional protocol to the European Convention on Human Rights, concerning the right to a healthy and safe environment' as a way of 'enhancing the human rights protection mechanisms vis-à-vis the challenges of climate change and environmental degradation processes'.[26] An Austrian-based study is focusing on the development of a comprehensive international treaty with a strong focus on responsibility-sharing mechanisms, as well as a guaranteed legal status for the displaced.[27]

There is variation among these proposals as to how the displaced are defined, whether they would be subject to individual status determination (like Convention refugees),[28] or whether protection would be extended prima facie on account of the objective country of origin conditions from which people flee.[29] While the underlying basis of each proposal is, presumably, to provide a rights-based framework for people forced to move when the impacts of climate change render life and livelihoods at home impossible, it is not self-evident that a treaty would presently best serve this end. There are three main reasons for this argument, examined below.

gradual environmental degradation that inexorably impacts their living conditions has the right to move within or outside of their home State. State Parties shall not in any way hinder, attempt to hinder or allow any hinderance with this displacement.'

[23] Bonnie Docherty and Tyler Giannini, 'Confronting a Rising Tide: A Proposal for a Convention on Climate Change Refugees' (2009) 33 Harvard Environmental Law Review 349, 350, 373.

[24] Sujatha Byravan and Sudhir Chella Rajan, 'The Ethical Implictions of Sea-Level Rise due to Climate Change' (2010) 24 Ethics and International Affairs 239, 253. They do not provide any details of how this treaty would operate.

[25] David Hodgkinson and others, '"The Hour When the Ship Comes In": A Convention for Persons Displaced by Climate Change' (2010) 36 Monash University Law Review 69. See generally, <http://www.ccdpconvention.com/> accessed 12 August 2011.

[26] The Council of Europe Parliamentary Assembly, Committee on Migration, Refugees and Population, 'Environmentally Induced Migration and Displacement: A 21st Century Challenge', Doc 11785 (23 December 2008) paras 6.3 and 121 respectively.

[27] Margit Ammer and others, 'Rechtsstellung und rechtliche Behandlung von Umweltflüchtlingen', Study 54/2010 commissioned by the Umweltbundesamt (Federal Ministry of the Environment 2010), referred to in Kälin and Schrepfer (n 9) 46. For the English version, see 'Legal Status and Legal Treatment of Environmental Refugees' <http://www.umweltbundesamt.de/umweltrecht-e/publikationen/texte_54_2010_kurzfassung_e.pdf> accessed 11 July 2011.

[28] For example, Maldives (n 16), Bangladesh (n 18), and Limoges (n 21) proposals.

[29] For example, Docherty and Giannini proposal (n 23), Hodgkinson and others proposal (n 25), and Biermann and Boas proposal (n 21). See also Chairperson's Summary, 'Nansen Conference on Climate Change and Displacement in the 21st Century' (Oslo, 6–7 June 2011) para 7.

III. Empirical Evidence on Movement

Treaty proposals are premised on certain assumptions about climate change and human movement that are not borne out in the empirical studies. First, although the '*displacement* dimension is most evident today in the context of sudden onset disasters',[30] the nature of that movement is likely to be predominantly *within* countries, not across international borders, and temporary in nature. Secondly, much climate-related movement is likely to be gradual, rather than in the nature of refugee 'flight'.[31] Thirdly, events that may lead to more permanent displacement may be triggered by geophysical rather than climate-related causes (eg earthquakes and volcanoes).[32] Finally, most cross-border displacement will occur within regions, rather than from the global south to the global north (an assumption that implicitly underpins a number of treaty proposals). It is important that debates about a universal treaty do not obscure the development of regional responses underscored by the human rights principles articulated in Chapter 9.[33]

As Chapter 6 on Bangladesh showed, even if initial movement from a disaster is rapid, it will almost always be *internal* movement. Even longer-term migration is predominantly within Bangladesh itself, largely because people's poor socio-economic circumstances preclude them from undertaking cross-border journeys. Some will move intra-rurally, while a large number will move from rural to urban areas within Bangladesh. Temporary and circular migration is a common longer-term survival strategy.[34] Since the displaced remain citizens of Bangladesh and therefore entitled to the protections that flow from that status, they will likely be treated as a domestic concern and not within the purview of international attention.

Similarly, in Kiribati and Tuvalu, people living on the outer islands will initially move internally to the main atoll, not directly overseas.[35] This accounts for some of the environmental and population stress on the main atolls. As examined in Chapter 5 and further below, *international* movement from Pacific islands such as Kiribati and Tuvalu is likely to be pre-emptive and planned in response to slow-onset changes that gradually degrade the islands, rather than in the nature of

[30] Volker Türk, Director, Division of International Protection, UNHCR, 'Can Protection of Environmentally Displaced Persons Be Found in Existing Protection Regimes? What Are the Next Steps from a Protection Perspective?' (Nansen Conference on Climate Change and Displacement in the 21st Century, Oslo, 7 June 2011) 4.

[31] See, eg, Frank Laczko and Christine Aghazarm (eds), *Migration, Environment and Climate Change: Assessing the Evidence* (IOM, 2009); EACH-FOR, 'Synthesis Report' (14 May 2009). See further, Ch 1.

[32] See the discussion of Montserrat in Ch 4.

[33] Kälin and Schrepfer (n 9) 8; on regional approaches, see 38–41.

[34] Chowdhury R Abrar and S Nurullah Azad, *Coping with Displacement: Riverbank Erosion in North-West Bangladesh* (RDRS Bangladesh, North Bengal Institute, and Refugee and Migratory Movements Research Unit, 2004) 113.

[35] 'I think most who have gone abroad live here in Tarawa. They live here. But I haven't heard of their relatives that they go straight from the outer island to go direct to somewhere else. No, I haven't heard of that': interview with an official from the Ministry of Environment, Lands and Agricultural Development (Tarawa, Kiribati, 14 May 2009).

sudden flight.[36] Such movement is, of course, conditional on people being eligible for existing visa categories (employment, education, family, etc). In this sense, its link to climate change will be 'invisible': those who can move through existing migration categories will do so, and will be counted among the general cohort of immigrants to a new country (not as migrants 'on account of' climate change).[37] Indeed, very few of those interviewed in Kiribati and Tuvalu articulated climate change as a predominant motivation for taking up these kinds of migration opportunities, although when asked, many acknowledged that it was an underlying consideration.[38] It is likely that the *perception* of risk of change is a far more significant factor in decision-making than actual biophysical change.[39] According to Barnett and Webber, 'social processes that create poverty and marginality are more important determinants of migration outcomes than environmental changes per se'.[40]

A particular challenge for any new treaty is adequately accounting for slow-onset movement brought about by gradual environmental deterioration, as opposed to flight from sudden disasters. The refugee and complementary protection paradigm, which premises protection needs on imminent danger, does not capture the need for safety from longer-term processes of climate change that may ultimately render a person's home uninhabitable. Indeed, even in the area of disaster management, experts lament that too often action comes as a response to an emergency, rather than as part of longer-term adaptation and risk-reduction strategies.[41] While some treaty proposals build in protection for those likely to be displaced as a result of gradual changes,[42] this is unlikely to get political traction. First, affected communities reject representations that cast them as victims in need of assistance, as opposed to people with skills to offer. As discussed in Chapter 2, they express considerable discomfort with a refugee-like paradigm. Secondly, this

[36] For background, see Jane McAdam and Maryanne Loughry, 'We Aren't Refugees', *Inside Story* (30 June 2009) <http://inside.org.au/we-arent-refugees/> accessed 1 July 2009.

[37] See Ch 1.

[38] See also Mortreux and Barnett (n 14). Compare Shawn Shen and François Gemenne, 'Contrasted Views on Environmental Change and Migration: The Case of Tuvaluan Migration to New Zealand' (2011) 49 International Migration e224 with Kathryn L Paton, 'At Home or Abroad: Tuvaluans Shaping a Tuvaluan Future' (Thesis submitted for Master of Development Studies, School of Geography, Environment and Earth Sciences, Victoria University of Wellington, 2009).

[39] Mortreux and Barnett (n 14) 107, citing Torsten Grothmann and Anthony Patt, 'Adaptive Capacity and Human Cognition: The Process of Individual Adaptation to Climate Change' (2005) 15 Global Environmental Change 199.

[40] Jon Barnett and Michael Webber, 'Migration as Adaptation: Opportunities and Limits' in McAdam (n 3) 38. 'If climate change were a major driver of migration from Funafuti, one would expect it be mentioned far more often than it was by respondents. Instead, employment and better opportunities were cited as the major reasons to migrate. Of the nine people planning to migrate, only one cited climate change as a primary factor influencing their decision to leave, with another mentioning climate change as a secondary factor. These two respondents also cited other factors that informed their decisions—namely family unification, employment and a desire for greater opportunities. Given this, it is clear that climate change itself is not a significant factor informing respondents' decisions to leave Funafuti': Mortreux and Barnett (n 14) 109.

[41] Charles Ehrhart and others, *Humanitarian Implications of Climate Change: Mapping Emerging Trends and Risk Hotspots* (2nd edn, OCHA and CARE, 2009) 3. On disaster management, see Ch 9 below.

[42] See, eg, Hodgkinson and others proposal (n 25).

would represent a radical reconceptualization of 'protection'. Of course, while it is open to States to craft any kind of protection instrument they see fit,[43] it seems highly unlikely that they would agree to such an open-ended instrument. Though refugee law is itself built on predictions about the future, the timeframes involved in predicting the slow-onset impacts of climate change, and especially the role that adaptation and resilience may play in mitigating against them, would likely be considered by States to be too uncertain for them to assume binding obligations towards affected populations. They would view such movement as more akin to migration—which remains squarely within the realm of State discretion—than displacement. This is why the potential of existing *non-refoulement* obligations was examined in Chapter 3, since it is likely that any expansion of these duties will be developed first through jurisprudential interpretation of their scope and content, rather than by the adoption of new treaty obligations. Finally, as the UN High Commissioner for Refugees (UNHCR) has noted, in 'slow onset disasters, people are not displaced or obliged to move as the result of a single event. Instead, an accumulation of factors leads to a tipping point at which people's lives and livelihoods come under such serious threat that they have no choice but to leave their homes.'[44] The tipping point will vary for different individuals, and it is important that any protection instrument is sufficiently nuanced to account for this. As has been observed by UNHCR in the refugee context:

> Persecution cannot and should not be defined solely on the basis of serious human rights violations. Severe discrimination or the cumulative effect of various measures not in themselves alone amounting to persecution, as well as their combination with other adverse factors, can give rise to a well-founded fear of persecution, or, otherwise said: make life in the country of origin so insecure from many perspectives for the individual concerned, that the only way out of this predicament is to leave the country of origin.[45]

Since the rationale behind a treaty is to address international movement, there is a risk that vast amounts of energy, time, and resources will be channelled into treaty-related advocacy at the expense of other, perhaps more appropriate and community-attuned, responses. As the International Organization for Migration (IOM) has observed, comments by the Intergovernmental Panel on Climate Change (IPCC) about the potential for mass displacement appear to be driving claims for a treaty, rather than an appreciation of the reality that much movement will be internal, gradual, and not necessarily suited to an international treaty response.[46]

[43] This is limited only by the requirement that they do not breach *jus cogens* norms.

[44] António Guterres, United Nations High Commissioner for Refugees, 'Statement' (Nansen Conference on Climate Change and Displacement in the 21st Century, Oslo, 6 June 2011) 2.

[45] Erika Feller, 'Statement by Ms Erika Feller, Director, Department of International Protection, UNHCR' (Strategic Committee on Immigration, Frontiers and Asylum, Brussels, 6 November 2002) 3. See also UNHCR, *Handbook on Procedures and Criteria for Determining Refugee Status under the 1951 Convention and the 1967 Protocol relating to the Status of Refugees* UN Doc HCR/IP/4/Eng/REV.1 (2nd edn, Geneva, 1992) para 5.

[46] Interview with Rabab Fatima and Anita Wadud, IOM (Dhaka, Bangladesh, 17 June 2010). See also International Organization for Migration (IOM), *Assessing the Evidence: Environment, Climate Change and Migration in Bangladesh* (IOM, 2010) 18. See also Ch 1.

From an advocacy perspective, one can appreciate that lobbying for a 'climate refugee' treaty may successfully generate attention and place climate change-related movement on the international agenda. Placing the 'maximalist' option of a treaty on the table may paradoxically encourage States at least to negotiate more minimalist responses, as a compromise or fallback position. However, it is imperative that advocacy is well-informed. If empirical evidence does not support claims being made,[47] it will not achieve its ends and could ultimately backfire.[48] Indeed, messy or alarmist work may lead to attempts to discredit the phenomenon of climate change-related movement altogether.[49] While ambiguous or multiple causality may complicate the establishment of parameters for dealing with climate-related movement, this is not unique to displacement situations generally and is a poor reason to overstate the role of climatic factors.

IV. Multicausality

Earlier chapters in this book have shown the difficulties of establishing climate change as *the* cause of movement, and the possible arbitrariness of linking solutions to particular drivers of movement rather than people's needs.[50]

Multicausality is not, of itself, a sticking point for devising a treaty. For example, under international law, the Refugee Convention does not require 'persecution' to be the predominant reason for flight. Indeed, other motivations are irrelevant provided a well-founded fear of persecution exists. The standard of proof in refugee law—a 'well-founded fear of persecution'—can be less than a 50 per cent chance.[51] Thus, the assessment of the intensity, severity, and nature of *future harm*, based on the individual's circumstances, is the relevant consideration that leads to refugee status being granted. That assessment is not a prediction, but rather a *supposition*, based on the available evidence.[52]

However, a climate change displacement treaty would necessarily require *a* link to climate change. This would require a number of causation hurdles to be jumped. First, the decision-maker would need to assess the nature of the alleged harm feared, for example, lack of food due to salt-water intrusion on agricultural

[47] See, eg, Christian Aid, *Human Tide: The Real Migration Crisis* (Christian Aid, 2007).

[48] For example, there is some discussion that the 'sinking' of the Carteret Islands is not being caused by sea-level rise attributable to climate change, but rather to subsidence: John Campbell, 'Climate-Induced Community Relocation in the Pacific: The Meaning and Importance of Land' in McAdam (n 3) 68.

[49] See, eg, reports in 2002 that a climate change lobby group in Washington DC was insisting that all Tuvaluans would evacuate the island in 2003: Islands Business Pacific (April 2002) 10 cited in John Connell, 'Losing Ground? Tuvalu, the Greenhouse Effect and the Garbage Can' (2003) 44 Asia Pacific Viewpoint 89, 90. See also the discussion in Ch 1.

[50] See Ch 1.

[51] *INS v Cardoza-Fonseca*, 480 US 421 (1987) 431; *Chan v Minister for Immigration and Ethnic Affairs* (1989) 169 CLR 379. The UNHCR *Handbook* (n 45) para 42 states: 'In general, the applicant's fear should be considered well-founded if he can establish, to a reasonable degree, that his continued stay in his country of origin has become intolerable to him for the reasons stated in the definition, or would for the same reasons be intolerable if he returned there.'

[52] See further, Chs 2 and 3.

land. Secondly, some of the treaty proposals suggest establishing an expert scientific body to determine whether the source of that harm (salt-water intrusion from king tides or sea-level rise) is attributable to climate change in each case.[53] However, as climate scientists explain, asking whether climate change has 'caused' a particular event is a nonsensical question, since it is impossible to conclusively attribute individual events to it.[54] Climate change is about assessing the likelihood of certain types of events or processes occurring, rather than making definitive findings about particular occurrences. One might say, therefore, that linking protection to cause in such a case is 'artificial to the point of flippancy'.[55] Thirdly, the decision-maker would need to assess whether that harm amounts to a sufficiently severe violation so as to attract the protection of the treaty. Given that climate change is bound up with underlying socio-economic circumstances, the degree to which climate change can—and needs to—be singled out as a factor would need careful consideration. These levels of complexity could lead to considerable difficulty and inconsistency in decision-making. And again, they raise the question: why is 'climate change' the key? In my view, this is inappropriate in the protection context. It misplaces the real focus of the inquiry, which should be on the nature of harm feared if a person is returned home.

V. Political Obstacles to a New Treaty

These conceptual critiques are linked to a more pragmatic one: the fact that States presently seem to lack the political will to negotiate a new instrument requiring them to provide international protection to additional groups of people. As the President of Kiribati observed, it would be very difficult for an elected politician to say, ' "We will accept people from Kiribati as a response to climate change." No politician's going to do that.'[56] The UNHCR has similarly remarked on the 'general recognition that in the current context, it will not at all be easy to establish a binding new international instrument relating to the rights of such people'.[57] This is in part 'because of largely incompatible interests of potential countries of origin and countries of destination of such movements. While the former would probably push for a maximum of rights regarding admission and status, the latter, taking into account their restrictive attitude towards refugees and asylum-seekers, are not likely to accept more than minimal obligations.'[58]

[53] See, eg, Biermann and Boas, Global Environmental Politics (n 21) 77; Hodgkinson proposal (n 25) 86; Docherty and Giannini (n 23) 389–90. Insurers use a similar mechanism to determine whether a particular event was covered by insurance (eg was there a 'flood' or 'stormwater runoff'?). Given that the damage sustained is the same, this has been criticized as an expensive, overly technical way of assessing claims. The difference in that context is that there is a direct cost incurred if all events are covered—premiums will rise to reflect this.

[54] Daniel G Huber and Jay Gulledge, 'Extreme Weather and Climate Change: Understanding the Link and Managing the Risk' (Centre for Climate and Energy Solutions, December 2011).

[55] Hersch Lauterpacht, *International Law and Human Rights* (Praeger, 1950) 422.

[56] Interview with President Anote Tong (n 10).

[57] Guterres (n 44) 4. [58] Kälin and Schrepfer (n 9) 57.

One Bangladeshi Parliamentarian observed:

> I think the first thing, before you go into the protocols and structures, what I think is needed is political weight, whether the appetite is there for governments, especially in the developed world, the Annex I countries to address the issue in Bangladesh, because I think if you have that will, if you have that willingness, that acceptance... then you can always work something out. I think one of the problems is that we're getting too involved in discussions on what sort of a structure we should have without first actually having the political will.... So I think the Bangladeshi position is that first the countries have to accept the concept, and once they accept it, then I'm sure we can find some sort of an adjustment.[59]

The reluctance of I-Kiribati and Tuvaluans to be called 'refugees', discussed in Chapter 2, highlights some of the central failures of the international protection system which are relevant in the present context as well. Most notably, their fear about languishing in camps is a real one, given that an absence of political will to implement the principle of burden-sharing currently leaves millions of refugees in protracted situations with no hope of durable solutions.[60] In light of the legal obligations that States already have towards Convention refugees, and the fact that some 15.4 million refugees today, not to mention other displaced people numbering some 43.7 million in total,[61] have no durable solution in sight, why would States be willing to commit to, and realize protection for, people displaced by climate change?[62]

In responding to the first of these questions, it could be argued that States might be prepared to adopt such an instrument precisely to call for shared responsibility. For example, an individual State might perceive a need to respond to potential arrivals of 'climate refugees', but be unwilling to unilaterally create legal avenues for their protection. Were it to elicit the support of other States in adopting a treaty, however, then its humanitarian impulse could be coupled with mutual self-interest, in that it could call on other States to share the responsibility of caring for such people.[63]

This is illustrated by the response of the Australian Labor Party, which (in Opposition) had proposed the creation of a Pacific Rim coalition to accept climate change 'refugees', and to lobby the UN to 'ensure appropriate recognition of climate change refugees in existing conventions, or through the establishment of a new convention on climate change refugees'.[64] When a Greens Senator proposed

[59] Interview with Saber Chowdhury MP, Member of the All Parliamentary Committee on Climate Change, Bangladesh (Dhaka, Bangladesh, 21 June 2010).

[60] On burden sharing, see Agnès Hurwitz, *The Collective Responsibility of States to Protect Refugees* (Oxford University Press, 2009) 138–71.

[61] UNHCR, *Global Trends 2010* (UNHCR, 2010) 5. The refugee figure includes 4.82 million Palestinian refugees.

[62] UNHCR, among others, argues that there is a risk that if the Refugee Convention is opened up for renegotiation there could be a reduction in protection overall: see quotes in Grant, Randerson, and Vidal (n 17). However, this could be avoided by creating a protocol rather than renegotiating the existing treaty text.

[63] That said, some of the States that host the largest numbers of refugees are not parties to the Refugee Convention or Protocol.

[64] Australian Labor Party (ALP), *Our Drowning Neighbours: Labor's Policy Discussion Paper on Climate Change in the Pacific* (ALP Policy Discussion Paper, 2006) 10.

the extension of protection visas to 'climate refugees' in June 2007,[65] the Labor party was quick to note that without a collaborative approach with other countries, assuming such an obligation would be a unilateral act and therefore inconsistent with its idea of international action.[66] Indeed, this is one of the strongest incentives for States to ratify such a treaty: it provides a basis on which they may take the moral high-ground, agitate for multilateral action, but also shift the 'burden' away from themselves alone on to the international community as a whole.

However, it is in response to the second question, why States would be willing to *realize* such protection, that real difficulties arise. As we see with the present refugee regime, problems of implementation—and durable solutions—stem predominantly from a lack of political will, rather than an absence of law. Despite the 148 States parties to the Refugee Convention and/or Protocol, the plethora of soft law relating to refugees, and an international agency (UNHCR) with a strong field as well as institutional presence, the displacement of millions remains unresolved.

In other words, a treaty per se does not 'solve' the problem. This is not an argument against the development of the law, but it does highlight one of the key obstacles in achieving treaty-based solutions (at least in the short- to medium-term), as well as the limits of a treaty even if negotiated. Indeed, difficulties in reaching consensus may mean that even if an instrument were adopted, certain States might refuse to ratify it. The end result would be a weakened, and possibly even undermined, instrument, leaving those in need of protection in 'a legal vacuum'.[67]

All these points relate back to the question how best to protect and promote the human rights of affected communities. While international human rights law principles should inform any decisions relating to movement, a protection-like response may not respond to communities' human rights concerns, especially those relating to cultural integrity, self-determination, and statehood.[68] It may also obscure other human rights that need attention.

Together, these concerns suggest that the focus on a multilateral treaty to extend States' international protection obligations may not presently be the most appropriate mechanism for achieving outcomes for populations severely affected by the impacts of climate change. There is a risk that legally defining a 'climate refugee' category may lead to a hardening of the concept, simultaneously defining groups 'in' or 'out' of protection needs.

Focusing attention on culturally sensitive outcomes for people in particular contexts, which respond to the nature, timing, and location of predicted movement within, from, and to particular States, and their own views about how they want to be perceived,[69] may ultimately better facilitate a human rights approach to the phenomenon.

[65] See Migration (Climate Refugees) Amendment Bill 2007.

[66] See debates following Second Reading Speech in Parliament of Australia: Commonwealth of Australia, *Senate: Official Hansard* (9 August 2007) 95ff. For the Labor government's present position with respect to climate change-related displacement from the Pacific, see Senate Foreign Affairs, Defence and Trade References Committee, *Economic Challenges facing Papua New Guinea and the Island States of the Southwest Pacific* (Commonwealth of Australia, 2009) paras 6.60–6.62.

[67] Kälin and Schrepfer (n 9) 57.

[68] On which, see Ch 5.

[69] See McAdam and Loughry (n 36).

VI. Government Advocacy for International Responses

States themselves have been divided on whether they wish to see climate-related displacement addressed in a multilateral instrument.[70] With the exception perhaps of Bangladesh, no government appears still to embrace the 'climate refugee' terminology or conceptualization.[71] At least two proponents of the treaty approach acknowledged in May 2011 that they had not consulted with affected populations about their proposals, and that they would obviously be of little utility were they premised on a model that such groups rejected.[72] Interestingly, neither Kiribati nor Bangladesh has identified migration or relocation as a priority issue under their National Adaptation Programme of Action (NAPA). Tuvalu's NAPA refers to the need to develop a post-disaster resettlement and rescue plan,[73] noting that '[a] last resort to adaptation would be migration and resettlement should the worse case scenario occur'.[74] Some interviewees in Kiribati and Tuvalu expressed concern about the international attention that the idea had garnered, intimating that its prominence was possibly complicating efforts to resolve the issue through bilateral or regional approaches on migration and humanitarian assistance.

The government of Kiribati is keen to secure international agreements in which other States recognize that climate change has contributed to their predicament and acknowledge 'relocation' as part of their obligations to assist (in a compensatory way).[75] By contrast, the governments of Tuvalu and the Federated States of Micronesia have resisted the inclusion of 'relocation' in international agreements because of a fear that if they do, industrialized States may simply think that they can 'solve' problems like rising sea levels by relocating affected populations, instead of by reducing carbon emissions—something which would not bode well for the world as a whole. According to the (now former) Solicitor-General of Kiribati, they say it is giving up—saying to the rest of the world, 'pollute all you like'.[76] In his view, these States are in a radically different position

[70] For discussion of the rejection of the 'refugee' label by Pacific communities, see Ch 2.

[71] The Bangladeshi Prime Minister has made some public comments using this terminology, however all government officials and NGOs I interviewed in Bangladesh rejected this term. See, eg, comments by Her Excellency Sheikh Hasina, Prime Minister of Bangladesh, that 'in Bangladesh, climate refugees are increasing every day': 'Statement to the Plenary' (United Nations Climate Change Conference (COP15), Copenhagen, 16 December 2009) <http://unfccc2.meta- fusion.com/kongresse/cop15_hls/templ/play.php?id_kongresssession=4125> accessed 13 August 2011.

[72] See comments by David Hodgkinson and Michel Prieur (Threatened Island Nations: Legal Implications of Rising Seas and a Changing Climate Conference, New York, 23–25 May 2011) <http://www.law.columbia.edu/centers/climatechange/resources/threatened-island-nations/tin_video_archives> accessed 10 August 2011.

[73] Government of Tuvalu, *Tuvalu's National Adaptation Programme of Action* (May 2007) 51–2.

[74] ibid, 25.

[75] Interview with President Anote Tong (n 10). See also the remarks of the Bangladeshi finance minister, Abul Maal Abdul Muhith, who prior to the 2009 Copenhagen climate conference stated: 'We are asking our development partners to honour the natural right of persons to migrate. We can't accommodate all these people': cited in John Vidal, 'Migration is the Only Escape from Rising Tides of Climate Change in Bangladesh', *The Guardian* (London, 4 December 2009) <http://www.guardian.co.uk/environment/2009/dec/04/bangladesh-climate-refugees/print> accessed 8 December 2009.

[76] Interview with Kiribati Solicitor-General David Lambourne (Tarawa, Kiribati, 8 May 2009).

from Kiribati: Tuvalu has only 10,000 inhabitants (a tenth of Kiribati's population) and close Polynesian ties to New Zealand, while citizens of the Federated States of Micronesia have rights to enter and work in the United States (and some internal migration options as well).[77] Kiribati wants 'to create that same safety net' for itself.[78]

A relatively common stance in Tuvalu, Kiribati, and Bangladesh is that officials would like to see an international framework acknowledging a moral (if not legal) responsibility to assist people displaced by climate change through a combination of strategies—funding for adaptation, offering migration pathways, committing humanitarian assistance if needed, and so on.[79] Most stop short of demanding admissions of liability or compensation per se, seeing the 'blame game' as futile and unproductive. Instead, they would like to see an overarching framework which (a) acknowledges the reality of climate change-related movement; (b) commits to funding adaptation to enable *in situ* responses for as long as this is feasible; and (c) commits in principle to facilitating movement through migration programmes, and responding to spontaneous movement where inevitable. These issues are explored below and in the next two chapters.

In my view, therefore, it is preferable to work with affected governments to try to reach solutions involving a combination of *in situ* adaptation and migration, with the acknowledgement that planned movement may be an adaptation strategy in some cases. This must be considered within a human rights framework, however. Adaptation cannot occur at all costs—at a bare minimum, it must be adaption with dignity.[80]

VII. Migration Options

Against this backdrop, some governments have turned their attention to planned migration options. Managed international migration can provide a relatively safe mechanism for enabling people to move away from the effects of climate change without artificially treating them as being in need of international 'protection' (from a persecutory or abusive State) in the traditional sense of refugee or human rights law. Managed migration pathways are also better suited to respond to slow-onset climate change impacts, which are unlikely to trigger existing (or future) temporary protection mechanisms designed for sudden disasters. A major reason for existing pressures on asylum systems in some industrialized countries is that

[77] See further, Ch 5. [78] Interview with David Lambourne (n 76).

[79] 'It's a global problem—globally induced problem—so within the solution it must be international': interview with President Anote Tong (n 10).

[80] Interview with Rizwana Hasan, Chief Executive of Bangladesh Environmental Lawyers Association (Dhaka, Bangladesh, 16 June 2010). On the importance of local knowledge, see Jon Barnett and John Campbell, *Climate Change and Small Island States: Power, Knowledge and the South Pacific* (Earthscan, 2010).

avenues for 'regular' economic or other independent migration are very restricted for poor people from developing countries.[81]

However, while migration can significantly increase communities' capacity to adapt to climate change, this is likely to be the case only where there is a significant degree of autonomy and choice in mobility decisions.[82] This can be maximized by ensuring that those who move are accorded the same rights as people in the host community (which, in domestic law, would ordinarily mean at least the rights of permanent residents); facilitating mutual understanding between these groups; providing assistance with relocation costs (hence calls for an international relocation fund); clarifying property rights; and strengthening emergency response systems.[83]

A. Kiribati and Tuvalu[84]

In Kiribati and Tuvalu, the development of labour, education, and family migration pathways—as opposed to international protection-like responses—are better attuned to (a) the desires of people in those countries; (b) the likely patterns of climate change on the environment (slow and gradual) and patterns of movement (pre-emptive and gradual, rather than in response to a sudden catastrophic event); and (c) the history of movement in the region.

The long-term strategy of the government of Kiribati is to secure 'merits-based migration' options to neighbouring countries like Australia and New Zealand, so that those who wish to move permanently have an early opportunity to do so.[85] In this way, the President hopes that 'pockets' of I-Kiribati communities will build up abroad and I-Kiribati culture and traditions will be kept alive.[86] This would enable the gradual, transitional resettlement of I-Kiribati in other countries, so that if and when the whole population has to move, there would be existing communities and extended family networks which those left behind could join.

[81] See also Kälin and Schrepfer (n 9) 50: 'Measures such as managed regular migration schemes, including immigration quota or targeted admission of migrants from particularly affected areas, would assist such adaptation efforts and, at the same time, reduce the risks of irregular migration.'

[82] See, eg, Jon Barnett and Michael Webber, *Accommodating Migration to Promote Adaptation to Climate Change* (Commission on Climate Change and Development, 2009) 2.

[83] See ibid. See also Alex Julca, 'Multidimensional Re-creation of Vulnerabilities and Potential for Resilience in International Migration' (2011) 49 International Migration 30, 34, who observes that while migration may enhance resilience by improving economic security (both for the migrant and the family/community at home), increase human and social capital, diversify wealth through remittances, increase economic growth, and reduce unemployment and associated social unrest, it can also disrupt family ties (often for indeterminate periods), lead to 'brain drain' (exacerbated when others follow) and a loss of possible community leaders, and result in remittance dependency.

[84] For background, see Ch 5.

[85] Interview with President Anote Tong (n 10); see also Goering (n 10) referring to remarks by I-Kiribati officials at Copenhagen.

[86] 'Once we have a significant pocket of our people in certain countries, it makes it easier for the rest to follow': interview with President Anote Tong (n 10).

We want to begin that [migration] now, and do it over the next twenty, thirty or forty years, rather than merely, in fifty to sixty years time, simply come looking for somewhere to settle our one hundred thousand people because they can no longer live in Kiribati, because they will either be dead or drown. We begin the process now, it's a win-win for all and very painless, but I think if we come as refugees, in fifty to sixty years time, I think they would become a football to be kicked around.[87]

The President makes clear, however, that his government would be lobbying neighbouring States like Australia and New Zealand for migration opportunities *even if* the climate change threat did not exist, given the other underlying pressures at home. However, the spectre of climate change makes those negotiations all the more pressing. As the (now former) Solicitor-General of Kiribati explained, the government would be negligent if it were to talk about migration purely as an economic issue.[88]

By contrast, the present approach of the Tuvaluan government is to reject a migration discourse in favour of *in situ* adaptation.[89] In interviews conducted during fieldwork, government officials would not be drawn on the extent to which migration opportunities were being pursued as a response to climate change. The shift in government policy in recent years may be a contributing factor to this.[90] However, it was clear that the notion of 'climate refugees' is resoundingly rejected and that a treaty constructed around this paradigm does not appeal.

Since the most devastating impacts of climate change in the Pacific will be slow-onset in nature, interim migration measures that permit temporary and circular movement—on the understanding that a permanent migration outcome will ultimately be possible once relocation is imperative—may appeal to affected and receiving countries alike.[91] Staggered or circular migration enables returning migrants to inform their friends and relatives at home about life elsewhere, which can in turn ease the transition for future groups of migrants. They may bring new skills and ideas as well.[92]

We don't want to mass migrate in 10, 20 years time when we find that we can no longer sustain the population here... We want to begin the process now. Not only would it be more gradual, less drastic on anybody, even the receiving country, but I think it allows us the opportunity to test it. And we can approach it in a more constructive manner. This is the strategy that we are advocating. It's primarily about skilling up. So 40,000 people are going to Australia every year—skilled from Asia. Okay, we don't want 40,000 moving

[87] President Anote Tong in Duncan Wilson, 'Climate Change: Nobody is Immune', *Islands Business* (2008) <http://www.islandsbusiness.com/islands_business/index_dynamic/containerNameToReplace=MiddleMiddle/focusModuleID=18087/overideSkinName=issueArticle-full.tpl> cited in Vikram Kolmannskog and Finn Myrstad, 'Environmental Displacement in European Asylum Law' (2009) 11 European Journal of Migration and Law 313, 325.

[88] Interview with David Lambourne (n 76).

[89] See Chs 1 and 5.

[90] See analysis of Tuvalu's approach in Ch 1.

[91] It may also be more palatable for governments to absorb some migrants in traditional labour categories, rather than to acknowledge the drivers behind the movement. Part of the challenge is to 'sell' the solution domestically, both within the country of origin and the host country.

[92] Jon Barnett and Natasha Chamberlain, 'Migration as Climate Change Adaptation: Implications for the Pacific' in Burson (n 4) 55.

from Kiribati in one year. But if we can achieve a couple of thousand a year, initially, I think that would go along with us. It's painless. [Some movement is] happening already.[93]

Of course, it is important that any such migration is reinforced by local adaptation measures, since the migration of skilled workers may further deplete local human resources. That said, 'skilling up' people can have direct benefits within Kiribati (eg provision of healthcare), as well as making a significant economic contribution through remittances, thereby increasing family resilience for those who remain.[94] As a former President of Kiribati acknowledged, migration is:

a double-edged sword, in the sense that it's good to have our people getting employment overseas and sending remittances back. But then you also have to see that the local economy, the local setup, also has to have enough skilled people. We don't want to just encourage all these skilled people to move out of the country and be left with nothing. That would be counterproductive on the local scene.[95]

By contrast, the current President argues that: 'we have an oversupply of brains; we can't employ our graduates. We want to move—that's exactly what we want to do—get our brains out. But keep educating our people—that is the strategy. So we need the responses—to [enable them to] educate up, skill up—so they can go.'[96] Another official sees this approach as having a local benefit as well, because 'people will have a part in the process, not just waiting for government to arrange everything for them. They have a part to play in the policy, in the system. And eventually, hopefully, they'll be part of the—of the relocation, you know, with less cost to the government.'[97]

By freeing up resources and alleviating stress on already fragile atoll environments, migration may enable at least a percentage of the population to remain in Kiribati and Tuvalu for longer, supported by remittances and extended family networks abroad. Remittances can help to fund climate change adaptation, among other things, and may support the subsequent migration of other family members over time. There is an emerging consensus that these States cannot carry populations of their present size if they are to approach any semblance of sustainable development.[98] Migration can help to relieve population pressure and fill skills shortages in other countries, thus providing a win–win situation.[99] This demonstrates how migration can be a form of adaptation. This process would also see new diaspora communities forming abroad, and receiving States adapting to their

[93] Interview with President Anote Tong (n 10).

[94] Mark Pelling and Juha I Uitto, 'Small Island Developing States: Natural Disaster Vulnerability' (2001) 3 Environmental Hazards 49; Julca (n 83) 38–41.

[95] Interview with Teburoro Tito, Former President of Kiribati (Tarawa, Kiribati, 18 May 2009). On 'brain drain', see Ch 6.

[96] Interview with President Anote Tong (n 10). On the positive elements of migration, see, eg, Barnett and Webber (n 82) 21ff.

[97] Interview with Betarim Rimon, Office of the President (Tarawa, Kiribati, 12 May 2009).

[98] See John Connell, 'Environmental Change, Economic Development and Emigration in Tuvalu' (1999) 22 Pacific Studies 1. Interview with David Lambourne (n 76): 'An atoll of this size really only can sustain 5,000, not 50,000, people.'

[99] See comments of Graeme Hugo in 'Seven Billion and Counting', *Sydney Morning Herald* (Sydney, 30 April 2011) News Review 4.

presence over time.[100] As President Tong observes: 'In my view, it's actually better for the receiving country to have people from that country already there in well established [communities], so that when the rest follow, it won't be such a burden on the receiving country.'[101]

Barnett and Webber regard migration as a central and often benign way of increasing community access to capital, technology, and information.[102] At present, some of the most isolated, resource-dependent communities do not have access to migration options, and only a change in policy will enable migration to become a genuine development and/or adaptation strategy for them.[103] Indeed, the World Bank's World Development Report 2010 on 'Development in a Changing Climate' identified migration as an issue worthy of further exploration, in particular 'the pertinence of mobility as an adaptive strategy for the poor, the associated policy and institutional challenges, and... both the positive and negative aspects of mobility'.[104] Barnett and Chamberlain advocate establishing a series of trials in the Pacific, which would permit discrete groups to access migration options and have their social and environmental consequences measured.[105] They have also suggested creating targeted training packages for Pacific islanders to develop skills that are needed in countries like Australia and New Zealand (eg agricultural labour), and matching remittance contributions with aid money.[106]

Such an approach builds on the historical migration patterns between Pacific countries and New Zealand, and these might constructively be developed as part of broader bilateral partnerships[107] and regional cooperation agreements.[108] New Zealand has long had special concessionary schemes for citizenship or permanent residence to promote economic development in Pacific island States,[109] one of the most recent iterations of which is the 2002 Pacific Access Category.[110] The rationale behind New Zealand's concessionary policies is to promote economic development in Pacific island States, although its original impetus came from a post-war period of industrial expansion. According to Stahl and Appleyard, such an approach is unique among developed States.[111]

[100] See Deborah McLeod, 'Potential Impacts of Climate Change Migration on Pacific Families Living in New Zealand' in Burson (n 4).

[101] Interview with President Anote Tong (n 10).

[102] Barnett and Webber (n 82).

[103] Barnett and Chamberlain (n 92) 55.

[104] They accordingly commissioned a report by Barnett and Webber (n 82) 3.

[105] Barnett and Chamberlain (n 92) 56.

[106] ibid, 57.

[107] Such as New Zealand's five year Strengthened Cooperation Programme with Niue from 2004 to 2009.

[108] Such as the Niue Declaration on Climate Change, 39th Pacific Islands Forum, Forum Communiqué, Annex B (19–20 August 2008) ('Niue Declaration').

[109] For descriptions of earlier categories, such as the special arrangements for Fiji, Tonga, and Western Samoa, and the South Pacific Work Permit Scheme (which included special schemes for Kiribati and Tuvalu), see 'Immigration and New Zealand: A Statement of Current Immigration Policy' (Immigration Division, Department of Labour, updated January 1986) (on file with author). For a very detailed overview, see Reginald T Appleyard and Charles W Stahl, *South Pacific Migration: New Zealand Experience and Implications for Australia* (Commonwealth of Australia, 1995), and see Appendix 1a and Appendix 1b of that publication for the Kiribati and Tuvalu work permit schemes.

[110] For discussion of this category, see Ch 4.

[111] Appleyard and Stahl (n 109) iv.

By contrast, Australia has insisted upon a 'non-discriminatory' policy that does not (formally) privilege any national group.[112] This is perhaps a reaction to its White Australia policy past and prior exploitation of Pacific labour, such as through 'blackbirding' (forced recruitment). However, since 2007 AusAID has funded the Kiribati–Australia Nursing Initiative (KANI), which offers around 30 young I-Kiribati the opportunity to train as nurses at Griffith University in Queensland, and, if successful, remain in Australia. In 2009, Australia implemented a three year Pacific Seasonal Workers Pilot Scheme, modelled in part on New Zealand's Recognised Seasonal Employer (RSE) scheme.[113] Over three years, up to 2,500 visas will be granted to people from Kiribati, Tonga, Vanuatu, and Papua New Guinea, to work in the Australian horticultural industry for between six to seven months in each 12 month period.[114] On 8 September 2011, the Australian government announced that the scheme would be expanded to include Nauru, Samoa, the Solomon Islands, and Tuvalu.[115] Anecdotal evidence suggests that the programme has been less successful than anticipated, not least because of a lack of job offers from Australian farmers who seem reluctant to provide labour conditions that can be avoided by relying on 'black market' labour.[116]

A 2009 inquiry by an Australian Senate committee revealed that '[w]hen asked about the possibility of forced re-location from Pacific island countries such as Kiribati and Tuvalu, DFAT [the Department of Foreign Affairs and Trade] informed the committee that it was not aware of any government consideration of this matter. Invited to comment again on whether these two islands were under consideration, DFAT replied no.'[117] The Committee recommended 'that the Australian

[112] ibid, v; interview with Sir Kamuta Latasi, Speaker of the Tuvaluan Parliament (and former Prime Minister) (Funafuti, Tuvalu, 27 May 2009) referring to Tuvalu's discussions in the mid-1990s with then Australian Prime Minister, Paul Keating.

[113] See, eg, Sankar Ramasamy and others, 'The Recognised Seasonal Employer Policy: Seeking the Elusive Triple Wins for Development through International Migration' (2008) 23 Pacific Economic Bulletin 171, available at <http://peb.anu.edu.au/pdf/PEB_23_3_Ramasamy%20et%20al_WEB.pdf> accessed 18 December 2009. The RSE scheme was introduced in April 2007 and currently includes Kiribati, Samoa, Tonga, Tuvalu, Vanuatu and, shortly, the Solomon Islands: see <http://www.dol.govt.nz/initiatives/strategy/rse/index.asp> accessed 19 December 2009.

[114] See <http://www.deewr.gov.au/Employment/Programs/PSWPS/Pages/default.aspx> accessed 12 December 2009.

[115] 'Expansion of Pacific Seasonal Workers Pilot Scheme' (Prime Minister of Australia, Press Release, 8 September 2011) <http://www.pm.gov.au/press-office/expansion-pacific-seasonal-workers-pilot-scheme> accessed 1 December 2011.

[116] See also Rochelle Ball, 'The Pacific Seasonal Workers Pilot Scheme (PSWPS) and Implications for Pacific Development' (2009) <http://peb.anu.edu.au/pdf/2009/PNG/ppp/Rochelle_Ball_Pacific_Update_081209.pdf> accessed 18 December 2009; Department of Education, Employment and Workplace Relations (Australia), 'Interim Evaluation of the Pacific Seasonal Worker Pilot Scheme: Executive Summary' (2010).

[117] Senate Foreign Affairs, Defence and Trade References Committee (n 66) para 6.60 (making reference to *Committee Hansard*, 21 November 2008, 28). On 21 November 2009, a spokesperson for the then Climate Change Minister, Penny Wong, was reported as acknowledging that permanent migration may eventually be the only option for some people, which will need to be dealt with by governments in the region: Adam Morton, 'Land of the Rising Sea', *Sydney Morning Herald* (Sydney, 21 November 2009) <http://www.smh.com.au/environment/land-of-the-rising-sea-20091120-iqub.html> accessed 27 November 2009. However, just three months earlier, the then Australian Prime Minister, Kevin Rudd, would not be drawn on this issue, reiterating only that Australia would

Government consider whether it may be necessary to review the legal and policy framework required in the event that regional communities may be forced to resettle as a consequence of changes in climate'.[118] It expressed its concern

> about the lack of government attention to formulating policy around the possibility that some Pacific island communities may have to re-locate because of rising sea levels or related environmental changes. The committee believes that the Australian Government should allow ample time to consider closely and carefully the legal and policy framework that may be required should such an eventuality arise. The committee believes that Australia could also make a valuable and significant contribution in practical ways to prepare those most at risk of having to resettle. It notes that the Government of Kiribati wants their people to be competitive and marketable. Australia could be a vital partner with countries such as Kiribati by helping with research, training, education and labour mobility arrangements to equip people, should they have to move, to take up productive positions in their new location. It believes that should migration be necessary from these Pacific Island countries, the basic principle underpinning the formulation of Australia's policy should be their 'migration with merit and dignity'.[119]

Moving into the future, migration schemes might be constructively developed as part of broader bilateral partnerships, such as New Zealand's five year Strengthened Cooperation Programme with Niue from 2004 to 2009, and through enhancing regional cooperation agreements, such as those adopted at the Pacific Islands Forum. Strategically, Australia and New Zealand would benefit from a more cooperative approach to migration, especially since many Pacific islanders view movement to New Zealand as the first step towards ultimately reaching Australia: once they obtain New Zealand citizenship,[120] they can freely travel to and work in Australia.[121]

B. Bangladesh

The previous chapter described in detail the nature of climate change-related movement in Bangladesh, noting that it is predominantly internal. In such circumstances, a treaty that did anything other than remind States of their existing human rights obligations towards those within its jurisdiction would be otiose. This section examines the role of international migration from Bangladesh.

provide practical support, such as helping to construct sea walls and protect against disease: Brendan Nicholson, 'PM Won't Be Drawn on Climate Change Refugees', *Sydney Morning Herald* (Sydney, 6 August 2009) <http://www.smh.com.au/environment/climate-change/pm-wont-be-drawn-on-climate-change-refugees-20090805-ea33.html> accessed 20 May 2011.

[118] Senate Foreign Affairs, Defence and Trade References Committee (n 66) para 6.62 (Recommendation 3).

[119] ibid, para 6.61 (referring to UN News Centre, 'Small Island Nations' Survival Threatened by Climate Change, UN Hears' (*UN News Service*, 25 September 2008) <http://www.un.org/apps/news/story.asp?NewsID=28265> accessed 29 April 2009).

[120] For people who obtained permanent residence post-21 April 2005, the waiting period is five years: Department of Internal Affairs, 'General Requirements for a Grant of New Zealand Citizenship' <http://www.dia.govt.nz/Services-Citizenship-General-Requirements-for-a-Grant-of-New-Zealand-Citizenship#six> accessed 1 December 2011.

[121] 1973 Trans-Tasman Travel Arrangement; see also the Special Category Visa (SCV) for New Zealand citizens since 1994. See generally, <http://www.immi.gov.au/media/fact-sheets/17nz.htm> accessed 20 October 2010.

International migration is a central pillar of Bangladesh's long-term economic growth strategy and there are understandably objectives to strengthen it.[122] There is already considerable international labour migration from Bangladesh to the Gulf States, Malaysia, and North Africa, where some 5.5 million Bangladeshis work.[123] Over time, certain 'migration corridors' have established themselves based on economic, geographic, and historical factors.[124] Kinship, friendship, and community networks are very important, since earlier migrants often 'act as conduits to channel later generations of movers to those destinations in an atmosphere of certainty'.[125] Not only do they provide assistance in terms of information about the market, but they also help with adjustment and settlement. This process is said to account for 60 per cent of labour migration from Bangladesh to South East Asia and the Middle East.[126] In 2009 alone, 475,000 Bangladeshis emigrated for work abroad and US$10.7 billion was remitted to Bangladesh.[127]

These emigrants are mainly in semi- and low-skilled jobs on temporary contracts that provide for little prospect of eventual integration. While they are not the most impoverished group within Bangladesh, they have very limited resources and choose migration as a livelihood strategy.[128] Overseas employment provides a way of possibly improving the economic condition and social status of the family, and in this regard may provide a short-term strategy to secure marriage or educational opportunities.[129] It is therefore a livelihood diversification and risk management tool, although it is vulnerable to shocks in the global economy.[130] However, domestic migration laws and bilateral agreements generally entrench low-skilled work as a temporary option, with return to the home country compelled once the contract ends. Long-term migration is therefore only an option for people of a high economic status. As Julca notes, 'highly skilled migrants generally move to improve

[122] IOM (n 46) 30 referring to Government of Bangladesh, 'Second National Strategy for Accelerated Poverty Reduction' (rev draft, 2009).

[123] IOM (n 46) 30 referring to 'Bangladesh Economic Review' (Finance Division, Ministry of Finance, Government of The People's Republic of Bangladesh, 2008); Syeda Rozana Rashid, 'Migration for Livelihood: Social Protection Issues of Rural Bangladeshis' in Chowdhury R Abrar and Janet Seeley (eds), *Social Protection and Livelihoods: Marginalised Migrant Workers of India and Bangladesh* (The University Press, 2009) 163.

[124] Laczko and Aghazarm (n 31) 42 referring to World Bank, *Migration and Remittances Factbook* (International Bank for Reconstruction and Development/World Bank, 2008).

[125] Tasneem Siddiqui and Mohammad Jalal Uddin Sikder, 'Rural to Urban Migration for Domestic Work in Bangladesh' in Abrar and Seeley (eds) (n 123) 58. This is the same in the Pacific context: 'The existence of social networks at the destination may contribute to an individual's decision since these provide important practical support in terms of accommodation and employment connections, as well as providing a sense of community': Mortreux and Barnett (n 14) 107.

[126] Siddiqui and Sikder (n 125) 58 referring to 'Ministry of Expatriates' Welfare and Overseas Employment: Bureau of Manpower Employment and Training (BMET)' (2006).

[127] BMET website (2010) <http://www.bmet.org.bd/BMET/index> cited in IOM (n 46) 30.

[128] Rashid (n 123) 165 and citations.

[129] ibid, 166. Although as Julca (n 83) 42 notes, 'immigrants working in the low ranked occupations often end up reproducing underlying labour market inequalities at destinations'.

[130] Interview with Dr Hameeda Hossain, Founder, human rights NGO Ain O Salish Kendra (ASK) (Dhaka, Bangladesh, 14 June 2010). At the time of the fieldwork in Bangladesh (June 2010), those interviewed reported that the fallout from the global economic crisis had not significantly affected Bangladesh.

income and quality of work, whilst low-skilled migration is typically driven by the expectation to reduce economic insecurity'.[131]

Bangladesh acknowledges the importance of international migration in contributing significantly to GDP through remittances, and its Climate Change Strategy and Action Plan proposes some ways of enhancing this.[132] Bangladeshis returning home in the wake of the global financial crisis have also taken advantage of new economic opportunities, while there is scope for investment in public/private partnerships by non-resident Bangladeshis following the successful Indian model.[133]

Currently, however, legal protections are lacking in the bilateral agreements between Bangladesh and destination countries, and there is considerable exploitation of Bangladeshi migrant workers, including cases of violent mistreatment and human trafficking.[134] Although migrant workers are protected by general human rights law, none of the destination countries mentioned above have ratified the Migrant Workers Convention, which expressly protects the rights of migrant workers and their families.[135] Bangladesh itself has not ratified this treaty either, which puts it in a weak bargaining position should it wish to lobby other countries on this issue.

Of course, international labour migration will not provide a mobility pathway for the poorest Bangladeshis affected by climate change, which is why a range of responses to climate change is essential. As one local NGO stated, 'those who will be affected the most unfortunately are not the skilled, so for them, the ability to move beyond the national boundary would be very difficult'.[136] However, the poor may benefit indirectly through remittances, which bring net wealth to the country, and as the better-resourced people take up opportunities overseas, so the capacity of urban centres to support internal migrants may gradually increase. Remittances provide a relatively stable income stream that is not undermined if another disaster hits, and this can assist families to recover. Following the Asian Tsunami in 2004, remittances to affected areas increased—a pattern replicated elsewhere.[137]

In addition, there may be some limited opportunities to expand seasonal labour migration into India for the poor. This could provide reciprocal benefits: both by

[131] Julca (n 83) 32. [132] Referred to in IOM (n 46) 8.

[133] Interview with Saber Chowdhury MP (n 59).

[134] See, eg, Tasneem Siddiqui, 'Introduction' in Tasneem Siddiqui (ed), *Migration and Development: Pro-Poor Policy Choice* (University Press Limited, 2005) 6.

[135] International Convention on the Protection of the Rights of All Migrant Workers and Members of Their Families (adopted 18 December 1990, entered into force 1 July 2003) 2220 UNTS 3; see UN Treaty Collection, Status of Ratifications as at 12 August 2011 <http://treaties.un.org/Pages/ViewDetails.aspx?src=TREATY&mtdsg_no=IV-13&chapter=4&lang=en> accessed on 12 August 2011.

[136] Interview with Rizwana Hasan (n 80); interview with SM Munjurul Hannan Khan, Deputy Secretary, Ministry of Environment and Forests and National Focal Point for the UNFCCC and IPCC, Bangladesh (Dhaka, Bangladesh, 15 June 2010).

[137] On the Asian Tsunami, see Frank Laczko and Elizabeth Collett, 'Assessing the Tsunami's Effect on Migration', Migration Information Source (Migration Policy Institute, April 2005) <http://www.migrationinformation.org/USfocus/display.cfm?id=299> accessed 13 August 2011. Paulson has shown that remittances to Samoan households increase significantly after cyclones: Deborah D Paulson, 'Hurricane Hazard in Western Samoa' (1993) 83 Geographical Review 43, cited in Barnett and Webber (n 82).

assisting India to meet labour shortages, as well as by opening up opportunities for Indian workers in Bangladesh (for example, it has been suggested that Bangladesh faces a shortage of skilled nurses which could be filled by Indians).[138] Greater cooperation towards a more mobile and flexible regional labour market could enhance prosperity in both countries. That cannot happen, however, until migration is depoliticized and can be openly discussed by both governments.

In this way, 'climate change migration' is likely to have a 'domino' effect. Those who move abroad may not necessarily be the ones most acutely or directly affected by climate change impacts, but move as an indirect result—as cities become overpopulated, resources become scarcer, and life becomes increasingly difficult. Thus, highly skilled, professional, or business migration from Bangladesh is likely to increase as internal rural–urban movement places acute pressure on the infrastructure of cities like Dhaka and 'pushes' the relatively wealthy—eligible for education and work visas—to move abroad. Enhancing migration options for the educated and well-resourced may, in turn, open up greater opportunities for those moving within Bangladesh. In this way, 'climate change migration' per se across international borders is likely to be a largely invisible phenomenon.

VIII. Conclusion

A refugee-like treaty is sometimes posited as *the* answer to climate change-related displacement, but it is dangerous to see it in this way. Any treaty is necessarily an instrument of compromise, and even once achieved, States must demonstrate sufficient political will to ratify, implement, and enforce it. While international law provides important benchmarks and standards to regulate State action, they must be supported by political will and action to be fully effective. As Aleinikoff argues, 'there can be no monolithic approach to migration management. Some areas might well benefit from norms adopted by way of an international convention; guiding principles might work best for areas in which a consensus is further away.'[139] Even if a treaty is thought desirable, it is important that it be viewed as one of a number of mechanisms that may respond to climate-induced displacement, rather than as the only solution.

Part of the problem may lie in the disciplinary constraints of international law and international relations. At their very core lies the objective to universalize—to create norms that take the 'particular' to a level of general applicability, that make individual rights 'human rights' at one and the same time. The risk, of course, is that if this is done without sufficient empirical understanding or foresight, we arrive at a level of generality that is too vague, and which cannot be translated into practical, rational policies and normative frameworks. It is clear that legal gaps

[138] Interview with Professor Chowdhury R Abrar, Refugee and Migratory Movements Research Unit, University of Dhaka (Dhaka Bangladesh, 14 June 2010).

[139] T Alexander Aleinikoff, 'International Legal Norms on Migration: Substance without Architecture' in Ryszard Cholewinski, Richard Perrechoud, and Euan MacDonald (eds), *International Migration Law: Developing Paradigms and Key Challenges* (TMC Asser Press, 2007) 476.

exist,[140] but they should be addressed first by a dispassionate, careful appraisal of the empirical evidence, rather than motivated by an assumption that existing frameworks should be extended. This is a risk of (prematurely) concentrating the diverse impacts of climate change on human movement into calls for treaties and the like. The local and the particular do not always speak well to an international law or governance agenda, where the 'cascading' effect requires broad, universalizing statements. A related critique might be the law's tendency to create rights-based frameworks, which cannot always respond directly or adroitly to primarily needs-based problems.

On the other hand, international law retains sufficient flexibility to respond to particular scenarios through bilateral and regional agreements. In my view, this is where attention would best be focused initially. Although national and regional responses may not seem as gratifying for some as securing a universal international treaty on climate-related movement, they may in fact be able to more swiftly and effectively provide targeted outcomes, which respond as particular scenarios in particular geographical areas unfold. Pursuing more bilateral and regional 'economic' migration opportunities would also help to address underlying problems relating to scarce resources, overcrowding, rapid urbanization, and environmental degradation.

At this stage, it seems more probable that the development of regional soft-law declarations, such as the Niue Declaration on Climate Change,[141] will provide a more effective springboard for developing responses, than will a new international instrument aiming to take into account the interests of all States in a wide variety of contexts. At the normative level, there are already clear frameworks to guide such actions—the human rights law regime is the most relevant and important, supplemented by overarching normative principles (discussed in the next chapter).

For these reasons, this chapter should not be interpreted as rejecting a treaty-based regime altogether, or the underlying basis of such a regime: that States ought to provide assistance to certain people who are unable to remain in their homes. International cooperation on climate-related movement is sorely needed.[142] Rather, the chapter's purpose is to caution against squeezing all forms of 'forced' movement into a protection paradigm, since this may not best address the patterns or needs of those who move.[143]

[140] Although as Kälin notes, perhaps fewer than some believe, given that a lot of movement will be internal: Walter Kälin, 'Conceptualising Climate-Induced Displacement' in McAdam (n 3).

[141] Niue Declaration (n 108). See also Betts (n 7).

[142] The UN High Commissioner for Refugees suggests that: 'A development-oriented approach is now required in response to displacement, emphasizing the inclusion of the most vulnerable and marginalized sections of society in efforts to ensure that they benefit from the livelihoods, services and security to which they are entitled': Antonio Guterres, 'Bracing for the Flood', *The New York Times* (New York, 10 December 2009) <http://www.nytimes.com/2009/12/11/opinion/11iht-edguterres.html?_r=1&emc=eta1> accessed 14 December 2009.

[143] As the UN High Commissioner for Refugees has noted, given that most displacement is predicted to be internal, primary legal responsibility for ensuring people's rights will lie with the States concerned: ibid.

8

Institutional Governance

I. Introduction

At the macro level, climate change-related movement can be categorized and responded to in a variety of ways—for instance, as a protection issue, a migration issue, a disaster issue, an environmental issue, a security issue, or a development issue. Each 'lens' contains an implicit set of assumptions that motivates different policy outcomes. As a *protection* issue, the assumption is that movement is forced and should be treated as refugee-like in nature, with binding protection obligations for States with respect to those displaced (hence calls for a new treaty). It might focus on the human rights deprivations were a person to remain in their country of origin, rather than leave in response to climate-related factors.

As a *migration* issue, movement is often cast as voluntary, and therefore as not compelling the 'international community' to respond. For instance, the absence of employment opportunities in the country of origin might be cast as the motivating factor for movement. The assumption here is that States can respond as and when they see fit through domestic immigration policy. As a *disaster* issue, assistance can be provided by *in situ* humanitarian relief and temporary relocation where needed.[1] As an *environmental* issue, the movement of 'climate refugees' from 'sinking islands' can be used as a potent political image in advocating for the reduction of carbon emissions and the protection of endangered ecosystems. Here, the 'refugee' terminology contributes to its dramatic effect. As a *development* issue, foreign aid and investment are seen as the tools that can fund adaptation measures and assist climate-affected countries to 'develop' their way out of poverty, poor governance, and so on and thereby enhance their capacity to adapt to climate change. In each of these conceptualizations, the extent to which climate change features as the key issue varies: it is predominant in the protection and environmental discourses; it is one of a number of relevant impacts in the migration, disaster, and development characterizations.

[1] There was an interesting suggestion made at the first session of the UN Human Rights Council's Advisory Committee that the Human Rights Council and the Secretary-General use their good offices to extend the principle of *non-refoulement* to 'hunger refugees': Human Rights Council (HRC), 'Report of the Advisory Committee on its First Session (Geneva, 4–15 August 2008)', UN Doc A/HRC/10/2, A/HRC/AC/2008/1/2 (3 November 2008) 15, Recommendation 1/6.

Governance matters, because how climate change-related movement is regulated may in fact shape how it evolves in practice. According to Warner, '[p]olicy interventions will largely shape the outcome'.[2] She argues that the efficacy of governance of climate-related movement will play 'a critical role in whether migrants will return, or whether they will stay away indefinitely'.[3] Similarly, the timing of interventions is crucial: 'even if people could technically return to hazard affected areas, they may not choose to return if rehabilitation does not take place soon enough to be in sync with life cycle or other developments (such as employment, or services like schooling for children)'.[4] Thus, policymakers may in fact exacerbate displacement if they do not carefully plan for responses to it at the local, national, and international levels. As Warner notes, the governance structures States put in place will 'play a leading role in determining the degree to which migration is a form of adaptation, or an indicator of a failure to adapt'.[5] If poverty and marginality are not reduced, and measures are not put in place for rehabilitation and recovery, then there is 'little doubt' that 'environmental change will continue to be an important proximate factor in migration decisions'.[6]

However, trying to identify which international agency is best equipped to deal with climate change-related movement is the wrong approach. Rather, the pertinent question is how the expertise of a number of relevant organizations can be most effectively utilized and integrated. The shortcomings of institutional responsibility in many ways mirror the conceptual and legal challenges identified in the earlier chapters of this book. While the 'climate change' element of movement may increase the number of relevant institutions, it may also overshadow the capacity of existing mechanisms and mandates to respond (eg natural disaster frameworks, refugee law where applicable, etc). Understanding the extent to which 'climate change' *matters* must therefore form part of the analysis.

The governance of climate change-related movement (like global migration governance more broadly) suffers from significant fragmentation, both vertically—with actors at the international, regional, and local levels—and horizontally—with

[2] Koko Warner, 'Assessing Institutional and Governance Needs Related to Environmental Change and Human Migration' (Study Team on Climate-Induced Migration, German Marshall Fund of the United States, June 2010) 4. As Mortreux and Barnett note, alarmist approaches to climate-related movement could be self-fulfilling if policymakers follow them, since 'large-scale migration may be an impact of climate change affected by policy responses in anticipation of climate impacts rather than by material changes in the environment per se': Colette Mortreux and Jon Barnett, 'Climate Change, Migration and Adaptation in Funafuti, Tuvalu' (2009) 19 Global Environmental Change 105, 111.

[3] Warner (n 2) 2.

[4] ibid, 3.

[5] ibid, 8. For example, heavy monsoonal rains in Bangladesh in 2007 reportedly led to 3,000 people a day moving to Dhaka owing to inadequate relief and livelihood support. By contrast, when aid has been effectively distributed and food markets supported, such migration has been minimal: see Richard Black, Dominic Kniveton, and Kerstin Schmidt-Verkerk, 'Migration and Climate Change: Towards an Integrated Assessment of Sensitivity' (2011) 43 Environment and Planning A 431, 443.

[6] Jon Barnett and Michael Webber, *Accommodating Migration to Promote Adaptation to Climate Change* (Commission on Climate Change and Development, 2009) 6. This report is also available as Jon Barnett and Michael Webber, 'Accommodating Migration to Promote Adaptation to Climate Change', Policy Research Working Paper 5270 (World Bank, 2010).

the phenomenon addressed in part or, more rarely, as a whole under the auspices of a range of other 'policy categories' and associated institutions. In part, this stems from slow recognition of the problem, confusion about how best to understand it (eg migration versus protection), and the multiple and diverse ways in which its impacts may be felt, which both impede and complicate its regulation. As a subset of environmental migration, climate change-related movement cuts across several areas of international governance—migration and asylum, the environment, development, human rights, and humanitarian aid and assistance—each of which is represented by a number of different United Nations (UN) and other bodies.[7] Yet, despite (or because of) the plethora of existing, as well as potential, governance mechanisms, processes, and institutions, no coherent multilateral governance framework exists for this purpose.

As discussed in previous chapters, it is important that responses at the global level do not overlook local knowledge bases for adaptation and resilience, and the cultural and livelihood needs of displaced communities. For example, an official from the Ministry for the Environment in Kiribati explained to me that while external donors want to fund the construction of hard sea walls, 'those structures are hindering the way nature moves'.[8] Sea walls need constant repairs and there is insufficient funding to maintain them. She argued that soft, natural structures (like plants) provided a better and more sustainable barrier: 'These are inexpensive; these are community-based, and hopefully the community will take on ownership and feel that they need to conserve their biodiversity'.[9]

Indeed, the UN Inter-Agency Standing Committee's (IASC) Task Force on Climate Change has noted the recent 'overall shift of focus away from global level negotiations towards regional and country level advocacy and implementation of fast track adaptation'.[10] It emphasizes the importance of listening to the experiences of national and local actors in implementing solutions, since they are 'fundamental to achieving real impact on the ground and improving resilience in the face of climate change'.[11] Local capacity will become critical as climate change worsens.[12]

[7] As Warner (n 2) 3 observes: 'Environmental change today blurs the mandates of humanitarian organizations'.

[8] Interview with an official from the Ministry of Environment, Lands and Agricultural Development, Kiribati (Tarawa, Kiribati, 14 May 2009).

[9] ibid.

[10] IASC Task Force on Climate Change, 'Mid-Term Report for 2010' (9 July 2010) 2.

[11] World Food Programme (WFP), International Federation of Red Cross and Red Crescent Societies (IFRC), and UN Office for the Coordination of Humanitarian Affairs (OCHA), 'Addressing the Humanitarian Challenges of Climate Change: Regional and National Perspectives: Preliminary Findings from the IASC Regional and National Level Consultations' (2009) 4.

[12] Capacity building is essential, as this remark illustrates: 'We are thinking very seriously about this. But when you ask us, "What can we do?", what can we do? We just observe and do whatever we want. We just plant the trees on the beaches just to prevent it. Those are the only things we can do at the moment': interview with Sir Kamuta Latasi, Speaker of the Tuvaluan Parliament (and former Prime Minister) (Funafuti, Tuvalu, 27 May 2009).

II. Spheres of Governance

Owing to the numerous cross-cutting and intersecting issues raised by climate change-related movement which relate to a variety of different institutional mandates, it risks being dealt with in an ad hoc and fragmented manner. At the same time, the specialist expertise of these different institutions is needed to ensure that climate change impacts are understood and addressed from a number of angles. The end goal should be a holistic appraisal of the needs of particular communities, and this necessarily requires information-sharing and coordination.

Partial efforts to respond to the phenomenon at the international level come from at least five traditional 'spheres of governance' and their corresponding institutions:[13]

- migration/asylum
(eg UN High Commissioner for Refugees (UNHCR), International Organization for Migration (IOM), Office of the High Commissioner for Human Rights (OHCHR), Special Rapporteur on the Human Rights of Migrants, Special Rapporteur on the Human Rights of Internally Displaced Persons,[14] International Labour Organization (ILO)—International Migration Programme, The Hague Process on Refugees and Migration, UN Population Fund (UNFPA), Internal Displacement Monitoring Centre (IDMC), IASC, Global Forum on Migration and Development);

- the environment
(eg UN Framework Convention on Climate Change (UNFCCC), UN Environment Programme (UNEP), International Institute for Sustainable Development (IISD));

- development
(eg UN Development Programme (UNDP), UNFPA, IISD, ILO, World Bank);

- disaster risk response and management[15]
(eg UN International Strategy for Disaster Reduction (UNISDR), Office for the Coordination of Humanitarian Affairs (OCHA), UN Special Representative

[13] This is a non-exhaustive list of institutions.

[14] The Special Rapporteur is an independent expert who reports to the Human Rights Council. Climate change, disasters, and internal displacement is a focus of the current Special Rapporteur, Chaloka Beyani, building upon on the excellent work of his predecessor, Walter Kälin. Following country visits, the Special Rapporteur has urged governments to establish national measures to deal with internal displacement relating to climate change and natural disasters through human rights-based preparedness measures and new laws on disaster risk reduction: 'UN Expert Urges Maldives to Tackle Displacement Caused by Climate Change' (*UN News Centre*, 21 July 2011) <http://www.un.org/apps/news/story.asp?NewsID=39111&Cr=displace&Cr1=> accessed 22 July 2011. The Brookings–LSE Project on Internal Displacement (which the Special Rapporteur co-directs) provides a research-oriented forum in which to develop initiatives such as the Regional Workshop on Internal Displacement caused by Natural Disasters and Climate Change in the Pacific, held in Fiji in May 2011: <http://www.brookings.edu/events/2011/0506_idp_fiji_workshop.aspx> accessed 1 July 2011.

[15] The role of disaster risk management in addressing climate-related displacement is discussed extensively in Ch 9.

of the Secretary-General for Disaster Risk Reduction, UNEP, UNDP, World Food Programme (WFP), Food and Agriculture Organization (FAO), the Red Cross (ICRC/IFRC), UNHCR);

- human rights/humanitarian aid agencies

(eg OHCHR, OCHA, IASC, ICRC/IFRC, UNHCR, IOM, Special Rapporteur on the Human Rights of Migrants, Special Rapporteur on the Human Rights of Internally Displaced Persons, UNFPA, UN Children's Fund (UNICEF), UN Women).

To this thematic list could be added 'security', especially given the links that national institutions, in particular, have sought to draw between climate-related movement and national (and human) security. This is discussed further below.[16]

Clearly, a number of these organizations cut across two or more of the policy areas, and a common entry point for humanitarian and development actors is the protection of vulnerable communities.[17] Each of them has a particular perspective and expertise to add to the context of climate change-related movement. The difficulty is that none of them provides a comprehensive and coherent multilateral framework regulating State responses to such movement. Moreover, institutions in the various policy fields may have overlapping or conflicting mandates, or alternatively such a limited/partial perspective that the phenomenon as a whole remains beyond their scope. Difficulties already exist within some of the sectors mentioned above: for example, the system of disaster management alone 'remains highly fragmented, increasingly specialised, and marred by institutional rivalries'.[18] There has sometimes been resistance to centralized control in the field, and there is considerable diversity between agencies in terms of the nature and timeliness of their responses. This was evident in reactions to the 2004 Asian Tsunami, where, notwithstanding significant resources and agency involvement, 'the basic needs of displaced people were compromised by difficulties in coordinating the delivery of the US$6.8 billion worth of assistance that was pledged, and the activities of the 16 UN agencies, 18 Red Cross response teams, 160 or more international NGOs, hundreds of private and civil-society groups, and 35 armed forces'.[19]

Indeed, whether a single organization could harness the interdisciplinary expertise required to address all aspects of the phenomenon from the science, to

[16] See section III.C. below. See also Jane McAdam and Ben Saul, 'Displacement with Dignity: International Law and Policy Responses to Climate Change Migration and Security in Bangladesh' (2010) 53 German Yearbook of International Law 233; Jane McAdam and Ben Saul, 'An Insecure Climate for Human Security? Climate-Induced Displacement and International Law' in Alice Edwards and Carla Ferstman (eds), *Human Security and Non-Citizens: Law, Policy and International Affairs* (Cambridge University Press, 2010).

[17] WFP and others (n 11) 4. For example, disaster risk reduction is 'the common agenda' for humanitarian and development agencies, even though at times it is 'not fully developed or integrated into either humanitarian action or development strategies'.

[18] François Gemenne, 'Environmental Changes and Migration Flows: Normative Frameworks and Policy Responses' (PhD thesis, Institut d'Etudes Politiques de Paris and University of Liège, 2009) 231.

[19] Barnett and Webber (n 6) 34 referring to UNHCR, *The State of the World's Refugees: Human Displacement in the New Millennium* (Oxford University Press, 2006) Ch 4.

mitigation and adaptation strategies, and precisely what its mandate would look like, remains unclear.[20] Presumably such an agency would still require the input and cooperation of other expert institutions in implementing policy, which might suggest that its role more appropriately would be in identifying risks and formulating possible responses, rather than operational. A multi-institutional approach would better respond to the many different types of movement that are encapsulated within the umbrella term of 'climate change-related movement', and particular agencies could be appointed as cluster leaders for particular types of mobility. Even if this were subsequently to transform into a new agency, it would necessarily require this broad range of expertise.

An inter-agency response with a central UN focal point or coordinator would be the more pragmatic, politically palatable, and resource-efficient approach. As needed, the additional expertise of inter-governmental agencies such as the IOM could be tasked with specific functions, subject always to appropriate supervision of the human rights implications of their activities. Much as the IDMC serves to monitor conflict-induced displacement worldwide, collating data and then advocating for durable solutions, such an agency could be tasked with identifying areas at risk and devising strategies about adaption and migration options—in consultation with local communities and receiving States.

In 2008, at a meeting of UN agencies and international and local NGOs in Fiji about climate change and human security in the Pacific, there was a strong sense that the cross-cutting and global nature of climate change impacts on human rights and human security required a multi-sectoral response.[21] One suggestion was that an agency structure like that of UNAIDS might be appropriate—namely, one that made full use of UN expertise across its associated organizations to create a coordinated global response. UNAIDS is guided by a Programme Coordinating Board, which comprises representatives from 22 governments, five NGOs, and 10 'cosponsors': UNHCR, UNICEF, WFP, UNDP, UNFPA, the UN Office on Drugs and Crime, ILO, UNESCO, WHO, and the World Bank, each of which has responsibility for at least one technical area (such as UNHCR's role in respect of HIV/AIDS infection among displaced populations). However, while the importance of strong institutional guidance and leadership was recognized, it was also considered essential to provide local communities with access to information, and

[20] For example, in terms of ascertaining the likelihood of migration as a spontaneous response, it is important to consider 'the socio-cultural-political-economic environment that communities exist in; the cognitive processes of the people experiencing the impact of climate change; the individual, household and community attitudes to migration and migration outcomes; and the type of climate stimulus that migration may be responding to': Dominic Kniveton and others, 'Climate Change and Migration: Improving Methodologies to Estimate Flows' (2008) IOM Migration Research Series No 33, 57. Understanding of this kind cannot come from a single discipline, but rather requires true interdisciplinarity.

[21] UN High Commissioner for Human Rights Regional Office for the Pacific, 'Roundtable Discussion on Climate Change and Human Security in the Pacific' (Suva, Fiji, 16 September 2008) (which the author attended).

to learn from local knowledge in order to implement adaptation strategies on the ground to empower local communities.[22]

Arguably, this is a role now assumed by the IASC Task Force on Climate Change, albeit in a less formal way. The IASC was created in June 1992 in response to a UN General Assembly resolution on the strengthening of humanitarian assistance,[23] and it is the primary mechanism for inter-agency coordination of humanitarian assistance.[24] It comprises representatives from the main UN humanitarian agencies, as well as standing invitees who include the Special Rapporteur on the Human Rights of Internally Displaced Persons (IDPs), the ICRC, Care International, the World Bank, the International Council of Voluntary Agencies, the UN High Commissioner for Human Rights, IOM, and InterAction.[25]

The IASC Task Force on Climate Change was established in June 2008 to raise awareness of, advocate for, and integrate climate change into various humanitarian agency programmes, and to encourage increased inter-agency analysis and cooperation.[26] Operationally, this means:

- Enhancing and scaling-up efforts to build the capacities of national governments (which bear primary responsibility for protecting their citizens), civil society and communities themselves to reduce disaster risk and to prepare for effective response to disasters.
- Increasing efforts to integrate climate risk information (eg forecasts) into programming.
- Forging closer partnerships between humanitarian and development actors, knowledge centres and governmental institutions to pursue common adaptation objectives.
- Advocating for greater disaster risk reduction (DRR) investments as a first line of defence in climate change adaptation.[27]

Additionally, some interesting coalitions have emerged between organizations whose interests are not traditionally linked. For example, in April 2008, the Climate Change, Environment and Migration Alliance (CCEMA) was formed as 'an informal framework for a global multi-stakeholder partnership on climate change, environment and migration',[28] comprised of IOM, Munich Re Foundation, Stockholm Environment Institute, UNEP, OCHA, the UN University Institute for Environment and Human Security, the University of Sussex Development Research Centre on Migration, Globalisation and Poverty, and the World Wildlife Fund.[29]

[22] See further Maryanne Loughry and Jane McAdam, 'Kiribati—Relocation and Adaptation' (2008) 31 Forced Migration Review 51, 51.

[23] UNGA res 46/182 (19 December 1991).

[24] UNGA res 48/57 (14 December 1993).

[25] IASC, 'Principals' (IASC, 2009) <http://www.humanitarianinfo.org/iasc/pageloader.aspx?page=content-princip-default&mainbodyid=3> accessed 18 May 2011.

[26] See, eg, IASC Task Force on Climate Change (n 10) 1.

[27] IASC, 'Quick Guide to Climate Change Adaptation' (IASC Task Force on Climate Change, 2010) 2.

[28] See 'About CCEMA' <http://www.ccema-portal.org/> accessed 1 July 2011.

[29] See ibid.

While this network has no separate international legal identity or mandate from governments, it provides a useful space for sharing ideas and developing common policies by 'mainstream[ing] climate change considerations into…migration management policies and practice'.[30]

To be comprehensive, any international institutional focal point will need to: (a) address the prevention of the causes of displacement; (b) coordinate multilateral humanitarian and emergency relief efforts to assist the displaced; (c) develop longer-term planning for relocation from slow-onset processes; (d) address the needs of those in post-emergency, return, and relocation phases; and (e) have a specific human rights protection-orientation, in addition to relief and assistance (like UNHCR).

III. Institutional Responses

Even though governance gaps arise from policy and institutional 'silos' at the national and international levels,[31] this does not mean that individual agencies cannot play an important role. If their processes are complementary and feed into each other, then the actions of individual bodies may help to shape and reinforce actions by the whole. For example, involving UNHCR is not intended to supplant *in situ* adaptation, but rather to provide assistance where that fails.[32]

Institutional interest in the relationship between climate change and migration has increased considerably in the past decade. Whereas climate change was predominantly conceived of as a scientific and environmental issue during the 1980s and 1990s, by the early 2000s the social and humanitarian consequences of climate change began to be more readily identified.[33] For example, the IFRC created a climate change centre in 2002 to 'better understand and address the risks of climate change, in particular in the context of disaster risk reduction, disaster management and health and care programs, with a focus on the most vulnerable people'.[34] Since then, the issue has gained further momentum, with an explosion of literature and increasing institutional and NGO engagement in the issue since the mid-2000s.[35]

[30] ibid.

[31] See Warner (n 2) 5–6. For a clear overview of governance gaps, see her Tables 1a and 1b (rapid-onset events); Table 2 (slow-onset events).

[32] For examples of what various agencies around the world are doing to implement climate change adaptation planning into their work, see WFP and others (n 11) 11.

[33] Nina Hall, 'Climate Change and Organizational Change in UNHCR' in Michelle Leighton, Xiomeng Shen and Koko Warner (eds), *Climate Change and Migration: Rethinking Policies for Adaptation and Disaster Risk Reduction* (UN University, 2011) citing telephone interview with IFRC staff member (16 April 2010).

[34] ibid, referring to the IFRC Climate Centre <http://www.climatecentre.org/site/about-us> accessed 10 May 2010. On these issues, see Ch 9.

[35] IOM and the Norwegian Red Cross have been particularly active, for example. See also recent reports by the UN Development Programme (UNDP), UN Environment Programme (UNEP),

In terms of the links between climate change, human rights, and displacement, a number of international institutions have initiated studies and adopted resolutions that are slowly beginning to shape the normative framework. At the very least, they demonstrate States' acknowledgement of the impact of climate change on human rights and settlements, and assist in identifying the relevant legal principles which need to underpin solutions. The following section provides an overview of some of this diverse activity.

A. UN Commission on Human Rights

In 2004, the UN Commission on Human Rights decided 'urgently to call upon the Sub-Commission on the Promotion and Protection of Human Rights to prepare a report on the legal implications of the disappearance of States for environmental reasons, including the implications for the human rights of their residents, with particular reference to the rights of indigenous people'.[36] Françoise Hampson, a member of the Sub-Commission, wrote two comprehensive working papers on the subject,[37] and the Sub-Commission appointed her as the Special Rapporteur on the issue to prepare 'a comprehensive study on the legal implications of the disappearance of States and other territories for environmental reasons, including the implications for the human rights of their residents, with particular reference to the rights of indigenous peoples',[38] and to send a questionnaire to affected States.[39] However, her appointment required the endorsement of the Commission on Human Rights and this did not occur before it was disbanded in 2006. When it was superseded that same year by the Human Rights Council, this working topic was not continued.[40]

OCHA, Oxfam, Christian Aid, World Vision, and so on, where climate change has become a focus.

[36] Decision 2004/122, UNCHR, 60th Session, UN Doc E/CN.4/2004/L.49 (2004). See also Sub-Commission on the Promotion and Protection of Human Rights, 'Prevention of Discrimination: Prevention of Discrimination and Protection of Indigenous Peoples: Report of the Working Group on Indigenous Populations on its Twenty-First Session', UN Doc E/CN.4/Sub.2/2003/22 (11 August 2003).

[37] Sub-Commission on the Promotion and Protection of Human Rights, 'The Human Rights Situation of Indigenous Peoples in States and Territories Threatened with Extinction for Environmental Reasons: Working Paper prepared by Mrs Françoise Hampson, Member of the Working Group on Indigenous Populations', UN Doc E/CN.4/Sub.2/AC.4/2004/CRP.1 (13 July 2005); Sub-Commission on the Promotion and Protection of Human Rights, 'Prevention of Discrimination: Prevention of Discrimination and Protection of Indigenous Peoples: Expanded Working Paper by Françoise Hampson on the Human Rights Situation of Indigenous Peoples in States and Other Territories Threatened with Extinction for Environmental Reasons', UN Doc E/CN.4/Sub.2/2005/28 (16 June 2005).

[38] Res 2005/20 (10 August 2005) para 2 (referred to the Commission on Human Rights—no decision was taken).

[39] For questionnaire, see UN Doc E/CN.4/Sub.2/AC.4/2006/CRP.2 (30 June 2006).

[40] See, eg, res 2006/16 (24 August 2006), seeking approval from the Human Rights Council for the appointment of a Special Rapporteur. See further UN Press Release, 'Human Rights Council Advisory Committee Discusses Studies prepared by the Sub-Commission on the Promotion and Protection of Human Rights' (13 August 2008) <https://hrea.org/lists2/display.php?language_id=1&id=9397> accessed 12 May 2011; 'Human Rights Council Working Groups' (OHCHR,

B. UN Human Rights Council and the Office of the High Commissioner for Human Rights

In December 2007, small island developing States adopted the Male Declaration on the Human Dimension of Global Climate Change.[41] It was the first international agreement to state expressly that 'climate change has clear and immediate implications for the full enjoyment of human rights',[42] and to suggest a process by which UN institutions could examine that relationship. It successfully called on the UN Human Rights Council to hold a dedicated debate on the relationship; for the OHCHR to conduct a detailed study on it; and for the process to feed into the UNFCCC negotiation process.[43]

In March 2008, the relationship between human rights and climate change was discussed at the UN Human Rights Council, the peak UN human rights institution comprised of government representatives. This led to the adoption of a resolution on 'Human Rights and Climate Change', which acknowledged that climate change 'poses an immediate and far-reaching threat to people and communities around the world and has implications for the full enjoyment of human rights'.[44] It tasked the OHCHR with preparing an analytical study of the topic.

In January 2009, the OHCHR presented its report on the links between human rights and climate change, which included a whole section on displacement.[45] While largely descriptive, it concluded with the view that 'dealing with such possible disasters and protecting the human rights of the people affected will first and

2006–11) <http://www2.ohchr.org/english/bodies/chr/workinggroups.htm> accessed 12 May 2011. Some general studies on the impact of climate change on indigenous peoples have been undertaken: see, eg, UN Permanent Forum on Indigenous Issues, 7th Session (21 April–2 May 2008) <http://www.un.org/esa/socdev/unpfii/en/session_seventh.html> accessed 12 May 2011.

[41] Adopted 14 November 2007. The Male Declaration was presented at the 13th meeting of the parties to the UNFCCC (COP13) in December 2007: see Maumoon Abdul Gayoom, President of the Republic of the Maldives, 'Opening Address' (Joint High-Level Segment of the 13th Session of the Conference of the Parties of the UNFCCC and the 3rd Session of the Meeting of the Parties to the Kyoto Protocol, Nusa Dua, 12 December 2007) <http://www.maldivesmission.ch/fileadmin/Pdf/Environment/President_at_Bali_Conference_2012122007_final_.pdf> accessed 8 August 2011.

[42] Male Declaration, Preambular para 12.

[43] Male Declaration, Arts 1, 4, 5. For a list of submissions made to the UNFCCC on the links between climate change and human mobility, see IOM, *Compendium of IOM's Activities in Migration, Climate Change and the Environment* (IOM, 2009). For analysis of this process, see Marc Limon, 'Human Rights Obligations and Accountability in the Face of Climate Change' (2010) 38 Georgia Journal of International and Comparative Law 543; John H Knox, 'Linking Human Rights and Climate Change at the United Nations' (2009) 33 Harvard Environmental Law Review 477.

[44] HRC res 7/23 (28 March 2008) Preambular para 1. A number of States argued quite vehemently against drawing such direct links: see discussion in Limon (n 43); Marc Limon, 'Human Rights and Climate Change: Constructing a Case for Political Action' (2009) 33 Harvard Environmental Law Review 439.

[45] 'Report of the Office of the United Nations High Commissioner for Human Rights on the Relationship between Climate Change and Human Rights', UN Doc A/HRC/10/61 (15 January 2009) paras 55–60, 73–4. See also the submissions made by States, UN organizations, regional intergovernmental organizations, national human rights institutions, and non-governmental organizations: OHCHR, 'OHCHR Study on the Relationship between Climate Change and Human Rights: Submissions and Reference Documents Received' (2008) <http://www2.ohchr.org/english/issues/climatechange/submissions.htm> accessed 12 May 2010. For an extensive analysis of the report, see Knox (n 43).

foremost require adequate long-term political solutions, rather than new legal instruments'.[46] Importantly, it recognized the disproportionate effects of climate change, stating that these would be 'felt most acutely by those segments of the population who are already in vulnerable situations due to factors such as poverty, gender, age, minority status, and disability'.[47]

Having received the report, the issue was considered further at the Human Rights Council's sessions in March and June 2009. This led to the adoption of a second resolution.[48] What is striking is its 'definitive acceptance' of a clear relationship between climate change and specific human rights—most notably the right to life and the right to self-determination—which some States had not been willing to acknowledge in the previous resolution.[49]

C. UN General Assembly and Security Council

Later in the same year, the UN General Assembly adopted a resolution on 'Climate Change and Its Possible Security Implications'.[50] This called on all relevant UN organs to 'intensify their efforts in considering and addressing climate change, including its possible security implications', and requested the Security Council to 'submit a comprehensive report to the General Assembly . . . on the possible security implications of climate change'.[51] Whereas the draft text of the resolution noted that the 'adverse impacts of climate change' could 'trigger population relocation and threaten the territorial integrity and sovereignty of some states',[52] this detail was omitted from the final resolution, suggesting the reluctance of some States to acknowledge these links.

The Security Council had first considered the impacts of climate change on international peace and security in April 2007. The President of the Security Council and United Kingdom Foreign Secretary, Margaret Beckett, stated that climate change was not merely a matter of national security but was about 'our collective security in a fragile and increasingly interdependent world'.[53] The UN Secretary-General stated that climate-induced movement could lead to major conflict and instability.[54]

The report prepared by the Security Council in response to the General Assembly's request contains a short section on population displacement and

[46] HRC (n 45) para 60 (fn omitted).

[47] ibid, para 42 (fn omitted).

[48] HRC res 10/4 (25 March 2009).

[49] See Limon (n 43) 564 and generally for an excellent overview of the process. See also resolutions on the links between human rights and the environment: res 9/1 (24 September 2008), res 12/18 (2 October 2009) res 16/11 (24 March 2011).

[50] UNGA res 63/281 (11 June 2009).

[51] ibid, operative, para 1.

[52] Draft res (2 September 2008) Preamble (copy on file with author).

[53] 'Security Council Holds First-Ever Debate on Impact of Climate Change on Peace, Security, Hearing over 50 Speakers' (5663rd Meeting, 17 April 2007), UN Department of Public Information News and Media Division.

[54] ibid. For an analysis of the debate, see Francesco Sindico, 'Climate Change: A Security (Council) Issue?' (2007) 1 Climate Change Law Review 29.

migration.[55] While recognizing the uncertainty about numbers, the fact that most displacement will occur within countries rather than across international borders, and that rural–urban movement will place enormous pressures on urban centres, the legal analysis is weak (and at times even inaccurate). Of additional concern is the under-theorized suggestion that climate change-related migration may result in increased conflict.[56]

In July 2011, the Security Council again debated whether climate change constitutes a threat to international peace and security. The President of Nauru appealed to the Security Council to recognize that climate change poses as great a threat as nuclear proliferation or terrorism, with the potential to 'destabilize our societies and political institutions'. He also argued that '[t]erritory loss could disrupt traditional systems of land ownership and spark conflicts over this and other increasingly scarce resources', as well as lead to the loss of thousands of years of cultural heritage. Loss of territory 'would force large numbers of our citizens to relocate; first internally, then across borders'.[57] However, on account of objections by China, Russia, and India, States could not agree to even a non-binding statement about the links between climate change and international peace and security, and suggested that the issue would be more appropriately dealt with by the lower levels of the UN.[58] Accordingly, the Security Council used very weak language in noting only 'its concern that *possible* adverse effects of climate change *may*, in the *long run*, *aggravate* certain existing threats to international peace and security', but also observed that 'possible security implications of loss of territory of some States caused by sea-level-rise may arise, in particular in small low-lying island States'.[59]

While there are certainly real human security concerns relating to climate change,[60] there remains considerable controversy about the extent to which climate change can be said to *cause* conflict (especially cross-border conflict).[61] A number of recent reports have 'flipped' the security analysis away from the 'human

[55] 'Climate Change and Its Possible Security Implications: Report of the Secretary-General', UN Doc A/64/350 (11 September 2009) paras 54–63, 71–3 (small island States).

[56] ibid, paras 61, 62, 70.

[57] Margaret Besheer, 'UN: Climate Change Could Cause Instability for Some Nations' (*Voice of America*, 20 July 2011) <http://www.voanews.com/english/news/environment/UN-Climate-Change-Could-Cause-Instability-for-Some-Nations-125915354.html> accessed 21 July 2011. See also Marcus Stephen (President of the Republic of Nauru), 'On Nauru, A Sinking Feeling', *The New York Times* (New York, 18 July 2011) <http://www.nytimes.com/2011/07/19/opinion/19stephen.html?_r=1> accessed 22 July 2011.

[58] 'Climate Change Not a Threat to Peace: UN Security Council' *AM* (ABC Radio National, Australia, 21 July 2011) <http://www.abc.net.au/am/content/2011/s3274205.htm> accessed 1 August 2011.

[59] UNSC Presidential Statement 15 (2011), UN Doc S/PRST/2011/15 (emphasis added).

[60] See McAdam and Saul, 'An Insecure Climate for Human Security?' (n 16); The President of Kiribati has called climate change '"eco-terrorism": this is terrorism in the gradual form but surely as deadly': interview with President Anote Tong (Tarawa, Kiribati, 12 May 2009).

[61] See McAdam and Saul, 'Displacement with Dignity' (n 16); Betsy Hartmann, 'Rethinking Climate Refugees and Climate Conflict: Rhetoric, Reality and the Politics of Policy Discourse' (2010) 22 Journal of International Development 233; Nils Petter Gleditsch, 'Armed Conflict and the Environment: A Critique of the Literature' (1998) 35 Journal of Peace Research 381; Jon Barnett and W Neil Adger, 'Climate Change, Human Security and Violent Conflict' (2007) 26 Political Geography 639.

security' of the most vulnerable communities affected by climate change impacts, to focus instead on the security of citizens of developed States that may receive 'climate migrants'. Gemenne characterizes this as part of the 'alarmist' approach to climate change and migration: movement is perceived as a security threat 'exacerbated *and brought to the policy level* by climate change'.[62] Just as the causal link between climate change and migration is complex and multifaceted,[63] so, too, is the connection between climate change impacts and conflict. Yet, advocates of the security approach seize on the speculative forecasts of scholars like Norman Myers and reports by NGOs such as Christian Aid,[64] which are pitched at a level of simplistic generality rather than grounded in solid empirical research,[65] to generate a sense of fear and uncertainty.

For example, in March 2008, the Council of the European Union (EU) released a report on 'Climate Change and International Security', which included a section on environmental migration.[66] Although the report documented the predicted security impacts on regions directly affected by climate change, it also highlighted the potential economic, political, and social impacts of this for the EU itself. The German Advisory Council on Global Change (WBGU) suggested that an intensification of migration would make the risk of conflict 'considerable'.[67] A highly controversial 2003 Pentagon-sponsored report described the national security implications of climate change as including border management, global conflict, and economic malaise,[68] warning of a potential 'flood of refugees to southeast U.S. and Mexico from Caribbean islands' by 2012.[69] A report based on the advice of a team of retired United States (US) generals and admirals concluded that climate change creates new security challenges for the US, such as 'increased immigration', with 'the potential to disrupt our way of life and force changes in how we keep ourselves safe and secure by adding a new hostile and stressing factor into the national and international security environment'.[70] In December 2008,

62 Gemenne (n 18) 122 (emphasis added).

63 See Ch 1.

64 Christian Aid, *Human Tide: The Real Migration Crisis* (Christian Aid, 2007); see also Friends of the Earth Australia, *A Citizen's Guide to Climate Refugees* (rev edn, Friends of the Earth Australia, 2007); IFRC, *World Disasters Report: Focus on Recovery* (IFRC and Red Crescent Societies, 2001) cited in Gemenne (n 18) 124, discussed in Ch 1.

65 See William B Wood, 'Ecomigration: Linkages between Environmental Change and Migration', in Aristide R Zolberg and Peter M Benda (eds), *Global Migrants, Global Refugees: Problems and Solutions* (Berghahn Books, 2001); Camillo Boano, Roger Zetter, and Tim Morris, 'Environmentally Displaced People: Understanding the Linkages between Environmental Change, Livelihoods and Forced Migration' (2008) Forced Migration Policy Briefing 1 (Refugee Studies Centre, University of Oxford); James Morrissey, 'Environmental Change and Forced Migration: A State of the Art Review' (2009) Refugee Studies Centre, University of Oxford, Background Paper <http://www.rsc.ox.ac.uk/events/environmental-change-and-migration/EnvChangeandFmReviewWS.pdf> accessed 4 August 2011.

66 Council of the European Union, 'Climate Change and International Security', (Report from the Commission and the Secretary-General/High-Representative to the European Council) Doc 7249/08 (3 March 2008).

67 German Advisory Council on Global Change, 'World in Transition: Climate Change as a Security Risk: Summary for Policy Makers' (WBGU Secretariat, 2007) 6.

68 Peter Schwartz and Doug Randall, 'An Abrupt Climate Change Scenario and its Implications for United States National Security' (October 2003) 3.

69 ibid, 17.

70 CNA Corporation, 'National Security and the Threat of Climate Change' (2007) 44.

the Australian government appointed a National Security Adviser, part of whose mandate encompassed the security implications of climate change for Australia, including 'unregulated population movements'.[71] That the threats to security in each of these examples are directed towards apparent threats to citizens of the EU Member States, US, and Australia may encourage a politics of fear, such as that which has dominated asylum debates in the industrialized world for the past decade. Within an already overhyped security climate, this approach easily gains political traction and may lead to States adopting restrictive responses to climate change-related movement.

D. The role of UNHCR

The current UN High Commissioner for Refugees, António Guterres, has taken a keen personal interest in climate change-related displacement and has driven the institution's engagement with the issue.[72] Although UNHCR's legal mandate does not extend to climate change-related displacement per se,[73] the High Commissioner perceives UNHCR as having a 'duty to alert states to these problems and help find answers to the new challenges they represent'.[74] He has identified climate change as 'one of the main drivers of forced displacement, both directly through impact on environment—not allowing people to live any more in the areas where they were traditionally living—and as a trigger of extreme poverty and conflict'.[75] At the meeting of States at UNHCR's Executive Committee in 2007, the High Commissioner told States that: 'We see more and more people forced to move because of extreme deprivation, environmental degradation and climate change', noting that:

> natural disasters occur more frequently and are of greater magnitude and devastating impact. Almost every model of the long-term effects of climate change predicts a continued expansion of desertification, to the point of destroying livelihood prospects in many

[71] Prime Minister Kevin Rudd, 'The First National Security Statement to the Australian Parliament', *The Australian* (Australia, 4 December 2008) 26.

[72] For an overview of UNHCR's work in this area, see <http://www.unhcr.org/pages/49e4a5096.html> accessed 15 May 2011.

[73] Indeed, this remains a key obstacle to UNHCR's formal involvement. Hall's interviews reflect the disquiet among some within UNHCR about taking a stance on the issue: Hall (n 33). Note, however, ECOSOC's praise of UNHCR's activities in 'the context of man-made disasters': ECOSOC res 2011 (LXI) (2 August 1976).

[74] António Guterres, UN High Commissioner for Refugees, 'Opening Statement' (58th Session of the Executive Committee of the High Commissioner's Programme, Geneva, 1 October 2007). However, a Working Group on Solutions and Protection within the Executive Committee of the UNHCR reported as early as 1991 that there was 'a need to provide international protection to persons outside the current international legal definition of refugee [where they were] forced to leave or prevented from returning to their homes because of human-made disasters, natural or ecological disasters': in Michelle Schwartz, 'International Legal Protection for Victims of Environmental Abuse' (1993) 18 Yale Journal of International Law 355, 379.

[75] Cited in Julian Borger, 'Conflicts Fuelled by Climate Change Causing New Refugee Crisis, Warns UN', *The Guardian* (London, 17 June 2008) <http://www.guardian.co.uk/environment/2008/jun/17/climatechange.food> accessed 20 July 2008. See also 'A Critical Time for the Environment' (2002) 127 Refugees 2.

parts of the globe. And for each centimeter the sea level rises, there will be one million more displaced. The international community seems no more adept at dealing with these causes than it is at preventing conflict and persecution.[76]

In late 2008, UNHCR produced its first policy paper on this subject, entitled 'Climate Change, Natural Disasters and Human Displacement: A UNHCR Perspective', which it revised (albeit without many substantive changes) in 2009.[77] In 2009, it also issued a factsheet outlining its involvement in the issue. While it recognized that it has no authority to assist or protect displaced people who fall outside its mandate, it stated that it was 'keen to influence the necessary dialogue on new or enhanced modalities of international cooperation to develop the capacity of States to respond to the challenges of forced displacement in the context of climate change. National, regional and international legal frameworks may need to adapt to these newly arising challenges'.[78] Since then, it has become more actively engaged through networks such as the IASC;[79] in commissioning research on climate change-related movement;[80] and in raising it as a normative protection gap at the 2010 High Commissioner's Dialogue on Protection Challenges and subsequent fora.[81] Climate-change-related displacement is a distinct focal point within UNHCR's Division of International Protection in Geneva.

UNHCR's mandate for the prevention of statelessness and IDPs may also be relevant in the climate-change displacement context.[82] UNHCR itself has argued that in cases where States are threatened in the long term by rising sea levels, and when 'their populations would be likely to find themselves largely in a situation that would be similar to if not the same as if statehood had ceased',[83] its mandate

[76] Guterres (n 74).

[77] UNHCR, 'Climate Change, Natural Disasters and Human Displacement: A UNHCR Perspective' (23 October 2008, revised 14 August 2009). See analysis in Hall (n 33).

[78] UNHCR, 'UNHCR Factsheet' (October 2009).

[79] Note also the Climate Change, Environment and Migration Alliance (CCEMA), formed in April 2008, which is comprised of IOM, Munich Re Foundation, Stockholm Environment Institute, UNEP, OCHA, UN University Institute for Environment and Human Security, University of Sussex Development Research Centre on Migration, Globalisation and Poverty, and the World Wildlife Fund. It was formed as 'an informal framework for a global multi-stakeholder partnership on climate change, environment and migration': see 'About CCEMA' (n 28).

[80] Roger Zetter, 'Protecting Environmentally Displaced People: Developing the Capacity of Legal and Normative Frameworks' (2011) Refugee Studies Centre Research Report <http://www.rsc.ox.ac.uk/pdfs/Zetter-%20EnvDispRep%2015022011.pdf> accessed 8 August 2011; see also the papers commissioned for the UNHCR Expert Roundtable on Climate Change and Displacement, Bellagio, February 2011: <http://www.unhcr.org/cgi-bin/texis/vtx/search?page=&comid=4e01e63f2&keywords=Bellagio-meeting> accessed 1 July 2011.

[81] See UNHCR, 'High Commissioner's Dialogue on Protection Challenges: Breakout Session 1: Gaps in the International Protection Framework and its Implementation: Report by the Co-Chairs' (8–9 December 2010) 3.

[82] Susin Park, 'Climate Change and the Risk of Statelessness: The Situation of Low-Lying Island States', UNHCR Legal and Protection Policy Research Series, PPLA/2011/04 (May 2011); UNHCR, supported by the International Organization for Migration and the Norwegian Refugee Council, 'Climate Change and Statelessness: An Overview' (Submission to the 6th Session of the Ad Hoc Working Group on Long-Term Cooperative Action (AWG-LCA 6) under the UN Framework Convention on Climate Change (UNFCCC), 1–12 June 2009) <http://unfccc.int/resource/docs/2009/smsn/igo/048.pdf> accessed 1 August 2011.

[83] UNHCR (n 81) 2. See also UNHCR (n 77); Ch 5.

to prevent and reduce statelessness would be triggered.[84] With respect to IDPs, UNHCR's institutional responsibility is limited to those displaced by conflict.[85] This is considerably narrower than the operational definition of 'IDPs' in the Guiding Principles on Internal Displacement, which expressly encompasses people who have fled their homes due to natural or human-made disasters.

Nevertheless, in recent years, UNHCR has responded on an operational level to disasters such as the 2004 Asian Tsunami, the 2005 Sri Lankan earthquake, flooding in Kenya in 2006, Cyclone Nargis in Burma in 2008, and the 2010 Haiti earthquake.[86] This is partly because it is seen as the institution with the most relevant experience, including on account of its emergency response capacity—quickly marshalling staff and providing shelter and urgent assistance to displaced people. Although some have suggested that UNHCR's willingness to assist in the cases above was linked not only to the scale of the disaster, but 'internal motives, which related to the agency's presence in the region and strategic considerations about its future role in the UN system',[87] in each case the agency was clear to stress that its work was purely humanitarian and not formally within its protection mandate.

The UN Emergency Relief Coordinator recently asked UNHCR to take on a more formal role for the protection of persons displaced by natural disasters—initially on a pilot basis. Despite considerable in-principle support, this was ultimately rejected by UNHCR's Standing Committee on the grounds that 'outstanding questions' relating to issues of State sovereignty, mandate implications, resources and capacity, and disengagement strategies remained unresolved.[88]

[84] See, eg, UNGA res 50/152 (9 February 1996) reiterated in UNGA res 61/137 (25 January 2007), UNGA res 62/124 (24 January 2008), UNGA res 63/148 (27 January 2009). The work of UNHCR extends in some cases to situations of *de facto* statelessness, such as in trying to get 'States to cooperate in the establishment of identity and nationality status of victims of trafficking, many of whom, especially women and children, are rendered effectively stateless due to an inability to establish such status, so as to facilitate appropriate solutions to their situations, respecting the internationally recognized human rights of the victims': Executive Committee of the High Commissioner's Programme, 'Statelessness: Prevention and Reduction of Statelessness and Protection of Stateless Persons', UN Doc EC/57/SC/CRP.6 (14 February 2006) para 7. See also, UNGA res 50/152 (9 February 1996) paras 14–15; UNGA res 3274 (XXIX) (10 December 1974); UNGA res 31/36 (30 November 1976). See further, Ch 5.

[85] There is debate in the literature about the extent to which resource scarcity could itself lead to conflict. While causation is complex, it would nevertheless be a sad irony if UNHCR's mandate were triggered due to inaction, as a non-violent situation escalated to one of conflict: see, eg, Lorraine Elliott, 'Climate Migration and Climate Migrants: What Threat, Whose Security?' in Jane McAdam (ed), *Climate Change and Displacement: Multidisciplinary Perspectives* (Hart Publishing, 2010); Kurt M Campbell and others, *The Age of Consequences: The Foreign Policy and National Security Implications of Global Climate Change* (Center for Strategic and International Studies and Center for a New American Security, 2007).

[86] See UNHCR, 'Emergency Preparedness and Response' <http://www.unhcr.org/pages/49c3646cc2.html> accessed 14 October 2011; Executive Committee of the High Commissioner's Programme, 'UNHCR's Role in Support of an Enhanced Humanitarian Response for the Protection of Persons affected by Natural Disasters', UN Doc EC/62/SC/CRP.19 (6 June 2011) para 4.

[87] Gemenne (n 18) 239.

[88] Executive Committee of the High Commissioner's Programme, 'Draft Report of the 51st Meeting of the Standing Committee (21–23 June 2011)', UN Doc EC/62/SC/CRP.25 (16 September 2011) para 32. For the background to the proposal, see Executive Committee (n 86). The conditions would be that there were relevant expertise, sufficient resources, and States asked for assistance.

Ultimately, greater engagement by UNHCR with natural disasters may be an indirect way in which it comes to deal with climate-related displacement, since disasters are predicted to increase in frequency and severity as a result of climate change. Thus, while responses would not be predicated on the climate change element per se, assistance and protection would be provided. Although a number of States have welcomed UNHCR's engagement in this area,[89] some have clearly limited their support to an increased *operational* presence rather than a formal extension of UNHCR's protection mandate.[90]

The High Commissioner has sought to position UNHCR's work within a broader framework of human movement, noting that the effectiveness of its existing mandate of protecting, assisting, and finding solutions for refugees and reducing statelessness is dependant on its 'ability to understand the broader patterns of people on the move in today's world'.[91] Certainly, UNHCR is uniquely placed to address the *protection* dimension of movement, and to assist the international debate through its expertise on forced migration and the nature of population movement. However, as explained throughout this book, not all climate change-related movement will amount to displacement requiring an international protection response, and approaching it from this perspective alone will be inadequate for addressing the needs of all affected groups. In particular, UNHCR may not be well-equipped to deal with longer-term movement away from slow-onset climate change impacts, although its experience relating to refugee resettlement may be instructive. As Warner notes, humanitarian organizations like UNHCR tend to focus on crisis and disaster management,[92] 'often with a short-term perspective and not with the goal (or capacity) to maintain long-term guidance, support and protection'.[93]

Finally, from a purely pragmatic perspective, UNHCR is already responsible for over 20 million refugees and other people of concern (including asylum seekers, returnees, stateless persons, and IDPs). Each year it relies on donations and the goodwill of States to provide it with funds to carry out its work in over 100 countries, and it has experienced significant budgetary crises over the years. It would require considerable additional resources to take on a formal role in relation to climate change-related displacement. As the government of France has

[89] Executive Committee of UNHCR, 'Summary Record of the 629th Meeting (Geneva, 28 September 2009)', UN Doc A/AC.96/SR.629 (12 October 2009) para 30 (Denmark); Executive Committee of UNHCR, 'Summary Record of the 633rd Meeting (Geneva, 30 September 2009)', UN Doc A/AC.96/SR.633 (12 October 2009) para 45 (Australia); Executive Committee of UNHCR, 'Summary Record of the 610th Meeting (Geneva, 1 October 2007)', UN Doc A/AC.96/SR.610 (19 October 2007) para 33 (Norway).

[90] eg Executive Committee of UNHCR, 'Summary Record of the 618th Meeting (Geneva, 6 October 2008)', UN Doc A/AC.96/SR.618 (7 November 2008) para 53 (Switzerland).

[91] Guterres (n 74).

[92] Arguably, UNHCR's role in relation to protracted refugee situations or long-term statelessness does not fit this model.

[93] Warner (n 2) 6.

observed, 'the impact of climate change challenge[s] UNHCR to adapt its mode of operation'.[94]

E. The role of IOM

In the context of climate change-related migration, IOM might also be regarded as a contender for a leading institutional role. It has been active since the early 1990s on the issue of environmental migration,[95] and more recently it has produced a number of significant research reports on the effects of climate-related movement, sponsored conferences and workshops, and run its own high-level intersessional dialogues on the topic.[96] In the past decade, IOM has received funding for more than 500 projects in this area.[97]

Unlike UNHCR, IOM is not part of the UN system and does not have a protection mandate. It does not oversee a treaty regime and has little normative vision of its own. Betts argues that it exists primarily 'as a service provider to individual states that pay for its services' in migration management.[98] While it certainly has relevant experience to contribute on migration, particularly through collaborative processes like the IASC, these other limitations make it an inappropriate principal body to oversee climate-related movement.[99]

However, IOM's recognition that the interactions between environment and mobility are much more intricate than displacement in the face of disaster, and that migration in the context of climate change can be a solution rather than a problem,[100] is vital in terms of *planning* for movement, rather than simply responding to it remedially. In conjunction with human rights and development agencies, its attention would be well focused on these aspects.

[94] Executive Committee of UNHCR, 'Summary Record of the 629th Meeting' (n 89) para 2 (France).

[95] See, eg, IOM, *Migration and the Environment* (IOM, 1992), which was the outcome of a conference on the subject. For IOM's other publications in this area, see <http://www.iom.int/jahia/webdav/shared/shared/mainsite/activities/env_degradation/flyer_env_series.pdf> accessed 15 May 2011.

[96] See generally, <http://www.iom.int/jahia/Jahia/activities/by-theme/migration-climate-change-environmental-degradation> accessed 15 May 2011. Its report on climate change-related movement in Bangladesh is a particularly important contribution to the field: IOM, *Assessing the Evidence: Environment, Climate Change and Migration in Bangladesh* (IOM, 2010).

[97] Laura Thompson, Deputy Director General of IOM, 'Opening Remarks' (delivered by Peter Schatzer) (IOM International Dialogue on Migration, Intersessional Workshop on Climate Change, Environmental Degradation and Migration, Geneva, 29 March 2011) 1 <http://www.iom.int/jahia/webdav/shared/shared/mainsite/microsites/IDM/workshops/climate-change-2011/IDM-Opening-Remarks-DDG.pdf> accessed 15 May 2011.

[98] Alexander Betts, 'Introduction: Global Migration Governance' in Alexander Betts (ed), *Global Migration Governance* (Oxford University Press, 2011) 8.

[99] For this reason, Gemenne's appraisal of IOM seems overly optimistic: (n 18) 241–3.

[100] Thompson (n 97) 3.

F. The role of the UNFCCC

The links between climate change, human rights, and displacement have also been taken up in the UNFCCC negotiations.[101] While the climate change regime provides a high-profile 'hook' for consideration of the protection and assistance concerns that arise from migration/displacement, it is not the appropriate forum to discuss the complexity of these issues in a structured or comprehensive way. Further, it does not have the appropriate expertise or operational capacity to oversee the issue of movement itself. However, the UNFCCC forum is a very important one for acknowledging the need for responsibility-sharing and financing of migration and other mobility strategies as a form of adaptation.

The first mention of migration in the UNFCCC context was in the assembly text of December 2008 at Poznan (COP14).[102] This was the precursor to the negotiating text drafted by the Ad Hoc Working Group on Long-Term Cooperative Action under the Convention, a group established pursuant to the 2007 Bali Action Plan to enhance action on adaptation. The text was revised in June 2009 and again in November 2009 just prior to the Copenhagen climate change conference (COP15). The final draft text considered at Copenhagen included two sections referring to human movement on which agreement had previously been unachievable. These called upon States to implement as part of their adaptation measures:

> [a]ctivities related to national, regional and international migration and displacement or planned relocation of persons affected by climate change, while acknowledging the need to identify modalities of inter-state cooperation to respond to the needs of affected populations who either cross an international frontier as a result of, or find themselves abroad and are unable to return owing to, the effects of climate change.[103]

They also called on States to 'jointly undertake action under the Convention to enhance adaptation at the international level', including through

> [a]ctivities related to migration and displacement or planned relocation of persons affected by climate change, while acknowledging the need to identify modalities of interstate cooperation to respond to the needs of affected populations who either cross an international frontier as a result of, or find themselves abroad and are unable to return owing to, the effects of climate change.[104]

[101] For a comprehensive overview, see Koko Warner, 'Climate Change Induced Displacement: Adaptation Policy in the Context of the UNFCCC Climate Negotiations' (2011) UNHCR Legal and Protection Policy Research Series, PPLA/2011/02.

[102] See UNFCCC Ad Hoc Working Group on Long-Term Cooperative Action under the Convention, 'Ideas and Proposals on Paragraph 1 of the Bali Action Plan: Revised Note by the Chair' (Fourth Session, Poznan, 1–10 December 2008) UN Doc FCCC/AWGLCA/2008/16/Rev.1 (15 January 2009) para 63(g) on measuring, verifying, and reporting of emissions reductions; paras 112(f), 112(h), 187(g) <http://unfccc.int/resource/docs/2008/awglca4/eng/16r01.pdf> accessed 1 July 2011. For a timeline of the process from 2007 to 2010, see Warner (n 101) 4.

[103] See Negotiating Text, UN Doc FCCC/AWGLCA/2009/14 (20 November 2009) 38, para 12(c) <http://maindb.unfccc.int/library/view_pdf.pl?url=http://unfccc.int/resource/docs/2009/awglca7/eng/14.pdf> accessed 14 December 2009.

[104] ibid, 56, para 13(b).

Although it was reported that there was 'no opposition from all the countries' to these provisions and they were 'most likely to go through',[105] it became clear that the draft negotiating text generally was not in a sufficiently finalized state to be discussed by the heads of State assembled in Copenhagen. Accordingly, the Copenhagen Accord[106]—the short, non-legally binding document adopted by States at Copenhagen—does not include any reference to migration.

However, the following wording was included in what was later accepted as the outcome text from the Copenhagen meeting, discussed in subsequent multilateral negotiations.[107] It invited States parties:

> to enhance adaptation action under the Copenhagen Adaptation Framework [for Implementation] taking into account their common but differentiated responsibilities and respective capabilities, and specific national and regional development priorities, objectives and circumstances, [and whereby developing country Parties shall be supported by developed country Parties and in accordance with paragraph 6 below], to undertake, inter alia:
>
> . . .
>
> (f) Measures to enhance understanding, coordination and cooperation related to national, regional and international climate change induced displacement, migration and planned relocation, where appropriate.

At the Tianjin climate change meeting in October 2010, paragraph 4(f) was slightly reworded to read: 'Measures to enhance understanding, coordination and cooperation with regard to climate change induced displacement, migration and planned relocation, where appropriate, at national, regional and international levels'.[108] As Warner notes: 'The new wording was important because it signalled to decision makers that migration and displacement have different forms and will require different types of policy approaches. The new order of wording also

[105] Bruno Sekoli, Chair of the Least Developed Countries (LDCs) group at Copenhagen cited in 'Good News for Climate Change Migrants' (*IRIN*, 11 December 2009) <http://www.irinnews.org/Report.aspx?ReportId=87405> accessed 30 December 2010. Reportedly, Australia was opposed to such references, the official reason being that it believed there was a need for further research on the links between climate change and movement: Lawton (n 106). Australia appeared wary of any text that might suggest it had legal obligations to assist people who might be displaced in the Asia-Pacific region. This is despite the fact that the total combined population of Kiribati and Tuvalu is around 110,000, which is less than Australia's annual immigration intake. As one government official in Kiribati observed: 'It's very small.... But for us it's very big. But for you guys it's very small. It's assistance you [would be] providing that is minimal, but yet monumental to us': interview with Tebao Awerika, Deputy Secretary, Ministry of Foreign Affairs (Tarawa, Kiribati, 12 May 2009).

[106] Copenhagen Accord (adopted 18 December 2009) <http://unfccc.int/files/meetings/cop_15/application/pdf/cop15_cph_auv.pdf> accessed 1 August 2011; Christopher Lawton, 'What about Climate Refugees? Efforts to Help the Displaced Bog Down in Copenhagen', *Spiegel Online* (17 December 2009) <http://www.spiegel.de/international/europe/0,1518,druck-667256,00.html> accessed 19 January 2010.

[107] UNFCCC, 'Work undertaken by the Conference of the Parties at Its Fifteenth Session on the Basis of the Report of the Ad Hoc Working Group on Long-Term Cooperative Action under the Convention', UN Doc FCCC/CP/2010/2 (11 February 2010) 14, para 4(f) <http://unfccc.int/resource/docs/2010/cop16/eng/02.pdf> accessed 1 July 2011; Negotiating Text, UN Doc FCCC/AWGLCA/2010/6 (17 May 2010) 17, para 4(f) <http://maindb.unfccc.int/library/view_pdf.pl?url=http://unfccc.int/resource/docs/2010/awglca10/eng/06.pdf> accessed 30 June 2010.

[108] Negotiating Text, UN Doc FCCC/AWGLCA/2010/14 (13 August 2010).

clarified that coordination and cooperation related to the topic could occur at national, regional, and international levels'.[109]

At the climate change conference in Cancún in December 2010 (COP16), this language was adopted as paragraph 14(f) of the 2010 Cancún Adaptation Framework. That provision 'invites' States to:

> enhance action on adaptation...taking into account their common but differentiated responsibilities and respective capabilities, and specific national and regional development priorities, objectives and circumstances, by undertaking, inter alia:
>
> ...
>
> (f) Measures to enhance understanding, coordination and cooperation with regard to climate change induced displacement, migration and planned relocation, where appropriate, at national, regional and international levels.[110]

Although this is phrased in non-mandatory language and thus does not *require* States to implement migration programmes or 'protect' people displaced by climate change, it provides an important recognition of the impacts of climate change on human movement, and the need for States to address this. Significantly, it recognizes migration as a form of adaptation, and this means that international adaptation funding may be directed towards preventing displacement and developing relocation and migration schemes.[111]

By contrast to its first iteration in December 2008, the provision reflects a more nuanced understanding of climate change-related movement,

> where different kinds of measures (research, coordination, cooperation), different types of mobility (displacement, migration, planned relocation), and levels (national, regional, international) of mobility were articulated. The very inclusion of a full sub-paragraph devoted to migration and displacement highlighted the importance for action, but the framing of the sub-paragraph gave Parties many sensible options for beginning to think about (and undertake activities to address) the issue.[112]

Although the Cancún Adaptation Framework is not legally binding, it has operational significance. There is anecdotal evidence that some governments are responding favourably to the idea of funding further work on 'migration as adaptation' as their contribution to this issue. The Framework also provides a space in which to deal with risk management in the climate change context, which is relevant to: (a) preventing displacement and having adequate response mechanisms in place when it occurs; (b) dealing with compensation for mitigation wrongs; (c) appealing to

[109] Warner (n 101) 12.

[110] See 'Draft Decision -/CP.16: Outcome of the Work of the Ad Hoc Working Group on Long-Term Cooperative Action under the Convention' (2010) <http://unfccc.int/files/meetings/cop_16/application/pdf/cop16_lca.pdf#page=3> accessed 15 March 2011.

[111] Current funding mechanisms include the Global Environment Facility (GEF) Trust Fund; the Least Developed Countries Fund (LDCF); the Special Climate Change Fund (SCCF); the Adaptation Fund (under the Kyoto Protocol); and a new Green Climate Fund (established at Cancún).

[112] Warner (n 101) 12.

ethical considerations around States' common but differentiated responsibilities; and (d) encouraging 'cross-ministry policy approaches'.[113] Warner notes that although the UNFCCC 'will not directly implement the array of issues mentioned under the Cancún Adaptation Framework', it will have a 'catalytic role'.[114]

IV. Regional Responses

There has been considerable activity at the regional level to address climate change-related movement, and this section highlights a number of examples. At the Pacific Islands Forum in 2008, Pacific leaders (including those of Australia and New Zealand) adopted the Niue Declaration on Climate Change. Relevantly, it recognized 'the importance of retaining the Pacific's social and cultural identity, and the desire of Pacific peoples to continue to live in their own countries, where possible'.[115] It also committed the Forum members 'to continue to advocate and support the recognition, in all international fora, of the urgent social, economic and security threats caused by the adverse impacts of climate change and sea level rise to our territorial integrity and continued existence as viable dynamic communities'.[116]

The non-binding, aspirational Anchorage Declaration of April 2009, adopted at the Indigenous Peoples' Global Summit on Climate Change in Alaska, called on countries to 'recognize, respect and implement the fundamental human rights of Indigenous Peoples', including ensuring that 'Indigenous Peoples have the right to mobility and are not forcibly removed or settled away from their traditional lands and territories'. It also noted that '[i]n the case of climate change migrants, appropriate programs and measures must address their rights, status, conditions, and vulnerabilities'.[117]

At a high-level climate change conference held in Kiribati in November 2010, leaders from 12 countries (Australia, Brazil, China, Cuba, Fiji, Japan, Kiribati, Maldives, Marshall Islands, New Zealand, Solomon Islands, and Tonga) adopted the Ambo Declaration. In that non-binding instrument, they expressed their '*[a]larm* at the impacts of the climate change crisis already being felt... especially the immediate threat to the livelihood and survival of the most vulnerable States',[118] and their support for 'consideration of the development and implementation of strategies and actions directed at protecting people displaced within or across borders as a result of adverse effects arising from climate change extreme events'.[119]

The Council of Europe Parliamentary Assembly's Committee on Migration, Refugees and Population compiled a report on environmentally induced migration and displacement in 2008. It suggested 'adding an additional protocol to the

113 ibid, 13.

114 ibid, 17. Interestingly, the Intergovernmental Panel on Climate Change (IPCC) is planning a sub-chapter on climate change and mobility in its fifth assessment report (2014), as well as addressing it in other chapters as a cross-cutting issue: Warner (n 101) 3.

115 Niue Declaration on Climate Change, 39th Pacific Islands Forum, Forum Communiqué, Annex B (19–20 August 2008) Preamble.

116 ibid, operative part.

117 Anchorage Declaration of the Indigenous Peoples' Global Summit on Climate Change (April 2009) para 11.

118 Ambo Declaration (November 2010) Art 1.

119 ibid, Art 15.

European Convention on Human Rights, concerning the right to a healthy and safe environment' as a way of 'enhancing the human rights protection mechanisms vis-à-vis the challenges of climate change and environmental degradation processes'.[120] So far, that proposal has not been pursued.

A number of international expert fora have noted that certain existing regional mechanisms may 'serve as vehicles for the design and implementation of programmes for the assistance and protection of persons whose livelihoods are affected by climate processes'.[121] They cite free movement agreements, such as those of the Economic Union of the Organisation of Eastern Caribbean States, the Caribbean Single Market and Economy, and the Economic, and Community of West African States (ECOWAS), which can strengthen sub-regional economic integration through increasingly freer movement of goods, capital, and people, and in turn promote the assimilation of migrants into receiving States.[122] As Türk notes, 'there is no reason to suppose their benefits should be denied to ECOWAS citizens crossing borders owing to environmental reasons'.[123]

It is important to ensure that there is an on-going dialogue between international, regional, and national organizations, so that normative frameworks speak to, and are informed by, particularized local needs. As UNHCR has observed,

> the international community does not always adhere to the principle of subsidiarity—whereby larger multilateral institutions do not take on tasks that can be adequately performed by local or regional organizations—during the initial stages of the humanitarian effort. International humanitarian organizations are expected to meet basic needs when governments cannot—or will not for political reasons. This calls for efforts to strengthen the preparedness of regional and sub-regional organizations, which can also operate as part of an effective early warning system.[124]

To facilitate this, the UNHCR-hosted Bellagio expert meeting on climate change and displacement recommended that international organizations:

- assist regional and sub-regional organizations to implement regional approaches, including the development of relevant expertise, the design and implementation of early warning systems, vulnerability assessments and adaptation strategies;
- coordinate regional efforts at the political and technical levels to build adaptive capacity within countries in the first instance;

[120] Council of Europe Parliamentary Assembly, Committee on Migration, Refugees and Population, 'Environmentally Induced Migration and Displacement: A 21st Century Challenge', COE Doc 11785 (23 December 2008) paras 6.3 and 121 respectively.

[121] 'Summary of Deliberations on Climate Change and Displacement' (UNHCR Expert Roundtable on Climate Change and Displacement, Bellagio, February 2011) (April 2011) ('Bellagio Deliberations') para 36; 'Chairperson's Summary' (Nansen Conference on Climate Change and Displacement in the 21st Century, Oslo, 6–7 June 2011) para 15.

[122] ibid.

[123] Volker Türk, Director, Division of International Protection, UNHCR, 'Can Protection of Environmentally Displaced Persons Be Found in Existing Protection Regimes? What Are the Next Steps from a Protection Perspective?' (Nansen Conference on Climate Change and Displacement in the 21st Century, Oslo, 7 June 2011) 5.

[124] UNHCR (n 19) 92 (fns omitted).

- assist countries in developing bilateral agreements that would ensure appropriate safeguards are in place for those individuals/communities that choose to migrate either in anticipation of, or due to, the manifestations of slow-onset climate related disasters; and
- support the design and implementation of migration programmes within and outside affected countries or regions.[125]

V. Conclusion

The dynamic and non-linear process of climate change-related mobility creates both opportunities and challenges for governance and institutional capacity.[126] The question is how to create policy packages that address the needs of those who move *and* those who remain, including livelihood issues, food security, social networks, and the maintenance of culture and community. Most authors agree that an early assessment and strengthening of a community's adaptive capacity is crucial. This requires educating communities about the situation, including providing access to information.

Recent deliberations at high-level international meetings suggest there is now an emerging consensus on the following issues:

- climate change-related migration is a multicausal phenomenon;
- climate-related displacement is likely to take different forms, and will require a variety of responses at the local, national, regional, and international levels;
- there is a need for further empirical studies on climate-related movement;
- migration is a rational adaptation strategy to climate-change processes, and should be supported as such;
- most people do not want to leave their homes, and sometimes the most vulnerable will not be able to move;
- there is a need to strengthen legal, policy, institutional, and administrative frameworks;
- planning and policy must be underpinned by a strong scientific and empirical basis;
- policy needs to be proactive, not just remedial;
- there is a need to strengthen operational and technical capacities, including through further funding;
- there must be sufficient budgetary support for long-term planning;
- affected populations should be included in decision-making through participatory processes;

[125] See Bellagio Deliberations (n 121) para 38.
[126] Warner (n 2) 8.

- comprehensive approaches are needed—for example, migration management should be linked with other policy objectives, including climate change adaptation, disaster risk reduction, humanitarian responses, and sustainable development;
- multi-stakeholder partnerships, involving public and private service actors, need to be developed; and
- all responses should be underpinned by basic human rights principles.[127]

As the developments discussed in this chapter indicate, most institutions are still grappling with the problem at a preliminary level and have yet to articulate clear policy responses. It is essential that multi-agency strategies are developed which complement and support each other. A failure to coordinate international responses to the problem may well result in ad hoc humanitarian assistance becoming the default response of the international community. This is unlikely to be adequate in terms of its scale, timeliness, durability, or comprehensiveness. It also risks imposing inequitable obligations on particular States, and may mean that the most vulnerable are not sufficiently well protected.

Accordingly, the need remains to develop an overarching framework for managing new flows of displaced persons, based on burden-sharing principles, a human-rights oriented approach, and the allocation of well-defined institutional responsibilities for humanitarian relief, legal protection, and the provision of permanent solutions. This is examined in the next chapter. In terms of rapid-onset disasters, there is now considerable recognition of the relevance of disaster risk reduction strategies and management mechanisms to the climate change context.[128] However, rational, planned migration in anticipation of the impacts of slower-onset processes, such as drought or sea-level rise, is not within the purview of the disaster mechanisms, which is why alternative legal pathways and institutional paradigms are also needed. It seems that the greatest challenge lies in creating institutional responsibility and governance arrangements for those whose movement is not triggered by a disaster per se, but rather by a disaster in slow motion: the slow-onset impacts of climate change over time.

[127] See, eg, Bellagio Deliberations (n 121); Thompson (n 97); Nansen Principles (appended to Chairperson's Summary) (n 121); Niue Declaration on Climate Change, Preamble. See also Fabrice Renaud and others, 'Control, Adapt or Flee: How to Face Environmental Migration?' (2007) 5 InterSecTions 33–5; Oli Brown, 'Migration and Climate Change' (2008) IOM Migration Research Series No 31; IOM, *World Migration Report 2010: The Future of Migration: Building Capacities for Change* (IOM, 2010) 7 which defines 'capacity-building' as '[s]trengthening knowledge, abilities, skills, resources, structures and processes needed to achieve goals effectively and sustainably and to adapt to change'.

[128] See, eg, IASC, 'Disaster Risk Reduction Strategies and Risk Management Practices: Critical Elements for Adaptation to Climate Change' (Submission to the UNFCCC Ad Hoc Working Group on Long Term Cooperative Action by the Informal Taskforce on Climate Change of the IASC and the International Strategy for Disaster Reduction, 11 November 2008) 7–8.

9

Overarching Normative Principles

Real and irresistible distress must be at all times a sufficient passport for human beings under any such application of human laws.[1]

I. Introduction

International protection frameworks, underscored by refugee and human rights law, provide important benchmarks for assessing needs and responses in other displacement contexts. They offer an existing body of rules and principles to guide and inform policymaking, with identifiable rights-bearers and duty-bearers. Though the scope for further 'activating'[2] human rights law may be limited in the climate change displacement context, at least at this point in time,[3] its normative principles provide a universal framework for guiding policy development, highlighting issues that might be obscured by a purely environmental or economic analysis, and helping to articulate claims about access, adaptation, participation, and balance.

As Chapters 3 and 4 canvassed, protection options under existing legal frameworks may be applicable in cases where there is an imminent, individualized risk that goes beyond generalized suffering (resulting in refugee status or complementary protection). There is also considerable State practice of providing at least temporary protection in response to displacement by sudden disasters,[4] premised by and large on the need for humanitarian assistance and to safeguard physical security. However, there remain normative gaps with respect to cross-border displacement in a number of other situations, and, as the previous chapter made clear, institutional direction.

In light of the difficulties inherent in achieving a new binding international agreement, discussed in Chapter 7, the United Nations High Commissioner for Refugees (UNHCR) has instead suggested that 'a more viable approach would be to at least develop a global guiding framework for situations of cross-border displacement resulting from climate change and natural disasters', noting that

[1] *The Eleanor* (1809) Edward's Admiralty Reports 135, 159–60.

[2] Stephen Humphreys, 'Competing Claims: Human Rights and Climate Harms' in Stephen Humphreys (ed), *Human Rights and Climate Change* (Cambridge University Press, 2010) 39.

[3] See Ch 3. See also 'Report of the Office of the United Nations High Commissioner for Human Rights on the Relationship between Climate Change and Human Rights', UN Doc A/HRC/10/61 (15 January 2009) para 70 (fn omitted), which notes that: 'While climate change has obvious implications for the enjoyment of human rights, it is less obvious whether, and to what extent, such effects can be qualified as human rights violations in a strict legal sense'.

[4] See Ch 4.

'UNHCR stands ready to support states in the development of such a framework, which could take the form of temporary or interim protection arrangements'.[5] Other high-level meetings have reiterated that 'States, in conjunction with UNHCR and other relevant stakeholders, could develop a guiding framework or instrument for the protection of people displaced externally due to sudden-onset natural disasters, including those related to climate change'.[6]

It has been suggested that guiding principles, similar in nature to the Guiding Principles on Internal Displacement,[7] may provide a useful template.[8] The advantage of a soft-law framework is that it is 'flexible and allows [States] to experiment with new ideas'.[9] States can maintain considerable freedom of action in defining standards of good conduct, while also improving the basis for international cooperation.[10] Based on existing refugee and human rights law principles, such an instrument would not require States to assume new obligations, but would clarify how those obligations might apply in the climate change displacement context. In Kälin's view, they would gain authority from the fact that they would reflect, and be consistent with, binding human rights law.[11] Indeed, because of this, they might best be described as a call to action. The Special Rapporteur on the Human Rights of Internally Displaced Persons (IDPs), Chaloka Beyani, has called them a basis for dialogue with States.[12] However, the extent to which they would go beyond clarifying the application of existing law, and instead start to create new principles, would ultimately be a political decision. As Kälin and Schrepfer observe, developing the law in this way 'would probably be difficult but, because of the non-binding character of soft law, [it] still would be easier than negotiating a treaty'.[13]

Even so, some States have already voiced concerns about the legitimacy of such a framework if they are not directly involved in its drafting (as was the case

[5] 'Statement by António Guterres, United Nations High Commissioner for Refugees' (Nansen Conference on Climate Change and Displacement in the 21st Century, Oslo, 6 June 2011) 4. The language of 'natural disasters' had become unpopular owing to the anthropogenic causes of disasters, but it is again being adopted as a means of distinguing predominantly environmental or climate change-related disasters from conflict-generated or political disasters.

[6] 'Chairperson's Summary' (Nansen Conference on Climate Change and Displacement in the 21st Century, Oslo, 6–7 June 2011) para 23.

[7] Guiding Principles on Internal Displacement, UN Doc E/CN.4/1998/53/Add.2 (11 February 1998).

[8] 'Summary of Deliberations on Climate Change and Displacement' (UNHCR Expert Roundtable on Climate Change and Displacement, Bellagio, 22–25 February 2011) (April 2011) ('Bellagio Deliberations') para 13.

[9] Walter Kälin and Nina Schrepfer, 'Protecting People Crossing Borders in the Context of Climate Change: Normative Gaps and Possible Approaches' (Study for the Swiss Ministry of Foreign Affairs, April 2011) 58, forthcoming as a paper in the UNHCR Legal and Protection Policy Research Series.

[10] Daniel Thürer, 'Soft Law', in *Max Planck Encyclopedia of Public International Law* (Oxford University Press, online edition at <http://www.mpepil.com>) para 6, cited in Kälin and Schrepfer (n 9) 58.

[11] Walter Kälin, 'Conceptualising Climate-Induced Displacement' in Jane McAdam (ed), *Climate Change and Displacement: Multidisciplinary Perspectives* (Hart Publishing, 2010) 93.

[12] Chaloka Beyani, 'A Human Rights Based Approach to Protection of Environmentally Displaced Persons' (Nansen Conference on Climate Change and Displacement in the 21st Century, Oslo, 7 June 2011) 1.

[13] Kälin and Schrepfer (n 9) 59.

with the Guiding Principles on Internal Displacement).[14] Indeed, the success of the Guiding Principles on Internal Displacement has been largely credited to the vision and tenacity of the two former Representatives of the UN Secretary-General on the Human Rights of IDPs, Francis Deng and Walter Kälin.[15] Without a similar institutional or individual champion it is unlikely that any new set of principles would achieve the same degree of support or implementation.[16] Presumably, some States also fear that a new soft-law framework could assume a life of its own. Over time, guiding principles may facilitate the implementation of such norms into domestic law, or inform, with the benefit of State practice, new multilateral instruments.[17] As Dupuy observed in relation to the Stockholm Declaration on the Environment: 'Although being only, from a formal point of view, a nonbinding resolution, many of its "principles" ... have been recalled by governments to justify their legal rights and duties. The subsequent state practice has been, no doubt, influenced by such a provision'.[18] Since a soft-law framework can exert influence on 'the interpretation, application and development of other rules of law',[19] this sense of 'creeping obligation' is no doubt what some States fear. For example, the 2009 Kampala Convention adopted by the African Union was not only modelled on the Guiding Principles, but built upon them by including an obligation to 'take measures to protect and assist persons who have been internally displaced due to natural or human made disasters, including climate change'.[20]

Nevertheless, each of the calls to action above has much to commend it, and provides one way of responding to paragraph 14(f) of the Cancún Adaptation Framework, which encourages States to take '[m]easures to enhance understanding, coordination and cooperation with regard to climate change induced displacement, migration and planned relocation, where appropriate, at national, regional and international levels'.[21] Having 'firmly put [climate-related movement] on the

[14] The Guiding Principles on Internal Displacement were drafted by a committee of experts, although soft law is usually negotiated by States: Kälin and Schrepfer (n 9) 59.

[15] See, eg, Alexander Betts, 'Towards a "Soft Law" Framework for the Protection of Vulnerable Irregular Migrants' (2010) 22 International Journal of Refugee Law 209, 224, 232.

[16] Even though many States have implemented the Guiding Principles in national law, and they were the genesis of the African Union Convention for the Protection and Assistance of Internally Displaced Persons in Africa (adopted 22 October 2009, not yet in force) ('Kampala Convention'), Kälin and Schrepfer consider support for the Guiding Principles to be relatively weak Kälin and Schrepfer (n 9) 25.

[17] See, eg, Kampala Convention, Art 5(4). See also Betts (n 15); Pierre Marie Dupuy, 'A Hard Look at Soft Law' (1990) 82 American Society of International Law Proceedings 371, 383.

[18] Dupuy (n 17) 383.

[19] Alan Boyle and Christine Chinkin, *The Making of International Law* (Oxford University Press, 2007) 223.

[20] Kampala Convention, Art 5(4).

[21] See 'Draft Decision -/CP.16: Outcome of the Work of the Ad Hoc Working Group on Long-Term Cooperative Action under the Convention' (2010) <http://unfccc.int/files/meetings/cop_16/application/pdf/cop16_lca.pdf#page=3> accessed 15 March 2011. This is discussed further in Ch 8.

agenda',[22] the paragraph evidences a limited political will by States to develop non-binding strategies to address the topic.[23]

The question remains, though, whether it is appropriate to create guiding principles focusing solely on climate change-related movement, or whether the scope should be broader, based on the *needs* and *rights* of the displaced irrespective of the cause.[24] The former suffers from the same conceptual issues canvassed in relation to a new treaty.[25] The latter would help to shift attention away from a (necessarily flawed) 'single cause' approach to instead acknowledge the interlocking and underlying socio-economic causes of movement, such as the 'structural problems of development'.[26] Indeed, there is a discernible shift in emphasis at the international level, in particular, away from 'climate change' displacement, towards viewing climate change as a subset of natural disasters (reflected in the calls to action above). In part, this may provide a way of elongating existing mandates and operational capacity, as well as acknowledging the complex causality of movement. However, if a framework responding to displacement by sudden-onset disasters is developed, it is vital that responses to slow-onset movement do not drop off the agenda.

Without endorsing any particular model, this chapter examines the normative principles that ought to inform any new framework—whether focused on protection (in the international refugee law sense); *in situ* adaptation to avoid movement; migration; or resettlement. It begins by assessing the premises of the two proposals above: first, that a framework should focus on natural disasters broadly conceived, rather than those linked only to climate change; secondly, that it should be limited to rapid-onset events; and thirdly, that temporary or interim protection may be sufficient. Next, it examines the importance of identifying protection needs according to the nature and phases of movement, using the Guiding Principles on Internal Displacement as a model. Finally, it identifies overarching normative principles that should provide the foundation for all policy interventions, whether by States, institutions, or other actors.

[22] Kälin and Schrepfer (n 9) 47.

[23] The paragraph itself does not mandate States to take action, and furthermore refers only to enhancing 'understanding, coordination and cooperation', which is relatively weak language.

[24] Betts (n 15) 215 advocates the creation of Guiding Principles on the Protection of Vulnerable Irregular Migrants, based on the Guiding Principles on Internal Displacement. At 226, he suggests that a series of instruments could be developed, for example for stranded migrants, survival migrants, forcibly deported people, or people who develop vulnerabilities in transit.

[25] See Ch 7.

[26] UNGA, 'International Co-operation to Avert New Flows of Refugees: Note by the Secretary-General', UN Doc A/41/324 (13 May 1986) para 38. At para 43, it notes that natural disasters are more likely to result in displacement in developing countries because they 'often lack the economic resources, infrastructure and service base (health, sanitary service, water supply, medical service, etc.) needed in order to deal fully with the crisis caused by the natural disasters'.

II. Protection from Natural Disasters

A. Introduction

The UN Emergency Relief Coordinator has suggested that more frequent and severe disasters may be 'the new normal'.[27] In 2010, insurance company Munich Re reported that there had been a marked increase in the number of extreme weather-related events in the past 30 years, citing climate change as the 'only plausible explanation' for this.[28] A study released in June 2011 by the Norwegian Refugee Council's Internal Displacement Monitoring Centre (IDMC) sought to calculate the numbers of people displaced by sudden-onset natural disasters in 2009 and 2010. It found that in 2010, 42.3 million people had been displaced by a sudden-onset natural disaster,[29] whether internally or across an international border. Of these, 38.3 million were displaced by a climate-related disaster (primarily flooding and storms) and 4 million by a geophysical one.[30] In the past decade, 2.4 billion people were affected by such disasters, compared to 1.7 billion in the previous decade.[31]

These calculations must be approached with caution given the methodological difficulties they necessarily entail.[32] Some researchers have suggested that statistics in part reflect increased populations in vulnerable regions, rather than a higher number of natural hazards. Nevertheless, a rising trend in the occurrence of natural disasters suggests that existing patterns of displacement are likely to continue, and possibly increase. This will mean an increase predominantly in *internal* displacement, although there may also be enhanced cross-border displacement (most likely where there are existing patterns of cross-border movement, with family and social networks abroad).[33]

B. Disaster risk reduction and assistance

The relevance of operational frameworks on disaster risk reduction to climate change-related displacement has been a recent but influential contribution to the debate, particularly at the international level. The Inter-Agency Standing

[27] Under-Secretary-General for Humanitarian Affairs and Emergency Relief Coordinator, 'Opening Remarks at the Dubai International Humanitarian Aid and Development Conference and Exhibition "DIHAD 2008 Conference"' (Dubai, 8 April 2008) <http://www.reliefweb.int/rw/rwb.nsf/db900sid/YSAR-7DHL88?OpenDocument> accessed 18 January 2010.

[28] Munich Re, 'Press Release: Two Months to Cancún Climate Summit/Large Number of Weather Extremes as Strong Indication of Climate Change' (27 September 2010) <http://www.munichre.com/en/media_relations/press_releases/2010/2010_09_27_press_release.aspx> accessed 27 July 2011.

[29] Norwegian Refugee Council, *Displacement due to Natural Hazard-Induced Disasters: Global Estimates for 2009 and 2010* (June 2011). The study used the term 'natural disaster' as shorthand for the human impact of extreme natural hazard events: 6.

[30] ibid, 4.

[31] IASC Task Force on Climate Change, 'Quick Guide to Climate Change Adaptation' (2010) 2, referring to <http://ochaonline.un.org/OCHAHome/InFocus/ClimateChangeHumanitarianImpact/TheThreatofClimateChange/tabid/5932/language/en-US/Default.aspx> accessed 22 November 2010.

[32] See Norwegian Refugee Council (n 29) 17–18; Ch 1.

[33] See Ch 6.

Committee (IASC)—which is the primary mechanism for the coordination, policy development, and decision-making of the key UN and non-UN humanitarian agencies—has explained that disaster risk reduction frameworks offer 'some of the most practical actions that support the goals of climate change adaptation'.[34] They provide a useful starting point for addressing the impacts of climate change because they are already applicable to national and local adaptation efforts and there are existing operational guidelines on which to draw.[35] While their focus is not on displacement per se, their lengthy elaboration of how best to protect the human rights of those affected by disasters is relevant to that context as well.

Both climate change adaptation and disaster risk reduction have similar objectives and methods: averting damage where possible, lessening its negative impacts when it does occur, protecting the most susceptible through risk and vulnerability assessments, building people's resilience, and taking multi-sectoral approaches to creating national strategies. Indeed, disaster risk reduction is a key element of adaptation.[36] Whereas climate change adaptation is still in a developmental stage, disaster risk response and management is 'well-established, with recognized and accepted tools and approaches'.[37] Assistance to communities at risk of severe climate change impacts may best be provided by collaborative and coordinated action in the pursuit of common disaster risk reduction and climate adaptation goals,[38] and there is considerable scope for practical and technical input by humanitarian agencies involved in disaster risk reduction.

Climate change will not necessarily alter the basic foundations of disaster response readiness, but the scale and frequency of disasters may increase.[39] Significantly, the effectiveness of disaster risk reduction strategies may affect whether or not people are displaced, and if so, how quickly they can return.[40] This is not just about physical protection from the disaster, but also economic protection. The loss or interruption of livelihoods through the destruction of crops, livestock, or productive assets may motivate people to move.[41] Indeed, in a study of people displaced by severe floods in Nepal in 2008 and 2009, nearly 25 per cent of those displaced from one area stated that even though their land was still suitable for cultivation, they were unwilling to return there because of other on-going vulnerabilities at home.[42]

[34] IASC Task Force on Climate Change (n 31) 1.

[35] ibid. For priority areas, see IASC, 'Disaster Risk Reduction Strategies and Risk Management Practices: Critical Elements for Adaptation to Climate Change' (Submission to the UNFCCC Ad Hoc Working Group on Long Term Cooperative Action by the Informal Taskforce on Climate Change of the Inter-Agency Standing Committee and the International Strategy for Disaster Reduction, 11 November 2008) 7–8.

[36] IASC Task Force on Climate Change (n 31) 4.

[37] ibid, 3.

[38] ibid.

[39] Charles Ehrhart and others, *Humanitarian Implications of Climate Change: Mapping Emerging Trends and Risk Hotspots* (2nd edn, OCHA and CARE, 2009) 3.

[40] Koko Warner, 'Assessing Institutional and Governance Needs Related to Environmental Change and Human Migration' (Study Team on Climate-Induced Migration, German Marshall Fund of the United States, June 2010) 3.

[41] ibid, 2.

[42] United Nations Resident and Humanitarian Coordinator's Office, 'Longer-Term Disaster Displaced: A Forgotten Group' (Field Bulletin, Nepal, 11 July 2011) 1 (fn omitted) <http://reliefweb.int/sites/reliefweb.int/files/resources/Full_Report_1759.pdf> accessed 22 July 2011.

Recognizing the mutually supportive objectives of sustainable development, poverty reduction, and good governance,[43] the Hyogo Framework for Action 2005–2015: Building Resilience of Nations and Communities to Disaster calls for the integration of 'existing climate variability and future climate change into strategies for the reduction of disaster risk and adaptation to climate change'.[44] It stresses the importance of monitoring disaster risks and enhancing early warning systems, reducing underlying risk factors, and using 'knowledge, innovation and education to build a culture of safety and resilience at all levels'.[45] Although the Hyogo Framework is a non-binding instrument, adopted by 168 States at the 2005 World Conference on Disaster Reduction, it is complemented by binding human rights obligations, which include a duty to mitigate against known disaster risks.[46] As Chapter 3 demonstrated, courts are beginning to recognize that States' responsibilities to protect against known environmental risks are inherent in their human rights obligation to protect the right to life. Two human rights treaties expressly refer to disaster relief, focusing on the need to protect and assist particular groups affected by disasters.[47]

The IASC Operational Guidelines on the Protection of Persons in Situations of Natural Disasters and the Red Cross' Guidelines for the Domestic Facilitation and Regulation of International Disaster Relief and Initial Recovery Assistance provide additional, operationally focused support.[48] They are intended 'to ensure that disaster relief and recovery efforts are conducted within a framework that protects and furthers human rights of affected persons'.[49] Neither is legally binding on States, but a number of States have incorporated them into their national disaster plans.[50]

On account of the international community's increased involvement in disaster relief assistance, but the relatively underdeveloped state of international law regulating it, the issue was included in the work programme of the International Law

[43] Hyogo Framework for Action 2005–2015: Building the Resilience of Nations and Communities to Disasters (World Conference on Disaster Reduction, Hyogo, 18–22 January 2005) para 4.

[44] ibid, para 14. [45] ibid.

[46] See Kälin (n 11); Ch 3 on the right to life. 'International, regional, and local actors have a shared responsibility to implement the principles enshrined in the Hyogo Framework for Action 2005–2015: Building Resilience of Nations and Communities to Disaster': Nansen Principles (appended to Chairperson's Summary) (n 6)Principle V.

[47] Convention on the Rights of Persons with Disabilities, UNGA res 61/106, Annex 1 (adopted 13 December 2006, entered into force 3 May 2008) Art 11; African Charter on the Rights and Welfare of the Child, OAU Doc CAB/LEG/24.9/49 (adopted 11 July 1990, entered into force 29 November 1999) Arts 23, 25.

[48] IASC Operational Guidelines on the Protection of Persons in Situations of Natural Disasters (Brookings–Bern Project on Internal Displacement, 2011) <http://www2.ohchr.org/english/issues/idp/docs/OperationalGuidelines.pdf> accessed 14 July 2011 (originally issued as IASC Operational Guidelines on Human Rights and Natural Disasters (Brookings–Bern Project on Internal Displacement, 2006)); Guidelines for the Domestic Facilitation and Regulation of International Disaster Relief and Initial Recovery Assistance (IFRC, 2007).

[49] IASC Operational Guidelines 2011 (n 48) 7.

[50] François Gemenne, 'Environmental Changes and Migration Flows: Normative Frameworks and Policy Responses' (PhD thesis, Institut d'Etudes Politiques de Paris and University of Liège, 2009) 234.

Commission (ILC)[51] in 2006.[52] Since 2007, the ILC has been preparing Draft Articles on the Protection of Persons in the Event of Disasters to clarify the legal regulation of disaster-related assistance.[53] The immediate impetus for this was the 2004 Asian Tsunami, which killed approximately 240,000 people in 12 States and left over one million people displaced.[54]

The ILC's background memorandum on the topic noted that whereas there is an extensive body of law applying in situations of armed conflict, there is no universal treaty comprehensively covering the main aspects of disaster relief—prevention, response, and protection.[55] 'Protection', in this context, relates to relief, assistance, and 'all activities aimed at obtaining full respect for the rights of the individual in accordance with the letter and the spirit of the relevant bodies of law (i.e. HR law, IHL, refugee law)'.[56] It can be responsive (to prevent or stop violations), remedial (to provide redress), or environment-building (to prevent future violations by 'creating the necessary legal and institutional framework, capacity and awareness that is necessary to promote respect for human rights and prevent future violations').[57] Disasters bring a further 'specialized conceptualization of protection, including, for

[51] Established in accordance with Art 13(1)(a) of the Charter of the United Nations (adopted 26 June 1945, entered into force 24 October 1945) 1 UNTS XVI ('UN Charter'), the International Law Commission's (ILC's) mandate is to prepare 'draft conventions on subjects which have not yet been regulated by international law or in regard to which the law has not yet been sufficiently developed in the practice of States', and to codify 'the more precise formulation and systematization of rules of international law in fields where there already has been extensive State practice, precedent and doctrine': Statute of the International Law Commission (1947), UNGA res 174(II), as amended by resolutions 485 (V) (12 December 1950), 984 (X) (3 December 1955), 985 (X) (3 December 1955), and 36/39 (18 November 1981) Art 15. When drafting articles, the ILC generally considers it impracticable to determine into which category each provision falls, such that the Draft Articles as a whole are said to 'constitute both codification and progressive development of international law': ILC, 'Introduction' <http://www.un.org/law/ilc/> accessed 9 May 2011.

[52] UNGA res 62/66 (6 December 2007); ILC, 'Report on the Work of its Fifty-Eighth Session (1 May–9 June and 3 July–11 August 2006)', UN Doc A/61/10, para 257.

[53] For text and commentary to date, see ILC, 'Report on the Work of its Sixty-Second Session (3 May–4 June and 5 July–6 August 2010)', UN Doc A/65/10 ('ILC Report') 321ff. See also ILC, 'Protection of Persons in the Event of Disasters: Memorandum by the Secretariat', 60th Session, UN Doc A/CN.4/590 (11 December 2007) ('ILC Memorandum'); ILC, 'Protection of Persons in the Event of Disasters: Memorandum by the Secretariat: Addendum' 60th Session, UN Doc A/CN.4/590/Add.1 (26 February 2008) (which contains definitions from various international and national instruments); ILC, 'Preliminary Report on the Protection of Persons in the Event of Disasters by Mr Eduardo Valencia-Ospina, Special Rapporteur', 60th Session, UN Doc A/CN.4/598 (5 May 2008); ILC, 'Second Report on the Protection of Persons in the Event of Disasters by Mr Eduardo Valencia-Ospina, Special Rapporteur', 61st Session, UN Doc A/CN.4/615 (7 May 2009); ILC, 'Third Report on the Protection of Persons in the Event of Disasters by Mr Eduardo Valencia-Ospina, Special Rapporteur', 62nd Session, UN Doc A/CN.4/629 (31 March 2010); ILC, 'Fourth Report on the Protection of Persons in the Event of Disasters by Mr Eduardo Valencia-Ospina, Special Rapporteur' 63rd Session, UN Doc A/CN.4/643 (11 May 2011).

[54] ILC Memorandum (n 53) para 2. [55] ibid, para 3.

[56] IASC, 'Protection of Internally Displaced Persons: Inter-Agency Standing Committee Policy Paper' (December 1999) 5, citing ICRC, 'Third Workshop on Protection, Background Paper' (7 January 1999). 'HR law' means 'human rights law'; 'IHL' means 'international humanitarian law'.

[57] IASC Operational Guidelines 2011 (n 48) 5. These ideas are reflected in: IASC (n 56); Global Protection Cluster Working Group, *Handbook for the Protection of Internally Displaced Persons*, 7 <http://www.unhcr.org/refworld/pdfid/4790cbc02.pdf> accessed 29 July 2011.

example, humanitarian access to the victims, securing safe zones, the provision of adequate and prompt relief and ensuring respect for human rights'.[58]

Whereas 'protection' in international refugee law implies (at the very least) respect for the principle of *non-refoulement*—non-return to serious harm—international disaster frameworks are *not* about responses to displacement per se. Rather, as Article 2 of the ILC Draft Articles explains, their purpose is 'to facilitate an adequate and effective response to disasters that meets the essential needs of the persons concerned, with full respect for their rights', whether displaced or not. They emphasize the primary duty of the State affected by the disaster to provide such protection.[59] Caution therefore needs to be exercised when determining the content of 'protection' in these frameworks. For example, the IASC Operational Guidelines divide human rights into four groups, stating that the first two categories must be protected immediately, whereas the last two may be implemented at a later stage:

1. Protection of rights related to protection of life; security and physical integrity; and the protection of family ties in the context of evacuations;
2. Protection of rights related to the provision of food; health; shelter; and education;
3. Protection of rights related to housing, land and property; and livelihoods; and
4. Protection of rights related to documentation, free movement in the context of durable solutions for internally displaced persons; re-establishment of family ties, expression and opinion; and elections.[60]

The Commentary to the ILC Draft Articles similarly distinguishes between human rights that are particularly relevant in the 'immediate aftermath' of a disaster compared to those in the 'recovery phase'.[61] This is because both instruments assume that protection is being afforded within the country where the disaster has occurred, and derogations from certain human rights may be justified if the disaster constitutes a public emergency threatening the life of the nation.[62] Derogation is premised on the idea that the emergency hampers the State's ability to guarantee all human rights (although some rights are considered so fundamental that they

[58] ILC Memorandum (n 53) para 251, referring to Economic and Social Council, 'Further Promotion and Encouragement of Human Rights and Fundamental Freedoms, Including the Question of the Programme and Methods of Work of the Commission: Human Rights, Mass Exoduses and Displaced Persons', UN Doc E/CN.4/1995/50 (2 February 1995) para 182. The ILC Memorandum notes that '[m]oreover, the substantive limits of this protection will differ in "simple" cases of disaster —where the focus will be on ensuring basic rights to food and shelter —as compared to complex emergencies in which questions of access predominate': para 251, fn 763.

[59] Commentary, Draft Article 5, para 4: ILC Report (n 53) 329, referring to UNGA res 46/182 of 19 December 1991 Annex, para 4. See also ILC, 'Fourth Report' (n 53) and Draft Articles 10 and 11 in UN Doc A/CN.4/L.794 (20 July 2011); Hyogo Framework for Action.

[60] IASC Operational Guidelines 2011 (n 48) 9.

[61] See ILC Memorandum (n 53) para 253.

[62] eg International Covenant on Civil and Political Rights (adopted 16 December 1966, entered into force 23 March 1976) 999 UNTS 171 ('ICCPR') Art 4.

cannot be derogated from).[63] By contrast, if people are displaced by a disaster across an international border, then that same rationale does not apply.[64] Accordingly, the human rights set out in the disaster frameworks should not be viewed as a template for the cross-border displacement context.

However, the ILC Draft Articles are valuable for identifying other norms of general international law that would usefully underpin frameworks and solutions relating to climate change-related movement—namely, the duty to cooperate, the principle of humanity, and the principle of human dignity.[65] The ILC has emphasized that the principle of humanity is 'the cornerstone for the protection of persons in international law since it place[s] the affected person at the centre of the relief process and recognize[s] the importance of his or her rights and needs'.[66] Similarly, human dignity is 'a principle underlying all human rights',[67] which should guide legal and policy outcomes. These principles are discussed in section V. below.

C. A disaster-oriented framework

As noted at the outset of this chapter, successive international expert fora have endorsed the creation of a new legal framework dealing with the displacement of people by sudden-onset disasters. UNHCR has indicated that it plans to discuss the development of such a regime with States at its Ministerial Meeting in December 2011 and hopes to secure pledges towards this.[68] While its precise form remains uncertain, it has become increasingly clear that what is envisaged is not in the nature of a new 'refugee' treaty, but rather an instrument that underscores States' existing human rights obligations, including the duty to cooperate.[69] In terms of the class of beneficiary, the focus has shifted away from people displaced by 'climate change' towards people fleeing from natural disasters. As the Secretary-General of the Norwegian Refugee Council has observed, it makes little sense to try to isolate climate change-related disasters from those with geophysical causes.[70]

[63] See further Jaime Oraá, 'The Protection of Human Rights in Emergency Situations under Customary International Law' in Guy S Goodwin-Gill and Stephan Talmon (eds), *The Reality of International Law: Essays in Honour of Ian Brownlie* (Oxford University Press, 1999).

[64] Unless, of course, the other State is similarly affected.

[65] Bellagio Deliberations (n 8) para 15; Nansen Principles (n 46) Principle I; ILC Draft Articles in ILC Report (n 53).

[66] Commentary, Draft Art 3, para 4: ILC Report (n 53) para 310.

[67] ILC, 'Protection of Persons in the Event of Disasters, Statement of the Chairman of the Drafting Committee' (20 July 2010) 7 <http://untreaty.un.org/ilc/sessions/62/DCChairman_statement4th_protection.pdf> accessed 10 December 2010.

[68] See remarks by Guterres (n 5) 5, 6; Volker Türk, Director, Division of International Protection, UNHCR, 'Can Protection of Environmentally Displaced Persons Be Found in Existing Protection Regimes? What Are the Next Steps from a Protection Perspective?' (Nansen Conference on Climate Change and Displacement in the 21st Century, Oslo, 7 June 2011) 6. However, note that in mid-2011 UNHCR's Standing Committee rejected a proposal for a pilot scheme whereby UNHCR would become the lead agency for the protection of those affected by natural disasters: Executive Committee of the High Commissioner's Programme, 'Draft Report of the 51st Meeting of the Standing Committee (21–23 June 2011)', UN Doc EC/62/SC/CRP.25 (16 September 2011) para 31.

[69] The author has been involved in a number of these fora and discussions have crystallized around this approach.

[70] Remarks by Elizabeth Rasmusson (Nansen Conference on Climate Change and Displacement in the 21st Century, Oslo, 7 June 2011); Chairperson's Summary (n 6) para 5. See also Kälin and

A needs-based approach is more appropriate than one that rests on particular triggers of displacement. It is also more empirically sound from a scientific perspective, since it focuses on managing a heightened state of risk rather than individual events. At the same time, however, it is important not to overlook the role of climate change in assessing policy responses to disasters. If disasters are viewed as discrete, unrelated events, then the climatic trends that may contribute to their frequency and severity may be overlooked and risk management inappropriately targeted.

A framework focused on disasters, rather than climate change specifically, obviates some of the conceptual difficulties discussed in the preceding chapters. It is a nod to the multicausality of displacement drivers, and would provide an important complement to existing disaster frameworks, including those being developed by the ILC. From an institutional perspective, a protection framework for displacement by rapid-onset disasters sits relatively comfortably alongside UNHCR's existing protection mandate (flight from persecution and other serious harm) and its long-standing operational involvement in responding to natural disasters.[71] In conjunction with the UN's emergency response agencies, such as the UN International Strategy for Disaster Reduction (UNISDR), the Special Representative of the UN Secretary-General for Disaster Risk Reduction, and the Office for the Coordination of Humanitarian Affairs (OCHA), it is perhaps unsurprising that legal responses are coalescing around these issues.

On the one hand, a disaster-related framework responds to the call for differentiated responses for different types of climate-related movement. As has been emphasized throughout this book, climate change-related movement requires a variety of nuanced responses, and the speed with which movement occurs *may* be a relevant determinant when it comes to assessing protection and assistance needs. A guiding framework on disaster-induced displacement could therefore provide a helpful and conceptually defensible framework for responding to one type of climate-related movement. As Chapter 7 demonstrated, States are very unlikely to adopt a longitudinal approach to assessing harm, which means that people moving pre-emptively from the projected longer-term impacts of climate change would not be regarded as having a protection need under such an instrument.

On the other hand, there is a risk that a framework based on disaster-related displacement privileges one type of movement over others, especially if complementary policy options are not actively pursued. As Ferris has remarked, 'sudden-onset disasters—cyclones, hurricanes, earthquakes...are the "easy" events to identify'.[72] The greater challenge lies in responding to the impacts of slower processes, which

Schrepfer (n 9) 61: 'it is conceptually sounder to look at disasters as a cause of displacement, and not to limit the focus to those that may be triggered by global warming'.

[71] On which, see Ch 8. The International Organization for Migration (IOM) is the official cluster leader on camp coordination and camp management in relation to disaster-induced internal displacement, and informal conversations with senior IOM staff indicate that some perceive this as a mandate grab by UNHCR.

[72] Elizabeth Ferris, 'Humanitarian Perspectives on "Protection of Persons in the Event of Disasters"' (Roundtable on 'Canada and International Law', Department of Foreign Affairs and International Trade, 21 April 2009) <http://www.brookings.edu/speeches/2009/0421_natural_disasters_ferris.aspx> accessed 14 July 2011.

potentially pose a more permanent risk to the sustainability of certain human settlements in the longer term. She has pertinently highlighted the problems with creating a framework that lacks a coherent theoretical foundation:

> [D]o people who migrate because of slow-onset, persistent drought have a privileged claim on the international community, in comparison, to people who flee grinding urban poverty? Or simply demographic pressure? Is the international community's response to people fleeing 'natural' conditions in their homelands a function of the suffering they endure? The magnitude of the effects of the phenomenon? Or the suddenness of its impact? How sudden does a phenomenon need to be to trigger an emergency humanitarian response? The effects of an earthquake may be felt in a matter of minutes, a tsunami in matter of hours, hurricanes and cyclones in days, flooding sometimes over a period of weeks, droughts over a period of months or even years. Can a dividing line be drawn between sudden and slow-onset disasters? Or is the question of time even relevant in deciding whether a particular phenomenon constitutes a disaster?[73]

Certainly, if States are willing to provide at least temporary protection for people fleeing the impacts of a sudden-onset disaster, then this is a measure worth pursuing. The principle of *non-refoulement* does not require the granting of permanent asylum, although human rights (and refugee) law mandates certain minimum standards of treatment irrespective of length of stay.[74] UNHCR perceives that 'the time has come to work with states to develop an internationally agreed doctrine of temporary protection, which would ensure the availability of interim protection to people in temporary need'.[75] But this should neither foreclose, nor be at the expense of, parallel measures aimed at securing migration opportunities, enhanced adaptation assistance *in situ*, and other measures for communities at risk of slower-onset processes. This is especially so since they may also require more permanent solutions, and negotiating these will take time. It is therefore vital that responses to slow-onset movement are not overlooked.

Furthermore, temporary protection will not always be sufficient even for victims of disasters. Though their displacement may happen suddenly, their return will not necessarily be expedited. A report on disaster displacement in Nepal explains that the duration of displacement depends on the intensity and type of disaster, with displacement following inundation of rivers generally short term in nature, while displacement resulting from floods and erosion typically longer term.[76] The report describes those displaced for a prolonged period as the 'forgotten group', much like those in protracted refugee situations whose needs tend to be overlooked once the

[73] ibid.

[74] See, eg, Jane McAdam, 'The Refugee Convention as a Rights Blueprint for People in Need of International Protection' in Jane McAdam (ed) *Forced Migration, Human Rights and Security* (Hart Publishing, 2008); Erika Feller, Assistant High Commissioner (Protection), 'The Refugee Convention at 60: Still Fit for Its Purpose?' (Workshop on Refugees and the Refugee Convention 60 Years On: Protection and Identity, Prato, 2 May 2011) 5: 'The non-*refoulement* principle prevents—or should prevent—return to persecution, but non-return can be achieved in a number of ways short of durable entry and stay'.

[75] Feller (n 74) 9.

[76] UN Resident and Humanitarian Coordinator's Office (n 42) 1 (fn omitted).

initial emergency phase has passed.[77] Thus, even if an international framework is developed on sudden-onset disaster displacement, it is unrealistic to suggest that temporary protection alone will provide an adequate response. Attention needs to be given to longer-term needs as well.

Although proponents for a disaster-based protection framework do recognize the need to develop complementary strategies,[78] there is a risk that without a strong institutional advocate, responses to displacement related to slow-onset climate change could start to slip from the global agenda. This is ironic given their intersection with so many different policy areas,[79] but relates to how such movement is conceptualized. The displacement/migration policy actors tend to view slower-onset climate impacts on that mobility predominantly as a development issue requiring the expertise of a different set of agencies. Warner suggests this results from their own focus on crisis and disaster management, 'often with a short-term perspective and not with the goal (or capacity) to maintain long-term guidance, support and protection'.[80] Yet, on the flipside, development agencies do not automatically 'see' (potential) displacement when developing poverty-alleviation and humanitarian assistance strategies. While development interventions to enhance resilience are essential,[81] and climate change adaptation funding is a key component in building sustainable development in vulnerable communities, these measures also need to be buttressed by strategies that recognize the role of migration. Indeed, evidence shows that since development in fact enhances people's capabilities and aspirations, it tends to coincide with *increased* migration, at least in the short to medium term.[82]

The following sections therefore draw together principles that apply to movement in response to both rapid-onset events *and* slow-onset processes. They are by no means an attempt to construct a set of guiding principles per se; rather, they identify legal principles that should underpin all strategies to address climate-related movement. In this way, they address current international deliberations without promoting a particular approach.

[77] On which, see Jean-François Durieux and Jane McAdam, '*Non-Refoulement* through Time: The Case for a Derogation Clause to the Refugee Convention in Mass Influx Emergencies' (2004) 16 International Journal of Refugee Law 4; Warner (n 40) 2.

[78] 'There is, however, a need to further explore the range of issues that could arise as a result of slow-onset disasters and longer-term climate change impacts, such as planned relocation and migration management': Chairperson's Summary (n 6) para 7. The UN High Commissioner for Refugees has noted that whereas the traditional response of the international community has been in emergency mode, '[w]e must now reconsider our approach': Guterres (n 5) 4.

[79] See Ch 8.

[80] Warner (n 40) 6. And yet, UNHCR also has a protection mandate in relation to large populations of statelessness people, who may never 'flee' but experience lifelong legal deprivations within their country of habitual residence.

[81] Chairperson's Summary (n 6) para 11, see also para 7; Nansen Principles (n 46) Principle V; Guterres (n 5) 4.

[82] Hein de Haas, 'Turning the Tide? Why Development Will Not Stop Migration' (2007) 38 Development and Change 819.

III. Guiding Principles on Internal Displacement

The Guiding Principles on Internal Displacement are relevant to climate-related internal displacement, and arguably require little alteration for that context.[83] Indeed, Chapter 6 examined how their application may enhance the assistance and protection of people displaced by climate impacts within Bangladesh. The following analysis highlights elements of the Guiding Principles that would also usefully inform a framework relating to cross-border movement. Their identification of rights and needs pertinent to particular phases of displacement—the pre-displacement phase, the phase of actual displacement, and the resettlement or relocation phase—is especially helpful.[84]

The first section of the Guiding Principles sets out general principles that underpin the instrument, such as equality,[85] non-discrimination,[86] and the right to protection and humanitarian assistance.[87] The second section concerns the pre-displacement phase. Principle 7 states that where displacement occurs 'other than during the emergency stages of armed conflict and disasters'—in other words, where it is planned—individuals should have access to information about the reasons and procedures for their movement, and, where applicable, on compensation and relocation. They should also be able to participate in the planning and management of their movement, and have the right to life, dignity, liberty, and security respected.[88] Additionally, the needs and interests of host communities must also be respected and carefully balanced in this process. This will have particular relevance where relocation schemes are developed.

Participation requires access to information, especially since information barriers inhibit movement (as evidenced by the importance of social networks in facilitating migration and the nature of migration patterns).[89] Without careful planning, negotiation, and consultation with both moving and host communities, relocation

[83] See Bellagio Deliberations (n 8) para 19; Chairperson's Summary (n 6) para 19: 'Both the Guiding Principles on Internal Displacement and the African Union's 2009 Kampala Convention for the Protection and Assistance of Internally Displaced Persons (IDPs) in Africa cover internal displacement resulting from natural disasters, including those linked to climate change'.

[84] See also Khalid Koser, 'Climate Change and Internal Displacement: Challenges to the Normative Framework' in Etienne Piguet, Antoine Pécoud, and Paul de Guchteneire (eds), *Migration and Climate Change* (Cambridge University Press and UNESCO Publishing, 2011); Susan F Martin, 'Managing Environmentally Induced Migration' in Frank Laczko and Christine Aghazarm (eds), *Migration, Environment and Climate Change: Assessing the Evidence* (IOM, 2009).

[85] Guiding Principles on Internal Displacement, Principle 1.

[86] ibid, Principle 4.

[87] ibid, Principle 3.

[88] ibid, Principles 7–9. Facilitating access to information and engagement with decision-making processes is also part of international environmental law: Declaration of the United Nations Conference on the Human Environment (adopted 14 June 1992) UN Doc A/CONF.151/26, vol I (12 August 1992) Annex I ('Rio Declaration') Principle 10. See also *Saramaka People v Suriname* (IACtHR, 28 November 2007) Series C No 172; Convention on Access to Information, Public Participation in Decision-Making and Access to Justice in Environmental Matters, ECE/CEP/43 (adopted 25 June 1998 entered into force 30 October 2001) ('Aarhus Convention'), which has been influential on the decisions of the European Court of Human Rights (see, eg, *Taşkin v Turkey* App no 46117/99 (ECtHR, 10 November 2004) para 99).

[89] On meaningful consultation, see *Ilmari Lansman v Finland* (1996) HRC Comm No 511/1992, para 9.5; *Apirana Mahuika v New Zealand* (2000) HRC Comm No 547/1993, para 9.8.

is likely to be fraught—especially since the effects of dislocation from home can last for generations and have significant ramifications for the maintenance and enjoyment of cultural and social rights by resettled communities.[90] Indeed, procedural rights of access to information, public participation in decision-making, and access to justice in environmental matters have been strengthened considerably by a number of international and regional instruments, and the jurisprudence of a number of human rights bodies.[91] In some jurisdictions, the duty arises pursuant to the obligation to protect life, health, and property, rather than a broader concern with environmental governance or public participation.[92]

The third section of the Guiding Principles concerns protection during displacement. Drawing on existing legal norms, it sets out a human rights framework for ensuring that a wide range of civil, political, economic, social, and cultural rights are respected.[93] In effect, it is a human rights charter focused specifically on the needs of the displaced, including protection of the right to life,[94] dignity and integrity,[95] liberty and security of the person,[96] freedom of movement,[97] an adequate standard of living,[98] medical attention,[99] education,[100] recognition before the law,[101] respect for family life and unity,[102] and the right not to be arbitrarily deprived of property and possessions.[103]

The fourth section deals with the provision of humanitarian assistance to IDPs. Though noting that the primary responsibility for this rests with national authorities, Principle 25 explains that international assistance may be offered, and if it is, it should not be viewed as an unfriendly act and consent should not be arbitrarily withheld. National authorities must also ensure that the distribution of assistance is not impeded.[104]

The final section contains principles relating to return, resettlement, and integration. This section provides that the competent authorities should ensure that the displaced are able to return home voluntarily—or resettle elsewhere in the country—in safety and with dignity, and are able to participate fully in the planning and management of their return or resettlement and subsequent integration.[105] The competent authorities should facilitate access for international humanitarian organizations to

See participatory approaches outlined in Government of Tuvalu, *Tuvalu's National Adaptation Programme of Action* (May 2007) 10–11; see also 35 for identified limitations/obstacles.

[90] John R Campbell, Michael Goldsmith, and Kanyathu Koshy, *Community Relocation as an Option for Adaptation to the Effects of Climate Change and Climate Variability in Pacific Island Countries (PICs)* (Asia-Pacific Network for Global Change Research, 2005).

[91] See, eg, Aarhus Convention; Rio Declaration, Principle 10; Ch 3 on the right to life and the environment.

[92] Alan Boyle, 'Human Rights and the Environment: A Reassessment' (UNEP paper revised, 2010) 18 <http://www.unep.org/environmentalgovernance/LinkClick.aspx?fileticket=GccCLN-brmg%3D&tabid=2046&language=en-US> accessed 27 June 2011, referring to the ECHR jurisprudence: see Ch 3.

[93] Guiding Principles on Internal Displacement, Principles 10–27.

[94] ibid, Principle 10.

[95] ibid, Principle 11.

[96] ibid, Principle 12.

[97] ibid, Principle 14.

[98] ibid, Principle 18.

[99] ibid, Principle 19.

[100] ibid, Principle 23.

[101] ibid, Principle 20.

[102] ibid, Principle 17.

[103] ibid, Principle 21.

[104] ibid, Principle 25. The prompt provision of adequate assistance can reduce longer-term migration and avert the need for more disruptive resettlement elsewhere.

[105] ibid, Principle 28.

assist IDPs in this process.[106] The competent authorities are also obliged to assist IDPs to recover property and possessions or be compensated for their loss.[107]

Resettlement should not be forced, but voluntary, rights-respecting, and based on the provision of full information about options.[108] Resettlement locations need to be selected in full consultation with existing communities in those areas, as well as with the potential new settlers.[109] For relocation to have the optimum chance of success, it has to be 'owned' by the affected communities, not imposed from above. As articulated in the 2011 Nansen Principles:

> National and international policies and responses, including planned relocation, need to be implemented on the basis of non-discrimination, consent, empowerment, participation and partnerships with those directly affected, with due sensitivity to age, gender and diversity aspects. The voices of the displaced or those threatened with displacement, loss of home or livelihood must be heard and taken into account, without neglecting those who may choose to remain.[110]

IV. Suggested Elements of a Guiding Framework

To be most effective, any legal framework on cross-border movement related to climate change impacts should follow a similar template to the Guiding Principles in terms of addressing three phases of movement: pre-movement/prevention, move-

[106] ibid, Principle 30.

[107] ibid, Principle 29. See also Framework on Durable Solutions for Internally Displaced Persons, UN Doc A/HRC/13/21/Add.4 (9 February 2010) <http://www2.ohchr.org/english/bodies/hrcouncil/docs/13session/A.HRC.13.21.Add.4_en.pdf> accessed 28 July 2011. See also Graeme Hugo, 'Lessons from Past Forced Resettlement for Climate Change Migration' in Piguet, Pécoud, and de Guchteneire (n 84)

[108] Evidence shows that adaptation strategies will be most successful where they are based on meaningful community participation, such as in constructing embankments: see Habibullah and others, 'Participatory Water Management: A Strategy for Climate Change Adaptation in Coastal Bangladesh' (2009) cited in IOM, *Assessing the Evidence: Environment, Climate Change and Migration in Bangladesh* (IOM, 2010) 17.

[109] As the Inter-American Court of Human Rights observed, it is necessary to 'obtain their free, prior, and informed consent, according to their customs and traditions': *Saramaka People v Suriname* (n 87) para 134.

[110] Nansen Principles (n 46) Principle X. A practical example is provided by the Carteret Islands Integrated Relocation Plan (Papua New Guinea). It 'involves a well thought out 14-step process, which when completed, leads ultimately to successful resettlement by those moving to Bougainville. The fourteen steps are: 1) Scoping out available land; 2) Identifying traditional land owners; 3) Negotiating with land title holders; 4) Engaging with landowners; 5) Exchange programmes; 6) Entering into land negotiations; 7) Carrying out social and resource mapping; 8) Planting gardens; 9) Identify families using objective selection criteria; 10) Prepare families for relocation; 11) Prepare host families for relocatee arrivals; 12) Building homes; 13) Moving families to the new resettlement sites; and 14) Exchanging traditionally valuable items such as shell money': Displacement Solutions, *Climate Change Displaced Persons and Housing, Land and Property Rights: Preliminary Strategies for Rights-Based Planning and Programming to Resolve Climate-Induced Displacement* (Displacement Solutions, 2009) 19. Despite this, however, relocation has been a very complicated process: see John Campbell, 'Climate-Induced Community Relocation in the Pacific: The Meaning and Importance of Land' in McAdam (n 11) 68–71.

ment, and relocation/resettlement.[111] It should distinguish between the different needs that may arise from movement relating to slow-onset climate change impacts and that occasioned by rapid-onset (disaster) events.[112] This is important in terms of the projected timeframe over which movement may occur, and whether solutions entail permanent settlement elsewhere or more temporary assistance. A framework should recognize the protection needs of nationals already in a third State at the time a disaster strikes (and/or where return becomes impossible),[113] and provide not only for a stay on removal but also adequate rights and entitlements while they remain, and for regularization of permanent stay after a certain period of time.

Bearing in mind the caveats expressed in Chapter 7 about creating an instrument *specifically* on climate-related movement, and noting the diversity of circumstances a comprehensive instrument would need to cover, any new framework ought to be based on the following premises:

- migration is a normal form of adaptation;
- climate change is rarely, if ever, the sole cause of movement, and protection and assistance should not depend on it being isolated as *the* cause;
- migration can be forced even where a person is not moving in response to *imminent* harm ('staggered flight');
- pre-emptive movement is a rational adaptive response that may avoid creating or exacerbating a disaster situation later on;
- planned movement can avoid disruption, loss of life, and sudden influxes of displaced people;
- responses to forced migration can be planned (as refugee resettlement demonstrates);[114]
- for many people, movement will be a solution of last resort (so concluding a soft-law framework will not 'open the floodgates').

These premises could be situated within the broader context of climate change adaptation by acknowledging that:

- the affected State has a responsibility to implement adaptation programmes;
- during the pre-movement/prevention phase, other States should consider directing financing for adaptation to internal relocation and resettlement initiatives;

[111] This mimics the Guiding Principles on Internal Displacement. See also Jon Barnett and Michael Webber, *Accommodating Migration to Promote Adaptation to Climate Change* (Commission on Climate Change and Development, 2009) 42–4 for a tabular summary of important issues to consider in internal and international movement, categorized by labour migration, displacement from rapid-onset disasters, displacement by slow-onset processes, and planned relocation.

[112] See ibid 42ff (and prior discussion, 30ff).

[113] See, eg, Ch 4 on Temporary Protected Status in the United States.

[114] The Director of the Intergovernmental Committee established by the 1938 Evian Conference was charged with undertaking 'negotiations to improve the present conditions of exodus (of refugees from Germany and Austria) and to replace them by orderly emigration'. Planned departures were also part of the 1989 Comprehensive Plan of Action for Indo-Chinese Refugees.

- where cross-border movement is inevitable and no internal relocation is possible in a way that safeguards human rights, States should also direct financing for adaptation towards international migration options (including providing for a social security fund for those who move), including to possible host States (eg financing burden-sharing);[115]
- there is utility in creating an international institutional focal point for coordinating multilateral efforts to assist people displaced on account of climate change, with a mandate (a) for coordinating humanitarian and emergency relief efforts; (b) for longer-term planning for relocation from slow-onset processes; and (c) with a specific human rights protection-orientation, in addition to relief and assistance.[116]

In terms of substantive rights, the Guiding Principles on Internal Displacement again provide an important model. Although the spatial element (internal/cross-border) is relevant to which State bears what obligations, it does not affect the basic needs of those who move. However, whereas those who move internally remain citizens of their country, those who cross an international border without authorization will be regarded as illegal migrants unless they are accorded a legal status. Accordingly, there need to be additional provisions for cross-border migrants with respect to non-rejection at the frontier, protection against expulsion, and regularization of status where temporary stay becomes more permanent.[117]

Any guiding framework should therefore:

- reflect and be consistent with international human rights law, international refugee law, and international humanitarian law (noting, in particular, any elements of customary international law which apply to all States);[118]

[115] Rayfuse has suggested that if small island States were to merge with larger States, conditions of merger could include a requirement that any revenue generated from these territorial acquisitions (such as via their extensive Exclusive Economic Zones) be placed into a trust fund to pay for the resettlement of the merging State's population (including on-going costs that might normally be borne by the State, such as pensions): Rosemary Rayfuse, 'W(h)ither Tuvalu? International Law and Disappearing States' (2009) University of New South Wales Faculty of Law Research Series Working Paper No 9, 11 <http://law.bepress.com/unswwps/flrps09/art9/> accessed 11 August 2011.

[116] See Ch 9.

[117] See, eg, the *bare minimum* content for temporary status proposed in UNHCR Executive Committee Conclusion No 22 (XXXII) 'Protection of Asylum-Seekers in Situations of Large-Scale Influx' (1981), which was endorsed in UNHCR, 'Protection of Persons of Concern to UNHCR Who Fall Outside the 1951 Convention: A Discussion Note', UN Doc EC/1992/SCP.CRP.5 (2 April 1992) paras 17, 19–21. Note that this was designed for States that were not parties to the Refugee Convention. Given the significant development of socio-economic rights in the intervening 30 years, and the recognition by the Committee on Economic, Social and Cultural Rights (in UN Committee on Economic, Social and Cultural Rights, General Comment No 3 'The Nature of States Parties Obligations (Art 2 Para 1 of the Covenant)', UN Doc E/1991/23 (14 December 1990) para 10) that the International Covenant on Economic, Social and Cultural Rights (adopted 16 December 1966, entered into force 3 January 1976) 993 UNTS 3 ('ICESCR') imposes an immediate 'minimum core obligation to ensure the satisfaction of, at the very least, minimum essential levels of each of the rights is incumbent upon every State party' (such as housing, social assistance, and education), these rights necessarily also comprise part of the basic human rights framework today.

[118] See also Kaelin and Schrepfer (n 9) 50ff.

- reaffirm the principle of non-discrimination (noting statements by the UN Human Rights Committee about the application of human rights to all people, including irregular migrants);
- identify any special protection that may apply to particular groups (eg children, women, the disabled, and so on, on the basis of specialist human rights treaties);
- respect the cultural and self-determination rights of communities, especially indigenous groups, and note that these need to be fostered in the place of relocation;
- recognize that States' *non-refoulement* obligations under human rights law are broader than Article 33 of the Refugee Convention,[119] and affirm that the principle of *non-refoulement* entails non-rejection at the border;
- respect the principle of family unity, with due regard to different cultural conceptions of 'family', and with the best interests of the child as a primary consideration;[120]
- apply the principle of *force majeure* in cases where people seek assistance and protection from a sudden-onset disaster;[121]
- facilitate access to rights by providing displaced people with a legal status (limbo is in no-one's interests);
- if protection is envisaged as temporary, such as in the aftermath of a disaster, provide a period after which access to permanent residence is made available;
- note that any derogations must be compliant with States' obligations under human rights law;[122]
- provide people who may be displaced—and the communities into which they may move—with access to information about their (potential) movement;[123]
- provide people who may be displaced—and the communities into which they may move—with opportunities to participate in discussion about and management of potential movement, and have their deliberations inform policy outcomes;[124]

[119] Convention relating to the Status of Refugees (adopted 28 July 1951, entered into force 22 April 1954) 189 UNTS 137, Art 1A(2), read in conjunction with Protocol relating to the Status of Refugees (adopted 31 January 1967, entered into force 4 October 1967) 606 UNTS 267

[120] This is particularly important in the context of displacement and is not elucidated expressly in the general human rights treaties (but see the Convention on the Rights of the Child (adopted 20 November 1989, entered into force 2 September 1990) 1577 UNTS 3, Arts 8, 10, 16, 20, 22).

[121] Guy S Goodwin-Gill, '*Non-Refoulement* and the New Asylum Seekers' (1986) 26 Virginia Journal of International Law 897, 909. This is discussed further below at section V.B.

[122] In the emergency context, the priority is normally addressing people's immediate needs through short-term solutions, with attention given to longer-term solutions only once these immediate needs have been met. However, the rationale for a staggered approach, in terms of according rights, does not make sense in the context of movement in response to slow-onset climate change: see ibid, 904 for the problems with justifying a temporary-style approach in the slow-onset context.

[123] See, eg, Guiding Principles on Internal Displacement, Principle 7.

[124] ibid; see also *Saramaka People v Suriname* (n 87) para 134: consultation alone is insufficient: the State must 'obtain [a people's] free, prior, and informed consent, according to their customs and traditions'. See section III. above.

- with respect to relocation (whether internal or international), identify human rights-centric mechanisms for resolving potential disputes about land tenure and access to resources, as well as associated economic, social, cultural, and spiritual costs;[125]
- facilitate humanitarian assistance to displaced populations;[126]
- facilitate return, where possible,[127] but also acknowledge the possibility for permanent settlement;
- require the establishment of effective and accessible monitoring, complaint, and response mechanisms to ensure that conditions on the ground comply with international human rights law.[128]

V. Overarching Normative Principles

In addition to ensuring that specific human rights are safeguarded, it is important that all responses to climate change-related movement, however conceived, are steeped in broader humanitarian norms, such as the 'fundamental principles of humanity, human dignity, human rights and international cooperation'.[129] These provide an overarching normative framework to guide the way in which solutions are crafted and implemented.

While there may seem to be a disconnect between abstract norms promulgated in international instruments, and problems on the ground requiring concrete solutions, it is important to appreciate that those norms can inform and give direction to action at the local level. Soft law 'concretizes abstract notions embodied in hard-law provisions',[130] such that vaguely worded human rights in treaty law can be given a specific, operational meaning. In other words, rights should be articulated not only for normative clarity, but also so that they can be translated by local and national authorities into effective interventions locally. There is also a constant dialogue: universal norms shape local action, which in turn gives meaning and content to those norms through their implementation in specific contexts.

[125] Campbell, Goldsmith, and Koshy (n 90) 42–3. Human rights law already contains mechanisms designed to assist in balancing rights, but these are often used to weigh up the rights of particular individuals, or the rights of an individual vis-à-vis the broader public interest.

[126] Guiding Principles on Internal Displacement, Principles 24–7; ILC Draft Articles (n 53) Art 2.

[127] Guiding Principles on Internal Displacement, Principles 28–30. Mass influx is often solved by mass repatriation once conditions improve, but this may not be possible where climate change renders the former habitat uninhabitable.

[128] Framework on Durable Solutions (n 107) paras 44–7.

[129] Bellagio Deliberations (n 8) para 15; Nansen Principles (n 46) Principle I; ILC Draft Articles (n 53).

[130] Bruno Simma, 'A Hard Look at Soft Law' (1990) 82 American Society of International Law Proceedings 371, 380.

A. The duty to cooperate

(1) Climate change and human rights

The duty to cooperate is a fundamental principle of international law. It is included in the UN Charter as one of the objectives of the UN,[131] and is part of almost all environmental law agreements and a number of human rights instruments.[132] The Office of the High Commissioner for Human Rights (OHCHR) has specifically noted that in the context of climate change, 'international cooperation is not only expedient but also a human rights obligation and that its central objective is the realization of human rights'.[133] The Commentary to the ILC's Draft Articles on the Protection of Persons in the Event of Disasters describes it as 'indispensable' in protecting victims of disasters.[134] States have a duty to seek international assistance where the disaster exceeds national response capacity,[135] and '[c]onsent to external assistance shall not be withheld arbitrarily'.[136]

The precise meaning of the duty to cooperate is unclear,[137] but at a minimum it entails an obligation to take into account the interests of other States.[138] To be effective, it arguably requires States to establish mechanisms for consultation, exchange of information, and the development of rules and standards. Article 2(1)

[131] UN Charter, Arts 1(3), 55, 56.

[132] ICESCR, Arts 2(1), 11, 15, 22, 23; Declaration on Principles of International Law concerning Friendly Relations and Co-operation among States in accordance with the Charter of the United Nations, UNGA res 2625 (XXV) (24 October 1970) Annex, para 1; Rio Declaration, Principles 5, 7, 13, 24, 27; UN Framework Convention on Climate Change (adopted 9 May 1992, entered into force 21 March 1993) 1771 UNTS 107, Arts 4(1)(c), (d), (e), (g), (h), (i), 5(c), 6(b); Prevention of Transboundary Harm from Hazardous Activities (ILC, 1997) Arts 4, 14, 16; Declaration of the United Nations Conference on the Human Environment (adopted 16 June 1972), UN Doc A/CONF.48/14/Rev.1 (1972) ('Stockholm Declaration') Principles 22, 24; see also UN Committee on Economic, Social and Cultural Rights, General Comments No 2 'International Technical Assistance Measures (Art. 22 of the Covenant)', UN Doc E/1990/23 (2 February 1990), No 3 (n 115), No 7 'The Right to Adequate Housing (Art 11.1 of the Covenant): Forced Evictions', UN Doc E/1998/22 (20 May 1997), No 14, 'The Right to the Highest Attainable Standard of Health (Article 12 of the International Covenant on Economic, Social and Cultural Rights)', UN Doc E/C.12/2000/4 (11 August 2000), No 15 'The Right to Water (Arts 11 and 12 of the International Covenant on Economic, Social and Cultural Rights)', UN Doc E/C.12/2002/11 (20 January 2003). Most recently, the 2006 Disabilities Convention states that the right to cooperate applies 'in situations of risk, including situations of armed conflict, humanitarian emergencies and the occurrence of natural disasters' (Art 11). In the context of natural disasters specifically, see UNGA res 46/182, 'Strengthening the Coordination of Humanitarian Emergency Assistance of the United Nations' (19 December 1991) Annex, para 5; ILC Draft Articles (n 53) Art 5; Guiding Principles on Internal Displacement, Principle 3. See generally Horst Fischer, 'International Disaster Response Law Treaties: Trends, Patterns, and Lacunae' in IFRC, *International Disaster Response Laws, Principles and Practice: Reflections, Prospects and Challenges* (IFRC and Red Crescent Societies, 2003) 24–44.

[133] OHCHR Report (n 3) para 99.

[134] Commentary, Draft Art 5, para 1: ILC Report (n 53).

[135] ILC Draft Articles (n 53) Art 10.

[136] ibid, Art 11(2).

[137] Philip Alston and Gerard Quinn, 'The Nature and Scope of States Parties' Obligations under the International Covenant on Economic, Social and Cultural Rights' (1987) 9 Human Rights Quarterly 156, 187.

[138] *Lac Lanoux Arbitration (France v Spain)* (1957) 24 ILR 101, para 22. See also *Fisheries Jurisdiction case (United Kingdom v Iceland) Merits* (1974) ICJ Rep 3, para 71: the duty to cooperate required that '[d]ue recognition must be given to the rights of both parties'.

of the International Covenant on Economic, Social and Cultural Rights (ICESCR) obliges each State party to

> take steps, individually and through international assistance and co-operation, especially economic and technical, to the maximum of its available resources, with a view to achieving progressively the full realization of the rights recognized in the present Covenant by all appropriate means, including particularly the adoption of legislative measures.

In relation to this provision, the UN Committee on Economic, Social and Cultural Rights has explained that '[w]here a State party is clearly lacking in the financial resources and/or expertise required to "work out and adopt" a detailed plan, the international community has a clear obligation to assist'.[139] Such assistance seems limited to providing financial or technical capacity or concluding agreements, rather than compelling States to undertake specific substantive obligations.[140] For example, it would be difficult to find authority to support the proposition that the duty to cooperate imposes a responsibility on States to facilitate adaptation through migration where *in situ* adaptation to climate change cannot remedy the pressures on the local population.[141]

However, Limon argues that the specific link drawn by the OHCHR between the duty to cooperate and the realization of human rights suggests that:

> all states that are party to the ICESCR have a legal obligation through international cooperation (i.e., the UNFCCC process) to reduce emissions to levels consistent with the full enjoyment of human rights (i.e., safe levels) in all other countries (especially vulnerable countries), to fund adaptation measures in vulnerable countries (depending on the availability of resources), and to ensure that the international climate change agreement due to be penned at COP 15 in Copenhagen is consistent with those human rights obligations and, at the very least, does not adversely impact human rights.[142]

A number of developed States, at least, would strongly resist that interpretation—just as they rejected the OHCHR's characterization of international cooperation as a human rights obligation, instead accepting only that it is 'important'.[143] As

[139] UN Committee on Economic, Social and Cultural Rights, General Comment No 11 'Plans of Action for Primary Education (Article 14 of the International Covenant on Economic, Social and Cultural Rights)', UN Doc E/C.12/1999/4 (10 May 1999) para 9.

[140] See also ICESCR, Art 23, which states: 'The States Parties to the present Covenant agree that international action for the achievement of the rights recognized in the present Covenant includes such methods as the conclusion of conventions, the adoption of recommendations, the furnishing of technical assistance and the holding of regional meetings and technical meetings for the purpose of consultation and study organized in conjunction with the Governments concerned'.

[141] See below on burden-sharing in international refugee law: section V.A.(2).

[142] Marc Limon, 'Human Rights and Climate Change: Constructing a Case for Political Action' (2009) 33 Harvard Environmental Law Review 439, 455 (fn omitted). See also John H Knox, 'Linking Human Rights and Climate Change at the United Nations' (2009) 33 Harvard Environmental Law Review 477, 492–6; John H Knox, 'Climate Change and Human Rights Law' (2009) 50 Virginia Journal of International Law 163, 212–18.

[143] Marc Limon, 'Human Rights Obligations and Accountability in the Face of Climate Change' (2010) 38 Georgia Journal of International and Comparative Law 543, 565–6, referring to UN Human Rights Council res 10/4 'Human Rights and Climate Change' (25 March 2009) preambular para: '*Recognizing also* that climate change is a global problem requiring a global solution, and that effective international cooperation to enable the full, effective and sustained implementation of the

Knox has observed in the context of climate change and human rights, the duty to cooperate could require merely that 'states work together to protect the people of the world from the effects of climate change on their human rights'[144]—although he argues that its minimum content requires States 'not to cause the widespread destruction of the human rights of those most vulnerable to climate change'.[145] Alston and Quinn state that it is 'difficult, if not impossible' to characterize States' commitment to international cooperation in the ICESCR as 'a legally binding obligation upon any particular state to provide any particular form of assistance',[146] although they note it may be possible to identify mandatory cooperation measures in the context of specific rights.[147] In practice, at least, there is a significant degree of cooperation between States, 'even if material contributions and political or moral support for the displaced waver and formal obligations are elusive'.[148]

(2) Forced migration

The importance of international cooperation is well recognized in the forced migration context and has been endorsed repeatedly by the UN General Assembly and UNHCR's Executive Committee.[149] The notion of burden-sharing (or

United Nations Framework Convention on Climate Change in accordance with the provisions and principles of the Convention is important in order to support national efforts for the realization of human rights implicated by climate change-related impacts'.

144 Knox, 'Climate Change and Human Rights Law' (n 142) 213.

145 ibid, 218.

146 Alston and Quinn (n 137) 191. Although virtually all States involved in drafting the treaty recognized the need for international assistance in aiding developing States to recognize the rights in the ICESCR, they did so at a level of generality. For example, a number of States expressed the view that there was no *right* to financial assistance, but only a right to request it. The French delegate suggested that 'countries with substantial resources *should* lend their assistance internationally', but could not identify any formal legal basis on which this would be required: UN Doc E/CN.4/SR.233, 8 (Mr Cassin, France), cited in Alston and Quinn (n 134) 188. Similarly, when the concept of human dignity was introduced into the Universal Declaration of Human Rights (adopted 10 December 1948) UNGA res 217A (III) 'UDHR', its purpose was 'to enable those participating in the debate to insert their own theory. Everyone could agree that human dignity was central, but not why or how'. As McCrudden notes, this is not to imply that 'dignity' has no content, but rather that it had 'different content for different people': Christopher McCrudden, 'Human Dignity and Judicial Interpretation of Human Rights' (2008) 19 European Journal of International Law 655, 678.

147 Alston and Quinn (n 137) 191. For example, in relation to the right to food, the UN Committee on Economic, Social and Cultural Rights has stated that '[p]riority in food aid should be given to the most vulnerable populations': UN Committee on Economic, Social and Cultural Rights, General Comment No 12 'The Right to Adequate Food (art. 11)', UN Doc E/C.12/1999/5 (12 May 1999) para 38. In relation to the right to health, it has stated that '[p]riority in the provision of international medical aid, distribution and management of resources, such as safe and potable water, food and medical supplies, and financial aid should be given to the most vulnerable or marginalized groups of the population', and that since 'some diseases are easily transmissible beyond the frontiers of a State, the international community has a collective responsibility to address this problem': General Comment No 14 (n 132) para 40. For examples, see the regional and domestic frameworks outlined in ILC Memorandum (n 53) paras 35–45.

148 Guy S Goodwin-Gill and Jane McAdam, *The Refugee in International Law* (3rd edn, Oxford University Press, 2007) 504 (fn omitted); Alexander Betts, *Protection by Persuasion: International Cooperation in the Refugee Regime* (Cornell University Press, 2009).

149 See, eg, UNGA res 45/702 (14 December 1990), UNGA res 47/106 (16 December 1992), UNGA res 65/264 (28 January 2011); UNHCR, *A Thematic Compilation of Executive Committee Conclusions* (4th edn, UNHCR, April 2009) 38–55 (for an extensive list of relevant Executive Committee Conclusions). See also Cotonou Agreement (2nd rev, 2010) Art 60. For a recent overview

responsibility-sharing) is expressed as a basic tenet of the international refugee protection regime and is underscored in the human rights context as well.[150] The Preamble to the Refugee Convention recognizes that 'the grant of asylum may place unduly heavy burdens on certain countries, and that a satisfactory solution of a problem of which the United Nations has recognized the international scope and nature cannot therefore be achieved without international co-operation'. It accordingly expresses 'the wish that all States, recognizing the social and humanitarian nature of the problem of refugees, will do everything within their power to prevent this problem from becoming a cause of tension between States'.

The most common way in which burden-sharing is effected is through financial and technical assistance.[151] Physical burden-sharing, through the resettlement of refugees from other countries, is the 'most radical and thereby least popular manner to relieve first asylum States',[152] and of the 148 States parties to the Refugee Convention and/or Protocol, only 23 are resettlement countries.[153] As Betts and Durieux have remarked, 'the North's dislike of uncontrolled migration is not a sufficient incentive for them to accept the substantially larger numbers of resettlement places required to make physical burden-sharing look real to overburdened host countries in the South'.[154] However, the duty to cooperate should also underpin the development of preferential migration schemes, including through the provision of financial support to poor countries admitting such migrants 'in a manner that safeguards the human rights of migrants concerned and protects them against exploitation, discrimination and marginalization'.[155]

The inter-connectedness of assistance, protection, and solutions means that comprehensive plans of action to address refugee situations may provide a useful template for developing multifaceted responses in the present context.[156] Comprehensive plans of action are multilateral agreements between certain stakeholders—countries of origin, host countries, UNHCR,[157] and affected communi-

of the principle in the displacement context, see UNHCR, 'International Cooperation to Share Burden and Responsibilities', Discussion Paper (Expert Meeting in Amman, Jordan, 27–28 June 2011).

150 See, eg, General Comment No 14 (n 132) para 40: 'States parties have a joint and individual responsibility, in accordance with the Charter of the United Nations and relevant resolutions of the United Nations General Assembly and of the World Health Assembly, to cooperate in providing disaster relief and humanitarian assistance in times of emergency, including assistance to refugees and internally displaced persons'. See also General Comment 12 (n 147) para 38.

151 For detailed analysis, see Agnès Hurwitz, *The Collective Responsibility of States to Protect Refugees* (Oxford University Press, 2009) 147–50.

152 ibid, 146.

153 UNHCR Executive Committee, 'Progress Report on Resettlement', UN Doc EC/61/SC/CRP.11 (31 May 2010) 3. 'Financial transfers may appear to be a questionable substitute for the core obligation to provide direct physical protection to refugees, especially where such transfers take place between highly developed and lesser-developed states and resemble "burden shifting"': Joan Fitzpatrick, 'Temporary Protection of Refugees: Elements of a Formalized Regime' (2000) 94 American Journal of International Law 279, 291.

154 Alexander Betts and Jean-François Durieux, 'Convention Plus as a Norm-Setting Exercise' (2007) 20 Journal of Refugee Studies 509, 527.

155 Kälin and Schrepfer (n 9) 50.

156 This is, in part, what the now abandoned Convention Plus initiative was to develop.

157 There remains a question whether UNHCR is the only agency that should be involved in light of the cross-cutting nature of the arrangement, which is an even more pertinent question in the climate change context: Goodwin-Gill and McAdam (n 148) 501.

ties—that seek to resolve specific refugee problems through a variety of measures such as diplomacy, coordination mechanisms, financial assistance, capacity building, humanitarian transfer/evacuation, and various durable solutions.[158] Typically, they:

- focus on a region and/or a shared refugee/IDP problem;
- have clear protection and durable solutions objectives;
- involve close cooperation among countries of origin, host States, UNHCR and its humanitarian and development partners, as well as affected communities and refugees;
- establish effective linkages between relevant political, peace-keeping, humanitarian, human rights, and developmental initiatives;
- set out the role and commitments of all stakeholders; and
- establish effective monitoring and follow-up mechanisms.[159]

A comprehensive plan of action would be one way of addressing potential movement from specific countries or regions in a planned, holistic manner, involving a variety of stakeholders (including affected groups). However, since the end of the 1989 Comprehensive Plan of Action for Indo-Chinese Refugees on 30 June 1996, no further such agreements have been forged, despite a number of attempts. In November 2005, negotiations on a UNHCR-proposed multilateral framework to address the irregular movement of asylum seekers were suspended on account of a lack of consensus.[160] The proposal incorporated key burden-sharing features, such as capacity-building, development assistance, and resettlement. While these were not the only issues on which agreement could not be reached, it appears that the political appetite for comprehensive strategies has waned in recent years.

B. The principle of humanity and situations of distress

Under the law of the sea, States have clear obligations of assistance towards people in situations of distress,[161] irrespective of their legal status or the circumstances in which they are found.[162] Vessels have a right of access to any port if they are compelled to take refuge on account of severe weather or other circumstances of

[158] Hurwitz (n 151) 157.

[159] UNHCR High Commissioner's Forum, 'Making Comprehensive Approaches to Resolving Refugee Problems More Systematic', UN Doc FORUM/2004/7 (16 September 2004) para 6.

[160] UNHCR High Commissioner's Forum, 'Convention Plus Core Group on Addressing Irregular Secondary Movements of Refugees and Asylum-Seekers', UN Doc FORUM/2005/7 (8 November 2005); see further Hurwitz (n 151) 161.

[161] ILC Articles on Responsibility of States for Internationally Wrongful Acts in 'Report of the International Law Commission on the Work of its Fifty-Third Session', UN Doc A/56/10 (2001) Art 23: 'the occurrence of an irresistible force or of an unforeseen event, beyond the control of the State, making it materially impossible in the circumstances to perform the obligation'.

[162] United Nations Convention on the Law of the Sea (adopted 10 December 1982, entered into force 16 November 1994) 1833 UNTS 396, Arts 18(2), 98, which reflect customary international law. For support of the principle of rendering assistance at sea as customary international law, see references in Rolf Einar Fife, 'The Duty to Render Assistance at Sea: Some Reflections after *Tampa*' in

force majeure, or if they are rendering assistance to people in danger or distress.[163] UNHCR's Executive Committee—comprised of States—has recognized that those rescued at sea should be provided with at least temporary admission to a State.[164]

Goodwin-Gill has suggested that this principle could apply to cases where people cross an international border to seek assistance and protection from a sudden-onset disaster.[165] Analogously, he argues, their situation of 'urgent distress . . . something of grave necessity'[166] justifies a departure from the normal requirements of domestic immigration law.[167] This argument is supported by the fact that the duty to render assistance in cases of distress or *force majeure* is based on a general principle of international law: elementary considerations of humanity. This 'presum[es] a common humanity that may be judicially protected by the application of universally applicable norms derivable from humanity's shared existence'.[168] As a source of law in its own right, this principle complements treaty and customary law obligations and applies in all areas of international law.[169] It is, in effect, a way of 'placing . . . certain principles behind and beyond the State', to which are attributed 'a certain normative force'.[170]

'Elementary considerations of humanity' are reflected in Article 6 of the ILC's Draft Articles as the 'principle of humanity'. This principle is described as 'the cornerstone for the protection of persons in international law since it place[s] the affected person at the centre of the relief process and recognize[s] the importance of

Jarna Petman and Jan Klabbers (eds), *Nordic Cosmopolitanism: Essays in International Law for Martti Koskenniemi* (Martinus Nijhoff, 2003) 477, fn 30.

163 Fife (n 162) 477.

164 See UNHCR (n 149) 470–3 for details of relevant Executive Committee Conclusions.

165 Goodwin-Gill (n 118) 917. It could be argued that this extends to people fleeing in the immediate aftermath of the event(s), provided the risk of harm is still imminent. In the seminal case of *The Eleanor* (n 1) 159–60, the court noted that there does not have to be 'an actual physical necessity existing at the moment; a moral necessity would justify the act, where, for instance, the ship has previously sustained damage, so as to render it dangerous to the lives of the persons on board'.

166 Goodwin-Gill (n 121) citing *The Eleanor* (n 1).

167 The situation of distress would render the person immune from the exercise of the receiving State's jurisdiction (but only with respect to the circumstances giving rise to the claim, not from every aspect of local law): ibid, 909. See also ILC Draft Articles on State Responsibility (n 161) Art 24, which states that certain acts that would otherwise be wrongful are justified when: 'the author of the act . . . has no other reasonable way in a situation of distress, of saving the author's life or the lives of the other person entrusted in the author's care'.

168 Matthew Zagor, 'Elementary Considerations of Humanity' in Karine Bannelier, Theodore Christakis, and Sarah Heathcote (eds), *The ICJ and the Evolution of International Law: The Enduring Impact of the Corfu Channel Case* (Routledge, 2012).

169 See further *The MV 'Saiga' (No 2) (Saint Vincent and the Grenadines v Guinea)*, Judgment, International Tribunal on the Law of the Sea (1 July 1999) para 155; Myron H Nordquist, *The United Nations Convention on the Law of the Sea 1982: A Commentary* (Martinus Nijhoff, 1985) 193; Bernard H Oxman, 'Human Rights and the United Nations Convention on the Law of the Sea' in Jonathan I Charney, Donald K Anton, and Mary Ellen O'Connell (eds), *Politics, Values and Functions: International Law in the 21st Century: Essays in Honor of Professor Louis Henkin* (Kluwer Law International, 1997) 377; Bernard H Oxman, 'Human Rights and the United Nations Convention on the Law of the Sea' (1998) 36 Columbia Journal of Transnational Law 404, cited in Fife (n 162) 482.

170 Guy S Goodwin-Gill, 'Europe and the Right to be Granted Asylum' (Address at the Australian National University, Canberra, 29 September 2010) 10 (copy on file with author).

his or her rights and needs'.[171] Its development and meaning can be traced through the jurisprudence of the International Court of Justice (ICJ).[172] First, in the *Corfu Channel* case, the ICJ stated that well-recognized and general principles of international law include 'elementary considerations of humanity, [which are] even more exacting in peace than in war'.[173] In the *Barcelona Traction* case, the ICJ referred to 'the principles and rules concerning the basic rights of the human person', which are owed to the international community at large.[174] In the *Nuclear Weapons Advisory Opinion*, it noted the existence of many rules of humanitarian law, 'so fundamental to the dignity of the human person' that States must observe them regardless of whether they have ratified the treaties in which they appear.[175] This is because 'they constitute intransgressible principles of international customary law', 'indicate the normal conduct and behaviour expected of States', and incorporate 'obligations which are essentially of an *erga omnes* character'.[176] Jørgensen has suggested that 'the Court seems to have had in mind principles which are so fundamental that they do not require translation into customary law in order to be applicable'.[177] As general principles, they can 'fill gaps or weaknesses in the law' and also provide 'a background of legal principles in the light of which custom and treaties have to be applied'.[178] The utility of the principle of humanity as such a principle has already been raised in the context of intra- and inter-generational equity, the protection of cultural heritage, and the precautionary principle.[179]

Nevertheless, the precise content and weight of the principle of humanity remains contested. The ICJ has not examined its scope and status, nor 'the methodology for its future elaboration and application'.[180] This is not unusual when it comes to the ICJ's consideration of general principles of international law, which is rarely accompanied by an elucidation about their precise application.[181] Indeed, Merrills has argued that general principles of international law are not 'a finished product, but a process in which the resources of legal culture are constantly being scanned by the judicial mind in a search for new solutions'.[182] Others have described general

[171] ILC Report (n 53) para 310.

[172] The following summary draws on Goodwin-Gill (n 170) 7–8.

[173] *Corfu Channel case (United Kingdom v Albania)* (1949) ICJ Rep 4, 22.

[174] *Barcelona Traction Light and Power Company, Limited (Belgium v Spain)* (1970) ICJ Rep 3, para 34.

[175] *Nuclear Weapons Advisory Opinion* (1996) ICJ Rep 226, para 79.

[176] ibid, cited also in *Israeli Wall Advisory Opinion* (2004) ICJ Rep 136, para 157.

[177] Nina HB Jørgensen, *The Responsibility of States for International Crimes* (Oxford University Press, 2003) 128, cited in Zagor (n 168) 280.

[178] Robert Jennings and Arthur Watts (eds), *Oppenheim's International Law* (vol 1, 9th edn, Longman, 1992) 40. In this respect, a general principle can modify the application of treaty or custom, since it has the same status in Art 38 of the ICJ Statute as those two sources of international law: Statute of the International Court of Justice (adopted 26 June 1945, entered into force 24 October 1945) 1 UNTS XVI, Annex.

[179] See, eg, *Moiwana Village v Suriname* (Inter-American Court of Human Rights, Case No 145, 15 June 2005) para 23 (Separate Opinion of Judge Trindade), cited in Zagor (n 168) 287–8.

[180] Zagor (n 168) 266.

[181] ibid, 276, referring to Giorgio Gaja, 'General Principles of Law', *Max Planck Encyclopedia of Public International Law* (Oxford University Press, online edition at <http://www.mpepil.com>, last updated 2007) para 20. See generally Jennings and Watts (n 178) para 12.

[182] John G Merrills, *The Development of International Law by the European Court of Human Rights* (2nd edn, Manchester University Press, 1993) 200, cited in Fife (n 162) 482.

principles of international law variously as equitable principles; non-binding general principles from which other norms can be derived; rules of custom; soft-law norms; and mere rhetorical devices with little legal significance.[183]

In the present context, the principle of humanity at the very least provides an overarching, normative concept to guide the development of strategies to respond to climate-related movement. It is closely connected with the concept of human dignity, discussed below, and it has been argued that together, they 'provide a complete framework guaranteeing respect for the protection of human rights of persons affected by disasters'.[184]

C. Human dignity

On the one hand, it is important not to overstate the role of overarching principles, since their generality may mean they have little independent content or meaning beyond a basic minimum core. As McCrudden has observed, the notion of 'dignity' has become a synonym for human rights generally, with its minimum content being that every human has an intrinsic worth, which should be recognized and respected by others, and that the State exists for the sake of the individual human being and not vice versa.[185] How this is conceptualized varies considerably from jurisdiction to jurisdiction 'because there appears to be no consensus politically or philosophically on how any of the three claims that make up the core of the concept are best understood'.[186] In some jurisdictions, 'dignity' is a right in and of itself, while in others it is more of a central organizing principle. Indeed, its significance at the time of drafting the Universal Declaration of Human Rights (UDHR), which led to its subsequent incorporation in other human rights instruments,[187] 'was that

[183] See Zagor (n 168) 266–7 for references and analysis of the evolution of the concept in ICJ jurisprudence. For further elaboration of its norm-creating potential, see *Nuclear Weapons Advisory Opinion* (n 175) 407 (Dissenting Opinion of Judge Shahabuddeen).

[184] ILC Report (n 53) para 305.

[185] McCrudden (n 146) 679.

[186] ibid, 678–9.

[187] See, eg, UDHR, Preamble, Arts 1, 22, 23; Geneva Conventions, Preamble, Common Art 3(1); Protocol I Additional to the Geneva Conventions of 12 August 1949, and relating to the Protection of Victims of International Armed Conflicts (adopted 8 June 1977, entered into force 7 December 1978) 1125 UNTS 3, Arts 75, 85; Protocol II Additional to the Geneva Conventions of 12 August 1949, and relating to the Protection of the Victims of Non-International Armed Conflicts (adopted 8 June 1977, entered into force 7 December 1978) 1125 UNTS 609, Art 4; Rome Statute of the International Criminal Court (adopted 17 July 1998, entered into force 1 July 2002) 2187 UNTS 3, Art 8; ICCPR, Preamble, Art 10; ICESCR, Preamble, Art 13; International Convention on the Elimination of All Forms of Racial Discrimination (adopted 7 March 1966, entered into force 4 January 1969) 660 UNTS 195, Preamble; Convention on the Elimination of All Forms of Discrimination against Women (adopted 18 December 1979, entered into force 3 September 1981) 1249 UNTS 13, Preamble; Convention against Torture and Other Cruel, Inhuman or Degrading Treatment or Punishment (adopted 10 December 1984, entered into force 26 June 1987) 1465 UNTS 85, Preamble; Convention on the Rights of the Child, Preamble, Arts 23, 28, 37, 39, 40; International Convention on the Protection of the Rights of All Migrant Workers and Members of Their Families (adopted 18 December 1990, entered into force 1 July 2003) 2220 UNTS 3, Arts 17, 70; International Convention for the Protection of All Persons from Enforced Disappearance (adopted 20 September 2006, entered into force 23 December 2010) UNGA res 61/177, 14 IHRR 582, Art 19; Disabilities Convention, Preamble, Arts 1, 3, 8, 16, 24, 25; American Declaration of the Rights and Duties of Man, OAS res XXX (1948) Preamble, Art XXIII; American Convention on Human Rights (adopted 22 November 1969, entered into force 18 July 1978) 1144 UNTS 123,

it supplied a theoretical basis for the human rights movement in the absence of any other basis for consensus'.[188] McCrudden concludes that it is used predominantly as a means of expressing a moral viewpoint.[189]

On the other hand, normative principles such as humanity and dignity may serve as interpretative aids in the progressive development of human rights. Furthermore, even if 'dignity' is understood as being coterminous with 'human rights', it signals the relevance of that body of legal obligations in particular contexts. Carozza suggests that since dignity is a central pillar of the UDHR, the substantive rights it contains ought to be understood as 'specifying to another degree the content of the more general recognition of the status and basic principle of human dignity'.[190] In other words, dignity expresses 'a value unique to itself, on which human rights are built'.[191]

As 'a principle underlying all human rights',[192] dignity is:

> both a justification and a framework principle within which other rights are forged. As the ethical and philosophical foundation of fundamental rights, the principle of respect for human dignity provides the basis for all other individual rights.[193]

At the very least, 'dignity' provides a way of framing desired conduct. Given that courts have accepted that an affront to human dignity constitutes degrading treatment[194] or a breach of the right to life,[195] it may be possible in certain circumstances to argue that conditions in the country of origin are so 'undignified' that return is precluded as a matter of law.[196] More generally, all policy interventions

Preamble, Arts 5, 6, 11; Arab Charter on Human Rights (adopted 22 May 2004, entered into force 15 March 2008) Preamble, Arts 2, 3, 17, 20, 33, 40; African Charter on Human and Peoples' Rights (adopted 17 June 1981, entered into force 21 October 1986) 21 ILM 58 (1982) Preamble, Art 5; Protocol to the African Charter on Human and Peoples' Rights on the Establishment of an African Court on Human and Peoples' Rights (adopted 9 June 1998, entered into force 25 January 2004), OAU Doc OAU/LEG/EXP/AFCHPR/PROT (III) Preamble; Charter of Fundamental Rights of the European Union [2010] OJ C83/389; Protocol No 13 to the Convention for the Protection of Human Rights and Fundamental Freedoms, concerning the Abolition of the Death Penalty in All Circumstances (adopted 3 May 2002, entered into force 1 July 2003) Council of Europe, CETS No 187, Preamble; Protocol to the African Charter on Human and Peoples' Rights on the Rights of Women in Africa (adopted 11 July 2003) Preamble, Arts 1, 3, 22–4.

[188] McCrudden (n 146) 677.

[189] ibid, 675 (fn omitted).

[190] Paolo G Carozza, 'Human Dignity and Judicial Interpretation of Human Rights: A Reply' (2008) 19 European Journal of International Law 931, 937. See also Andrew Clapham, *Human Rights Obligations of Non-State Actors* (Oxford University Press, 2006) 545–6 (fns omitted), who identifies its importance in four key spheres: '(1) the prohibition of all types of inhuman treatment, humiliation, or degradation by one person over another; (2) the assurance of the possibility for individual choice and the conditions for "each individual's self-fulfilment", autonomy, or self-realization; (3) the recognition that the protection of group identity and culture may be essential for the protection of personal dignity; (4) the creation of the necessary conditions for each individual to have their essential needs satisfied'.

[191] McCrudden (n 146) 681.

[192] ILC (n 67).

[193] ILC, 'Fifth Report on the Expulsions of Aliens by Mr Maurice Kamto, Special Rapporteur', 61st Session, UN Doc A/CN.4/611 (27 March 2009) para 69.

[194] See, eg, *East African Asians v United Kingdom* (1973) 3 EHRR 76. See also *Pretty v United Kingdom* (2002) 35 EHRR 1, para 52; *Ireland v United Kingdom* (1979–80) 2 EHRR 25, para 167.

[195] *Yakye Axa v Paraguay* IACHR (Judgment of 17 June 2005); *Francis Coralie Mullin v The Administrator, Union Territory of Delhi* 1981 (1) SCC 608.

[196] See Ch 3.

relating to climate change-related movement, whether in response to rapid or slow-onset impacts, must ensure that people can maintain their dignity, which necessarily requires ensuring respect for their human rights. As Principle 1 of the Stockholm Declaration states:

> Man has the fundamental right to freedom, equality and adequate conditions of life, in an environment of a quality that permits a life of dignity and well-being, and he bears a solemn responsibility to protect and improve the environment for present and future generations.

VI. Conclusion

This chapter has surveyed a range of premises and principles drawn from international law that ought to underpin legal, policy, and institutional responses to climate change-related movement. While these principles are relevant to extant discussions in international fora about the creation of a guiding framework on disaster-related displacement, their applicability is not contingent on any particular type of instrument being adopted. Indeed, this book has consistently stressed the need for multifaceted solutions, noting that a one-size-fits-all approach to climate-related movement is inappropriate. This is because different groups will have different needs depending on the nature of their mobility: 'One group may need livelihood assistance, another may need resettlement assistance, another may need humanitarian assistance, and all may need some kind of differentiated legal protection'.[197] It is important that if consensus is achieved on a mechanism for dealing with one type of climate-related movement (such as guiding principles on displacement from sudden-onset disasters), this does not overshadow the parallel development of strategies to address other types of movement (such as that linked to slow-onset impacts).[198] Similarly, while temporary protection may be an adequate response in some cases, in circumstances where the long-term habitability of certain areas is threatened, more sustainable options will need to be found. In shaping responses by States and other actors, including international agencies, the principles of humanity and dignity may act as 'meta-principles' with 'interstitial normativity', helping to set the boundaries of acceptable conduct and the limits of other rules and principles when they conflict or overlap.[199] Fundamentally, they place the needs and rights of the affected individual in the centre, so that legal and policy responses are human rights-focused.

[197] Warner (n 40) 6.

[198] Of course, such strategies may not necessarily be framed in terms of 'climate change', but may be framed instead as responses to development, migration needs, demographic pressures, or education. See further Ch 7.

[199] A Vaughan Lowe, 'Sustainable Development and Unsustainable Arguments' in Alan E Boyle and David Freestone (eds), *International Law and Sustainable Development: Past Achievements and Future Challenges* (Oxford University Press, 1999) 31.

Conclusion

> Migration may be an adjustment mechanism of first resort or a survival mechanism of last resort.[1]

This book has argued that legal responses to climate change-related movement must be carefully informed by empirical evidence about its nature, timing, and underlying causes. If they are developed in the abstract, they risk being not only non-responsive, but possibly maladaptive, which may make already precarious situations worse.

Climate change impacts will certainly affect mobility in some parts of the world, but there is insufficient evidence to suggest that they will radically alter, rather than replicate or build upon, existing patterns of movement. Empirical evidence to date suggests that most increases in movement will be within countries, rather than across international borders, and temporary, rather than permanent.

Certainly, it is important to recognize the idiosyncratic elements that climate change brings to the analysis of human mobility. In particular, the relative speed with which climate change impacts are likely to be felt means that people's resilience and traditional coping methods may be challenged. More fundamentally, however, it is critical to appreciate the way in which climate change relates to and interacts with underlying political, economic, social, and environmental pressures. Climate change is best conceived of as a threat multiplier which exacerbates pre-existing vulnerabilities. In this way, it may provide a tipping point that would not otherwise have been reached. The tipping point will vary from individual to individual, since it depends on a range of variables which will interact in different ways at different times. These include individual tolerance thresholds, resources, support mechanisms, and the extent of family and kinship networks elsewhere to help facilitate movement and provide assistance on arrival. Deciding whether or not to move is therefore a complex process.[2]

The extent to which disaster risk reduction and climate change adaptation strategies are implemented will also affect people's resilience in the aftermath of a disaster—whether they can return, rebuild, and rehabilitate—and their ability to adapt in the face of anticipated future impacts. In certain contexts, the latter will require pre-emptive measures to facilitate movement, so as to alleviate demographic and resource pressures and obviate exposure to further risk.

[1] Gil Marvel Tabucanon and Brian Opeskin, 'The Resettlement of Nauruans in Australia: An Early Case of Failed Environmental Migration' (2011) 46 Journal of Pacific History (forthcoming), text available online <http://papers.ssrn.com/sol3/papers.cfm?abstract_id=1851910> accessed 17 July 2011.

[2] See Chs 1, 5–6.

While a refugee-like protection paradigm may be appropriate in certain circumstances, such as where people flee spontaneously across an international border in search of protection, in many situations it will not match the empirics of climate-related movement.[3] The international protection regime, based on refugee and human rights law,[4] is remedial in nature. It is 'triggered' only once a person has crossed an international border, and it provides protection from the prospect of relatively imminent harm on return. Indeed, in cases where the impacts of climate change are slower-onset in nature, the overarching objective should be to avoid the protection discourse needing to be engaged at all, by developing other methods for movement that give more choice to affected communities about if and when they wish to move. Paradoxically, however, the protection system—which is an imperfect fit in this context—may have to be resorted to if no action is taken in the interim to secure safe and early migration options for those who wish to move.

Relying on the courts to develop the meaning and scope of existing protection instruments to assist those who move in response to the impacts of climate change will be a slow, unpredictable, and jurisdictionally varied method for securing protection. Courts can play a vital role in ensuring that human rights treaties are interpreted as living instruments which can respond to changing social circumstances,[5] but they need good test cases to do this. Furthermore, some jurisdictions have very limited (if any) grounds on which such human rights-based claims may be brought. This is why some argue that climate-related displacement and migration is a policy issue that would be best addressed by the international community as a whole, rather than by the courts. Nevertheless, progressive interpretations by regional courts such as the European Court of Human Rights can compel State action at the domestic level, which may in turn spur legislative change. A judicial–legislative dialogue can sometimes yield the most fruitful results.

Clearly, there remains a need to clarify the application of existing legal and institutional frameworks to the context of climate change-related movement, and to develop responses where gaps exist.[6] However, because the concept of 'climate change-related movement' encompasses so many different forms of mobility—from displacement, at one end of the spectrum, to voluntary movement, at the other—a one-size-fits-all response will be woefully inadequate. Furthermore, unless local knowledge and preferences inform responses at the national, regional, and international levels, they will be ineffective and possibly ill-targeted. As Barnett and Campbell have observed:

> adaptations will not be implemented, nor be successful, unless they are consistent with the social and cultural mores particular to the community in which adaptation takes place.

[3] See Ch 7. [4] See Chs 2–4.

[5] For example, Birnie, Boyle, and Redgwell note that the environmental jurisprudence of the European Court of Human Rights is growing so extensively that proposals for an environmental protocol to the treaty have not been pursued: Patricia Birnie, Alan Boyle, and Catherine Redgwell, *International Law and the Environment* (3rd edn, Oxford University Press, 2009) 275. Nonetheless, the European Convention on Human Rights does not guarantee the right to a safe and healthy environment per se, and while the court's jurisprudence may foreshadow a progressive approach towards claims relating to non-return to environmental harm, it is unlikely that this is the context in which the most significant developments of the human rights–environment nexus will occur.

[6] See Chs 7–9.

Put another way, adaptation activities have to be aligned with what people consider to be their needs, rights and values, otherwise implementation of adaptation will fail. It is therefore important that adaptation strategies are not imposed by outsiders, and that local communities are carefully and deliberately involved and empowered in decision making about and implementation of adaptation activities.[7]

As the case studies of Kiribati, Tuvalu, and Bangladesh examined in this book have highlighted, legal and policy responses must involve a combination of strategies, rather than an either/or approach. Physical adaptation needs to be financed and developed, and migration options, including opportunities for economic, family, and educational migration, need to be accepted as a rational and normal adaptation strategy, rather than as an indication that adaptation has failed. While movement can be a sign of vulnerability, it can also be a means to achieve security and realize human rights, especially when it is able to be planned. Policies to restrict migration rarely succeed, are often self-defeating, and increase costs to migrants, communities of origin, and destination communities.[8]

While it may be possible to reach general international agreement on overarching normative principles, based on human rights law, the duty to cooperate, and general international law principles of humanity and dignity, the detail and implementation of responses will need to be crafted for particular circumstances. Localized and/or regional responses may be better able to respond to the particular needs of affected populations in determining who should move, when, in what fashion, and with what outcome. Staggered migration, circular migration, or the promise of a place to migrate to should it become necessary might be welcome measures that could appeal both to host and affected communities alike. Furthermore, by contrast to many other triggers of displacement, the slow onset of some climate change impacts, such as rising sea levels, provides a rare opportunity to *plan* for responses, rather than relying on a remedial instrument in the case of spontaneous (and desperate) flight.

Crucially, policy responses to climate-related movement must not operate in a vacuum. To be effective, interventions must be attuned to and complement policies relating to development, housing, family planning, and the 'carrying capacity' of particular environments. This requires a sustained dialogue between actors from a variety of policy and disciplinary spheres to ensure that responses are appropriately targeted. In particular, transformative, sustainable solutions require long-term development, which the humanitarian system alone cannot mobilize.[9] Accordingly, there must be a vertical dialogue between the different levels

[7] Jon Barnett and John Campbell, *Climate Change and Small Island States: Power, Knowledge and the South Pacific* (Earthscan, 2010) 17.

[8] Hein de Haas, 'Turning the Tide? Why Development Will Not Stop Migration' (2007) 38 Development and Change 819.

[9] Louis-Georges Arsenault (UNICEF), 'Adapting Humanitarian Disaster Response to Climate-Induced Displacement' Persons' (Nansen Conference on Climate Change and Displacement in the 21st Century, Oslo, 7 June 2011) 2 <http://d2530919.hosted213.servetheworld.no/expose/global/download.asp?id=2262&fk=1621&thumb=> accessed 13 August 2011.

of policymaking, as well as a horizontal dialogue across the different spheres of policymaking, so that responses are cross-cutting, complementary, and holistic.

International law does not contain all the answers to resolving climate change-related displacement and facilitating pre-emptive movement. Human rights concepts are often used in climate change discussions in a broad, ethical sense, rather than in a technical, legal sense, such that 'the assertion of universal human rights is not, at base, a legal assertion at all; it is first a moral or political assertion, and as such frequently carries greater weight and authority than its narrower legal cousin'.[10] However, the boundaries which international law sets for State conduct, and the rights it recognizes for individuals and groups, provide important, normative parameters to guide and shape legal and policy development in a human rights-centric manner.

[10] Stephen Humphreys, 'Introduction: Human Rights and Climate Change' in Stephen Humphreys (ed), *Human Rights and Climate Change* (Cambridge University Press, 2010) 11.

Appendix

References to migration/relocation as an adaptation strategy in submitted National Adaptation Programmes of Action (NAPAs) (as at 5 January 2012)

1. **Afghanistan, *National Capacity Needs Self-Assessment for Global Environmental Management (NCSA) and National Adaptation Programme of Action for Climate Change (NAPA), Final Joint Report* (February 2009)**
   - No relevant discussion of migration/relocation as an adaptation strategy.
   - There are many references to migration/displacement in Afghanistan's political context. Refugees are discussed in terms of the general development issues facing Afghanistan (see, eg, 'the disruption of traditional practices that has resulted from insecurity, migration, a breakdown of social structures, poverty and drought which has led to over exploitation of the natural resource base' (p 6)).
2. **Angola, *National Adaptation Programme of Action under the United Nations Framework Convention on Climate Change (UNFCCC)* (2011)**
   - No relevant discussion of migration/relocation as an adaptation strategy (although note p 31).
3. **Government of the People's Republic of Bangladesh, *National Adaptation Programme of Action (NAPA), Final Report* (November 2005)**
   - No relevant discussion of migration/relocation as an adaptation strategy (although see pp 35–7).
4. **Benin, *Programme d'Action National d'Adaptation aux Changements Climatiques du Benin (PANA-Benin)* (January 2008)**
   - No relevant discussion of migration/relocation as an adaptation strategy, although there is some discussion of migration as a past coping strategy (p 49) and the risk of future movement (p 60).
5. **Bhutan, *National Adaptation Programme of Action* (no date, submitted May 2006)**
   - Potential Adaptation Activities by Sector (Table):
     - Relocation/resettlement of affected towns (Natural Disaster and Infrastructure) (p 16).
   - Project Profile 1: 'Disaster Management Strategy—Planning for Food Security and Emergency Medicine to Vulnerable Communities':
     - Activities: '[i]dentification of potential areas for resettlement of villages' (p 30).
   - Project Profile 4: 'Landslide Management and Flood Prevention' (Pilot Schemes in Critical Areas):
     - Activities: '[i]dentification of potential areas for resettlement for vulnerable communities' (p 44).

- Project Profile 5: 'Flood Protection of Downstream Industrial and Agricultural Area':
    - Activities: '[i]dentification of potential areas for resettlement in case of such requirements' (p 48).

6. **Burkina Faso,** ***Programme d'Action National d'Adaptation à la Variabilité et aux Changements Climatiques (PANA du Burkina Faso)*** **(November 2007)**
    - No relevant discussion of migration/relocation as an adaptation strategy, although there is some discussion of migration as a past coping strategy (p 22) and the risk of future movement (p 33).
7. **Republique du Burundi,** ***National Adaptation Plan of Action to Climate Change 'NAPA'*** **(January 2007)**
    - No relevant discussion of migration/relocation as an adaptation strategy (but see p viii).
8. **Royal Government of Cambodia,** ***National Adaptation Programme of Action to Climate Change (NAPA)*** **(October 2006)**
    - No relevant discussion of migration/relocation as an adaptation strategy.
9. **Republic of Cape Verde,** ***National Adaptation Programme of Action on Climate Change 2008–2012, Executive Version*** **(December 2007)**
    - No relevant discussion of migration/relocation as an adaptation strategy, but it is noted that an impact of climate change is the migration of coastal populations to the interior of islands (p 9).
10. **Central African Republic,** ***Programme d'Action National d'Adaptation (PANA) aux Changements Climatiques*** **(May 2008)**
    - No relevant discussion of migration/relocation as an adaptation strategy (although see pp 23, 29).
11. **Republique du Tchad,** ***Programme d'Action National d'Adaptation aux Changements Climatiques (PANA-Tchad)*** **(February 2010)**
    - Focus on reducing migration generally; no relevant discussion of migration/relocation as an adaptation strategy (although see pp vii, xiii, 33).
12. **Union of the Comoros,** ***National Action Programme of Adaptation to Climate Change (NAPA)*** **(March 2006)**
    - No relevant discussion of migration/relocation as an adaptation strategy, although it is noted that 'emigration is the response strategy of the underprivileged populations and sometimes the only solution, but it is likely to lead to social disintegration' (p 10).
13. **Republique Democratique du Congo,** ***Programme d'Action National d'Adaptation au Changement Climatique de la République Démocratique du Congo*** **(September 2006)**
    - No relevant discussion of migration/relocation as an adaptation strategy (although see pp 54, 68).
14. **République de Djibouti,** ***Programme d'Action National d'Adaptation aux Changements Climatiques*** **(October 2006)**

- No relevant discussion of migration/relocation as an adaptation strategy (although see pp 20, 31, 41, 67, 68).

15. **State of Eritrea, *National Adaptation Programme of Action* (April 2007)**
    - Annex 1: 'Coping Strategies and Potential Adaptation Projects identified by Stakeholders' lists the following relevant potential adaptation strategies:
        - Livestock Sector: '[c]asual work in towns or permanent displacement of young people' is listed as an autonomous current coping strategy, but not as a project or programme response (p 24).
        - Marine and Coastal Zone sector: migration to highlands or a 'relatively suitable and cold area' is listed as an autonomous current coping strategy; 'assistance for relocation of island inhabitants' is listed as a potential adaptation activity (p 27).
        - 'Public Health' table: '[p]reparing programmes for resettlement involving socio-economic infrastructure' (p 28).
    - No resettlement/migration options are listed in the actual project profiles, but 'assistance for relocation of island inhabitants' is noted as a major adaptation need identified for coastal zones and marine ecosystems during stakeholder consultations (p 12).

16. **The Federal Democratic Republic of Ethiopia, *Climate Change National Adaptation Programme of Action (NAPA) of Ethiopia* (June 2007)**
    - Table 3.3: 'Summary of Potential Adaptation Options proposed by the Regional Consultative Workshops':
        - Proposed cross-sectoral adaptation options include '[p]roper land use policy, resettlement' (p 43).
    - It is noted that migration has been a past coping strategy (pp 5, 25, 29)

17. **Gambia, *National Adaptation Programme of Action (NAPA) on Climate Change* (November 2007)**
    - Project 9: 'Restoration/Protection of Coastal Environments':
        - 'Risks and Barriers' noted include '[t]here are parts of the Kotu stream that are already settled upon with residential properties, the rehabilitation of the stream will displace some people. Therefore the issue of resettlement and compensation will have to be considered' (p 82).
    - Table 3.4: 'Adaptation Options and Their Relationship to Vulnerability of Water Resources Systems'
        - 'Resettlement of people' is listed as an option (p 47).
    - Table 3.6: 'Adaptation Options and Their Relationship to Vulnerability in Sensitive Coastal Environments'
        - 'Resettlement of people' is listed as an option (p 48).
    - Stakeholder consultations reveal seasonal/permanent migration as a current strategy of fisherfolk (pp 36–7).

18. **République de Guinee, *Plan d'Action National d'Adaptation aux Changements Climatiques (PANA) de la République de Guinee* (July 2007)**
    - No relevant discussion of migration/relocation as an adaptation strategy (although see pp vii, 14).

19. **Republic of Guinea-Bissau, *National Programme of Adaptation to Climate Changes* (December 2006)**
    - No relevant discussion of migration/relocation as an adaptation strategy.
    - Some discussion of migration as a consequence of rainfall deterioration and other factors (see, eg, p 32).
20. **République d'Haïti, *Plan d'Action National d'Adaptation (PANA)* (October 2006)**
    - No relevant discussion of migration/relocation as an adaptation strategy (although see pp 12, 14, 16, 43).
21. **Republic of Kiribati, *National Adaptation Program of Action (NAPA)* (January 2007)**
    - 'Where flooding leads to erosion ... people have to relocate themselves or retreat. Few incidents of this form of adaptation have occurred' (p 12).
    - Notes that 'CCA Policy and CCA Strategy elaborate on the long term basis of adaptation planning and implementation. The CCAS stresses eight focal areas: including *population and resettlement'* (p 23).
    - Complementary Table on NAPA and KAP II lists 'support population and resettlement programs' under Policy Coordination and Planning (p 59).
    - Past resettlement in a foreign country on account of drought in the 1950–60s is noted (p 6).
22. **Lao People's Democratic Republic, *National Adaptation Programme of Action to Climate Change* (April 2009)**
    - No relevant discussion of migration/relocation as an adaptation strategy.
23. **Lesotho, *Lesotho's National Adaptation Programme of Action (NAPA) on Climate Change* (no date, submitted June 2007)**
    - Migration as a current coping strategy by farmers and pastoralists is noted (pp 6, 9).
24. **Liberia, *National Adaptation Programme of Action (NAPA)* (2008)**
    - No relevant discussion of migration/relocation as an adaptation strategy.
25. **République de Madagascar, *Programme d'Action National d'Adaptation au Changement Climatique* (no date, submitted December 2006)**
    - No relevant discussion of migration/relocation as an adaptation strategy.
26. **Malawi, *Malawi's National Adaptation Programme of Action (NAPA)* (March 2006)**
    - The NAPA selects priority activities through criteria selected by the Least Developed Countries Expert Group (see p 7). Reconstruction and resettlement programs are listed in tables evaluating the criteria for each adaptation option for Malawi, but score low against other adaptation options (pp 38–40).
27. **Republic of Maldives, *National Adaptation Programme of Action (NAPA)* (2007)**
    - Project 1: 'Integration of Future Climate Change Scenarios in the Safer Island Strategy to Adapt Sea Level Rise and Extreme Weather Risks Associated with Climate Change':
      - 'The primary aim of this project is to enhance people's safety from climate change impacts and natural disasters, and the path selected is to improve and

strengthen the Safer Island Strategy developed by the government to resettle communities from the smaller, more vulnerable islands into larger, better protected ones' (p 47).

- Project 2: 'Coastal Protection of Safer Islands to Reduce the Risk from Sea Induced Flooding and Predicted Sea Level Rise':
    - 'The location for implementing this project is to be selected from the islands identified to be developed as a safer island under the Safer Island Strategy (SIS) .... The SIS was developed following the Indian Ocean Tsunami of 2004, to resettle communities' (p 50).
- Project 10: 'Protection of Human Settlements by Coastal Protect Measures on Safer Islands' is an aspect of the SIS (pp 72–3).

28. République du Mali, *Programme d'Action National d'Adaptation aux Changements Climatiques* (July 2007)

- No relevant discussion of migration/relocation as an adaptation strategy (although see pp 13, 34–5, 57).

29. Islamic Republic of Mauritania, *National Adaptation Programme of Action to Climate Change NAPA-RIM* (November 2004)

- Project 7.6: 'Marine and Coastal Ecosystems':
    - 'The implementation of a safeguard plan for the town of Nouakchott and its infrastructures' lists as an objective '[t]o relocate by 2009 over 50% of the infrastructures established on the dunes and to orchestrate a planned removal of all the infrastructures established in the sectors susceptible to the effects of climate change' (p 71).
- Project title: 'The Reorganization of Populations Adversely Affected by Climate Change, Taking into Consideration the Options They Have Already Adopted':
    - 'The climate changes resulting from the deterioration of the rainfall level have affected the great majority of the rural and nomadic populations. This situation has driven these populations to adopt various strategies to adapt to the new conditions. These strategies include rural exodus, connection between tarred roads and permanent settlement in the favourable areas (wetlands)' (p 70). Activities include 'the redirection of options already adopted by the populations: population settlement along the tarred roads (adaptation to climate change) and disordered settlement' (p 70).

30. Mozambique, *National Adaptation Programme of Action (NAPA)* (December 2007)

- Strategies listed as emerging from the participatory evaluation process include 'Resettlement of people in areas not prone to floods' (p 13).
- Project 1: 'Strengthening of an Early Warning System'. Activities listed include 'Resettle crowded populations from flood and cyclone prone areas' (p 29).

31. Nepal, *National Adaptation Programme of Action (NAPA) to Climate Change* (September 2010)

- Annex 4: 'Local Perceptions of Climate Change Impacts':
    - Impacts under the heading 'Urban Settlements & Infrastructure' include '[i]mpact of disasters leading to greater migration to urban areas' (p 60).

- Annex 5: Gender Sensitivity Analysis of Climate Change Impacts:
  - Impacts listed under 'Urban Settlements' include: 'Migration and frequent movements due to temporary displacement related to flash floods pose risks of insecurity and sexual violence against women' (p 61).
- Annex 11: List of Priority Adaptation Options to Address Climate-Induced Disasters. Adaptation options include:
  - '[r]esettlement of vulnerable community' (Increases in Intense Rainfall/Floods; Increases in Intense Rainfall/Landslides) (p 73).
  - '[d]iscouraging and restricting settlements in high-risk areas' (Increases in Intense Rainfall/Floods) (p 73).
  - '[d]iscouraging and restricting people living in high-risk areas' (Increases in Intense Rainfall/Landslides) (p 73).

32. **Republic of Niger, *National Adaptation Programme of Action* (August 2006)**
    - No relevant discussion of migration/relocation as an adaptation strategy.

33. **Republic of Rwanda, *National Adaptation Programmes of Action to Climate Change—NAPA-Rwanda* (December 2006)**
    - No relevant discussion of migration/relocation as an adaptation strategy, although there is discussion of migration as an existing response to environmental vulnerability (pp 17, 37), and of attempts to reduce the need for migration as a result of climate change (pp 58, 69).

34. **Samoa, *National Adaptation Programme of Action* (2005)**
    - 'Summary List of Existing and Potential Adaptation Activities' notes the following *existing* adaptation activities:
      - 'Relocate families' (p 18).
    - 'Summary List of Existing and Potential Adaptation Activities' notes the following *potential* adaptation activities:
      - 'Assistance for relocation of communities inland' (p 18).
      - 'Relocation of roads further inland' (p 19).
    - 'Summary List of Key Adaptation Needs':
      - Communities: 'Assistance for relocation of communities inland' (p 20).
      - Works Transport and Infrastructure: 'Relocation of roads further inland' (p 20).
    - 'List of Priority Activities':
      - Implement Coastal Infrastructure Management Plans for Highly Vulnerable District, including assistance for relocation of roads further inland and assistance for relocation of communities inland (p 23).
    - Project Profile 7: Implement Coastal Infrastructure Management Plans for Highly Vulnerable Districts Project:
      - Expected outcomes include '[i]ncremental relocation of community and government assets outside the CHZs' [coastal hazard zones] (p 48).
      - Indicators/Expected Outputs include '[a] trend of inland retreat or relocation' (p 48).

- Suggested actions required include '[t]o identify a new site for relocation of districts hospitals and or clinics outside of the hazard zones' (p 49).

35. **Sao Tomé and Príncipe, *National Adaptation Programmes of Action on Climate Change* (December 2006)**
    - 'Project Title: Displacement of Local Communities' notes: '[f]or occasion of torrential rains and invasion of coast and beach at Santa Catarina e Malanza; consequently, it cause floods as the sea level rise. This community of fishermen, farmers that have to stop their activities, and put their family at risk of subsistence. … A construction of new homes as well as the communities' displacement, to face the Climate Changes phenomenon is necessary' (p 48). This priority project was ranked 9th overall (p 9).
    - 'Vulnerability Evaluation and Needs of Adaptation: Adapted Solutions for Climate Change':
        - Under 'Infrastructure, Public Works and Tourism Sectors', one solution is '[r]elocations of some communities or parts of them' (p 18).
        - Under 'Measures for the Varied Phase of an Occurrence: During the Occurrence', one measure is '[t]o proceed to the displacement, temporary lodging and relocation of populations that the emergency situation imposes' (p 19).
    - 'Characterization of Priority Options':
        - Includes as one priority option the '[r]elocation of some communities in risk or part of them' (p 21).
    - Table 1: 'Crossing among Impacts, Vulnerability Factors, Adaptation Measures and Proposed Projects for Public Works, Infrastructures and Tourism Sectors'
        - One verified impact is 'sea level rise', and both an adaptation measure and a proposed project in response is the relocation of local communities (p 40).
    - Table 5: 'Crossing among Impacts, Vulnerability Factors, Adaptation Measures and Proposed Projects for Fisheries Sector'
        - An adaptation measure in response to coastal erosion is '[i]mmigration for other beaches (communities)' (p 44).

36. **Republique du Senegal, *Plan d'Action National pour l'Adaptation aux Changements Climatiques* (2006)**
    - It is noted that: 'Les phénomènes d'érosion côtière affectent presque toutes les grandes villes côtières du Sénégal. Ils se traduisent par des destructions d'infrastructures et de bâtiments pouvant nécessiter des relocalisations' (p 22). (See also pp 55, 61).

37. **Sierra Leone, *National Adaptation Programme of Action (NAPA), Final Report* (December 2007)**
    - No relevant discussion of migration/relocation as an adaptation strategy.

38. **Solomon Islands, *National Adaptation Programme of Action* (November 2008)**
    - Component 3: 'Human Settlement':
        - '[o]ne of the key potential adaptation options for many of these communities and/or villages is to relocate. The question of relocation has serious political, economic and socio-cultural implications. Given the above, the focus of this

component is to enhance the capacity of such islands and communities to plan for adaptation' (p 80).

- Component 5.2: 'Climate Change Adaptation on Low-Lying and Artificially Built-Up Islands in Malaita and Temotu Provinces':
  - Outcome 1: To develop and implement plans to relocate as an adaptation measure (p 86). Discussed in depth at pp 86–7.
- Relocation and Resettlement objectives are listed in Annex 1: Adaptation Options under 'Water Sector' and 'Health Sector' and 'Human Settlement' (pp 121–3, 127).
- See also discussion of past and potential relocation at 2.3.4 ('Key Vulnerabilities—Human Settlements'): 'Securing land becomes a critical issue when considering relocation and resettlements schemes throughout the country' (p 39); 'Climate change and sea level rise is likely to displace a number of communities and/or villages. Of particular note is the limited potential for adaptation in low-lying atolls and artificially-built islands. These communities and/or villages often have very little or no land to move to and therefore are limited to living in high risk, disaster-prone and climate-sensitive environments' (p 40); 'The low-lying atolls and artificially-built islands do not have direct access to lands on larger islands and therefore these groups will be seriously affected by sea-level rise. It will be necessary for these groups of people to be resettled but will require and depend entirely on national and provincial government efforts' (p 42); 'The impacts of climate change and sea-level rise on human settlements pose serious questions regarding adaptation options. One of the key options will be through relocation and resettlement schemes. Thus, it will require legislative and structural changes to the land tenure and land management systems in the country to facilitate such actions' (p 42).

39. Republic of the Sudan, *National Adaptation Programme of Action* (July 2007)

- No relevant discussion of migration/relocation as a future adaptation strategy, although there is discussion of past movement (pp 4, 9, 30, 33, 36) and possible strategies to reduce future movement (pp 39, 43, 45).

40. United Republic of Tanzania, *National Adaptation Programme of Action (NAPA)* (January 2007)

- Selected project activities include the implementation of 'sustainable tourism activities in the coastal areas and relocation of vulnerable communities from low-lying areas' (ix).
- Relocation is considered as a Potential Adaptation Activity in these sectors:
  - Coastal and Marine Resources: 'relocation of services, properties and existing infrastructures due to sea level rise' (p 24).
  - Human Settlement Sector: 'relocation of vulnerable communities to other areas' (p 29).
  - Tourism Sector: 'relocation of people living in wildlife corridors' (p 30).
- In relation to preventing the need for movement, see 'Project Title: Shifting of Shallow Water Wells Affected by Inundation on the Coastal Regions of Tanzania Mainland and Zanzibar'.

- Due to salt-water inundation in coastal areas, 'there is an immediate need to take action to curb the situation, otherwise if unchecked; people living along the coast will be forced to migrate to other areas, something which may cause social conflicts and other environmental degradation due to overpopulation and utilization of resources' (p 42).

41. **Democratic Republic of Timor-Leste, *National Adaptation Programme of Action (NAPA) on Climate Change* (December 2010)**
    - 'Water Sector': Increased flooding resulting from climate change impacts will increase urban migration (p 28).
    - 'The Biodiversity (Forests, Freshwater Aquatic, Marine and Invasive Alien Species) Sector': Climate change impacts will include: 'Migration/displacement to locations with more suitable environmental conditions' (p 30).
    - 'Disasters—Cross-cutting': The interaction of climate change and natural disasters will result in '[d]isplacement due to flooding' (p 34).

42. **Togo, *Plan d'Action National d'Adaptation aux Changements Climatiques—PANA* (September 2009)**
    - Table 8: 'Synthèse des mesures adaptatives passées et actuelles mises en œuvre par les populations face à la variabilité et aux changements climatiques':
        - 'Déplacement' is listed several times (pp 46–8).
    - For general references to displacement and flight, see pp 2, 19, 29, 30, 36, 38, 46–7, 56–9, 64, 85, 88, 92, 102–3, 107.

43. **Tuvalu, *Tuvalu's National Adaptation Programme of Action* (May 2007)**
    - NAPA Project Profile 6: 'Strengthening of Community Disaster Preparedness and Response Potential'. Activities include:
        - '[d]evelop[ing] a post-disaster resettlement and rescue plan (p 51).
        - $15,000 earmarked in tentative budget for post-disaster resettlement and rescue plan (p 52).
    - 3.3.2: 'Potential Future Vulnerability of Tuvalu':
        - 'A last resort to adaptation would be migration and resettlement should the worse case scenario occur' (p 25).
    - General references to internal migration 'due to increasing changes in lifestyle and dependence on imported foods' (pp 6, 15); 'increasing population and internal migration to urban areas and the increasing changes in climate and variability resulting in erratic rainfall pattern changes caused the water problem in Tuvalu' (p 45).

44. **Uganda, *Climate Change—Uganda National Adaptation Programmes of Action* (2007)**
    - Project No 4: 'Community Water and Sanitation Project'. Activities include:
        - 'Relocate communities to safer areas/districts' (p 56).
    - Migration is also discussed as a current coping strategy to environmental degradation (pp 10, 44, 53, 60).

45. **Republic of Vanuatu, *National Adaptation Programme for Action (NAPA)* (no date, submitted December 2007)**

- 'Relocation of settlements and various infrastructures' is cited as an adaptation option for Torba Province (p 23), Sanma Province (p 24), Penama Province (24), Malampa Province (p 25), Shefa Province (p 25), and Tafea Province (p 26).
- Relocation of vulnerable settlements and infrastructure is listed as a general adaptation strategy (p 28).
- The NAPA also discusses the relocation of a settlement called Lateau on Tegua, an island in Torba province, as 'the first-ever global climate change adaptation project in Vanuatu' (pp 19–20).

46. **Republic of Yemen, *National Adaptation Programme of Action* (no date, submitted April 2009)**

- No relevant discussion of migration/relocation as an adaptation strategy.

47. **Republic of Zambia, *Formulation of the National Adaptation Programme of Action on Climate Change (Final Report)* (September 2007)**

- No relevant discussion of migration/relocation as an adaptation strategy.

Bibliography

1. BOOKS AND MONOGRAPHS

Abrar, Chowdhury R and S Nurullah Azad, *Coping with Displacement: Riverbank Erosion in North-West Bangladesh* (RDRS Bangladesh, North Bengal Institute, and Refugee and Migratory Movements Research Unit, 2004).

Abrar, Chowdhury R and Janet Seeley (eds), *Social Protection and Livelihoods: Marginalised Migrant Workers of India and Bangladesh* (The University Press, 2009).

Arnold, Denis G (ed), *The Ethics of Global Climate Change* (Cambridge University Press, 2011).

Barnett, Jon and John Campbell, *Climate Change and Small Island States: Power, Knowledge and the South Pacific* (Earthscan, 2010).

Betts, Alexander, *Protection by Persuasion: International Cooperation in the Refugee Regime* (Cornell University Press, 2009).

Birnie, Patricia, Alan Boyle, and Catherine Redgwell, *International Law and the Environment* (3rd edn, Oxford University Press, 2009).

Blake, Nicholas and Raza Hussain, *Immigration, Asylum and Human Rights* (Oxford University Press, 2003).

Boyle, Alan and Christine Chinkin, *The Making of International Law* (Oxford University Press, 2007).

Bratspies, Rebecca M and Russell A Miller (eds), *Transboundary Harm in International Law: Lessons from the Trail Smelter Arbitration* (Cambridge University Press, 2006).

Brownlie, Ian, *Principles of Public International Law* (6th edn, Oxford University Press, 2003).

Burson, Bruce (ed), *Climate Change and Migration: South Pacific Perspectives* (Institute of Policy Studies, Victoria University of Wellington, 2010).

Castoriadis, Cornelius, *The Imaginary Institution of Society* (Polity Press, 1987).

Clapham, Andrew, *Human Rights Obligations of Non-State Actors* (Oxford University Press, 2006).

Connell, John and John P Lea, *Urbanisation in the Island Pacific: Towards Sustainable Development* (Routledge, 2002).

Crawford, James, *The Creation of States in International Law* (2nd edn, Oxford University Press, 2006).

Duursma, Jorri C, *Fragmentation and the International Relations of Micro-States: Self-Determination and Statehood* (Cambridge University Press, 1996).

Faaniu, Simati and others, *Tuvalu: A History* (Institute of Pacific Studies and Extension Services, USP and the Ministry of Social Services, 1983).

Fischer, Steven R, *A History of the Pacific Islands* (Palgrave Macmillan, 2002).

Foster, Michelle, *International Refugee Law and Socio-Economic Rights: Refuge from Deprivation* (Cambridge University Press, 2007).

Gearty, Conor and Virginia Mantouvalou, *Debating Social Rights* (Hart Publishing, 2010).

Goodwin-Gill, Guy S, and Jane McAdam, *The Refugee in International Law* (3rd edn, Oxford University Press, 2007).

Harris, David J, *Cases and Materials on International Law* (6th edn, Sweet and Maxwell, 2004).
Hathaway, James C, *The Law of Refugee Status* (Butterworths, 1991).
Higgins, Rosalyn, *The Development of International Law through the Political Organs of the United Nations* (Oxford University Press, 1963).
Humphreys, Stephen (ed), *Human Rights and Climate Change* (Cambridge University Press, 2010).
Hurwitz, Agnès, *The Collective Responsibility of States to Protect Refugees* (Oxford University Press, 2009).
International Institute for Environment and Development, *Environmental Refugees* (Earthscan, 1984).
Jennings, Robert and Arthur Watts (eds), *Oppenheim's International Law* (vol 1, 9th edn, Longman, 1992).
Joseph, Sarah, Jenny Schultz and Melissa Castan, *The International Covenant on Civil and Political Rights: Cases, Materials, and Commentary* (2nd edn, Oxford University Press, 2004).
Kanjilal, Tushar, *Who Killed the Sundarbans?* (Tagore Society for Rural Development, 2000).
Khan, Irene, *The Unheard Truth: Poverty and Human Rights* (WW Norton and Co, 2009).
Laczko, Frank, and Christine Aghazarm (eds), *Migration, Environment and Climate Change: Assessing the Evidence* (IOM, 2009).
Lauterpacht, Hersch, *International Law and Human Rights* (Praeger, 1950).
Leighton, Michelle, Xiomeng Shen, and Koko Warner (eds), *Climate Change and Migration: Rethinking Policies for Adaptation and Disaster Risk Reduction* (UN University, 2011).
Lieber, Michael D (ed), *Exiles and Migrants in Oceania* (The University Press of Hawaii, 1977).
Lowe, Vaughan, *International Law* (Oxford University Press, 2007).
Marek, Krystyna, *Identity and Continuity of States in Public International Law* (Librairie E Droz, 1954).
McAdam, Jane, *Complementary Protection in International Refugee Law* (Oxford University Press, 2007).
——(ed), *Climate Change and Displacement: Multidisciplinary Perspectives* (Hart Publishing, 2010).
Myers, Norman and Jennifer Kent, *Environmental Exodus: An Emergent Crisis in the Global Arena* (Climate Institute, 1995).
Nordquist, Myron H, *The United Nations Convention on the Law of the Sea 1982: A Commentary* (Martinus Nijhoff, 1985).
O'Connell, Daniel P, *State Succession in Municipal Law and International Law* (Cambridge University Press, 1967).
Ollapally, Deepa M, *The Politics of Extremism in South Asia* (Cambridge University Press, 2009).
Parry, Martin L and others (eds), *Climate Change 2007: Impacts, Adaptation and Vulnerability: Contribution of Working Group II to the Fourth Assessment Report of the Intergovernmental Panel on Climate Change* (Cambridge University Press, 2007).
Piguet, Etienne, Antoine Pécoud, and Paul de Guchteneire (eds), *Migration and Climate Change* (Cambridge University Press and UNESCO Publishing, 2011).
Powles, Michael (ed), *Pacific Futures* (Pandanus Books, 2006)
Price, Matthew E, *Rethinking Asylum: History, Purpose, and Limits* (Cambridge University Press, 2009).

Saul, Ben and others, *Climate Change and Australia: Warming to the Global Challenge* (Federation Press, forthcoming, 2012).
Shain, Yossi, *Governments-in-Exile in Contemporary World Politics* (Routledge, 1991).
Shaw, Malcolm N, *International Law* (6th edn, Cambridge University Press, 2008).
Solomon, Susan and others (eds), *Climate Change 2007: The Physical Science Basis: Contribution of Working Group I to the Fourth Assessment Report of the Intergovernmental Panel on Climate Change* (Cambridge Universiy Press, 2007).
Stephens, Tim, *International Courts and Environmental Protection* (Cambridge University Press, 2009).
Stern, Nicholas, *The Economics of Climate Change: The Stern Review* (Cambridge University Press, 2007).
Talmon, Stefan, *Recognition of Governments in International Law: With Particular Reference to Governments in Exile* (Oxford University Press, 1998).
Tegart, William JM, Gordon W Sheldon, and D Colin Griffiths (eds), *Climate Change: The IPCC Impacts Assessment* (Australian Government Publishing Service, 1990).
William JM, Gordon W Sheldon, and D Colin Griffiths (eds), *Climate Change: The IPCC Impacts Assessment* (Australian Government Publishing Servies, 1990).
UNHCR, *The State of the World's Refugees: Human Displacement in the New Millennium* (Oxford University Press, 2006).
Verheyen, Roda, *Climate Change Damage and International Law: Prevention Duties and State Responsibility* (Martinus Nijhoff, 2005).
Watson, Robert T and others (eds), *IPCC Special Report: The Regional Impacts of Climate Change: An Assessment of Vulnerability: Summary for Policymakers* (IPCC, November 1997).
Weis, Paul, *Nationality and Statelessness in International Law* (2nd rev edn, Sijthoff and Noordhoff International Publishers, 1979).
Wisner, Ben and others, *At Risk: Natural Hazards, People's Vulnerability and Disasters* (2nd edn, Routledge, 2004).
Zolberg, Aristide R and others, *Escape from Violence: Conflict and the Refugee Crisis in the Developing World* (Oxford University Press, 1989).

2. ARTICLES, CHAPTERS, REPORTS, AND OCCASIONAL PAPERS

Advisory Opinion on the Legality of the Threat or Use of Nuclear Weapons [1996] ICJ Rep 226.
Afsar, Rita, 'Internal Migration and the Development Nexus: The Case of Bangladesh' (Refugee and Migratory Movements Research Unit, Dhaka and DFID, 2003)
Agnew, John, 'The Territorial Trap: The Geographical Assumptions of International Relations Theory' (1994) 1 Review of International Political Economy 53.
Ain o Salish Kendra (ASK), *Human Rights in Bangladesh 2008* (ASK, 2009).
Aleinikoff, T Alexander, 'International Legal Norms on Migration: Substance without Architecture' in Ryszard Cholewinski, Richard Perrechoud, and Euan MacDonald (eds), *International Migration Law: Developing Paradigms and Key Challenges* (TMC Asser Press, 2007).
Alston, Philip, 'Peoples' Rights: Their Rise and Fall' in Philip Alston (ed), *Peoples' Rights* (Oxford University Press, 2001).
Alston, Philip and Gerard Quinn, 'The Nature and Scope of States Parties' Obligations under the International Covenant on Economic, Social and Cultural Rights' (1987) 9 Human Rights Quarterly 156.

Ammer, Margit and others, 'Rechtsstellung und rechtliche Behandlung von Umweltflüchtlingen', Study 54/2010 commissioned by the Umweltbundesamt (Federal Ministry of the Environment, 2010).

Anderegg, William RL and others, 'Expert Credibility in Climate Change' (2010) 107 Proceedings of the National Academy of Sciences 12107.

Appleyard, Reginald T and Charles W Stahl, *South Pacific Migration: New Zealand Experience and Implications for Australia* (Commonwealth of Australia, 1995).

Aqorau, Transform, 'Rethinking Sovereignty in the Pacific' in Michael Powles (ed), *Pacific Futures* (Pandanus Books, 2006).

Arsenault, Louis-Georges (UNICEF), 'Adapting Humanitarian Disaster Response to Climate-Induced Displacement' Persons' (Nansen Conference on Climate Change and Displacement in the 21st Century, Oslo, 7 June 2011).

Asia Pacific Forum of National Human Rights Institutions, *Human Rights and the Environment: Reference Paper* (2007).

——*Human Rights and the Environment: Final Report and Recommendations* (2007).

Asian Development Bank, *Asian Development Outlook 2011: South–South Economic Links* (Asian Development Bank, 2011).

—— *Country Partnership Strategy: Tuvalu 2008–2012* (Asian Development Bank, 2008).

Australian Academy of Science, *The Science of Climate Change: Questions and Answers* (Australian Academy of Science, 2010),

Australian Labor Party, *Our Drowning Neighbours: Labor's Policy Discussion Paper on Climate Change in the Pacific* (ALP Policy Discussion Paper, 2006).

Ball, Rochelle, 'The Pacific Seasonal Workers Pilot Scheme (PSWPS) and Implications for Pacific Development' (2009) <http://peb.anu.edu.au/pdf/2009/PNG/ppp/Rochelle_Ball_Pacific_Update_081209.pdf>.

Barnett, Jon, 'The Effect of Aid on Capacity to Adapt to Climate Change: Insights from Niue' (2008) 60 Political Science 31.

Barnett, Jon and W Neil Adger, 'Climate Change, Human Security and Violent Conflict' (2007) 26 Political Geography 639.

—— 'Climate Dangers and Atoll Countries' (2003) 61 Climatic Change 321.

Barnett, Jon and Natasha Chamberlain, 'Migration as Climate Change Adaptation: Implications for the Pacific' in Bruce Burson (ed), *Climate Change and Migration: South Pacific Perspectives* (Institute of Policy Studies, Victoria University of Wellington, 2010).

Barnett, Jon and Michael Webber, 'Accommodating Migration to Promote Adaptation to Climate Change' Policy Research Working Paper 5270 (World Bank, 2010).

—— 'Migration as Adaptation: Opportunities and Limits' in Jane McAdam (ed), *Climate Change and Displacement: Multidisciplinary Perspectives* (Hart Publishing, 2010).

—— *Accommodating Migration to Promote Adaptation to Climate Change* (Commission on Climate Change and Development, 2009).

Bedford, Richard and Charlotte Bedford, 'International Migration and Climate Change: A Post-Copenhagen Perspective on Options for Kiribati and Tuvalu' in Bruce Burson (ed), *Climate Change and Migration: South Pacific Perspectives* (Institute of Policy Studies, Victoria University of Wellington, 2010).

Berman, Nathaniel, '"But the Alternative is Despair": European Nationalism and the Modernist Renewal of International Law' (1993) 106 Harvard Law Review 1792.

Betts, Alexander, 'Introduction: Global Migration Governance' in Alexander Betts (ed), *Global Migration Governance* (Oxford University Press, 2011).
—— 'Towards a "Soft Law" Framework for the Protection of Vulnerable Irregular Migrants' (2010) 22 International Journal of Refugee Law 209.
Betts, Alexander and Esra Kaytaz, 'National and International Responses to the Zimbabwean Exodus: Implications for the Refugee Protection Regime' (2009) UNHCR New Issues in Refugee Research, Research Paper No 175.
Betts, Alexander and Jean-François Durieux, 'Convention Plus as a Norm-Setting Exercise' (2007) 20 Journal of Refugee Studies 509.
Beyani, Chaloka, 'A Human Rights Based Approach to Protection of Environmentally Displaced Persons' (Nansen Conference on Climate Change and Displacement in the 21st Century, Oslo, 7 June 2011).
—— 'Preparing for a Warmer World: Towards a Global Governance System to Protect Climate Refugees' (2010) 10 Global Environmental Politics 60.
Biermann, Frank and Ingrid Boas, 'Protecting Climate Refugees: The Case for a Global Protocol' (2008) 50(6) Environment 8.
—— 'Preparing for a Warmer World: Towards a Global Governance System to Protect Climate Refugees' (November 2007) Global Governance Working Paper No 33.
Black, Richard, 'Environmental Refugees: Myth or Reality?' (2001) UNHCR New Issues in Refugee Research, Working Paper No 34.
Black, Richard, Dominic Kniveton, and Kerstin Schmidt-Verkerk, 'Migration and Climate Change: Towards an Integrated Assessment of Sensitivity' (2011) 43 Environment and Planning A 431.
Blackman, Jeffrey L, 'State Successions and Statelessness: The Emerging Right to an Effective Nationality under International Law' (1998) 19 Michigan Journal of International Law 1141.
Boano, Camillo, Roger Zetter, and Tim Morris, 'Environmentally Displaced People: Understanding the Linkages between Environmental Change, Livelihoods and Forced Migration' (2008) Forced Migration Policy Briefing 1 (Refugee Studies Centre, University of Oxford).
Boyle, Alan, 'Human Rights and the Environment: A Reassessment' (UNEP paper revised, 2010) <http://www.unep.org/environmentalgovernance/LinkClick.aspx?fileticket=GccCLN-brmg%3D&tabid=2046&language=en-US>.
Bridgeman, Natalie L, 'Human Rights Litigation under the ATCA as a Proxy for Environmental Claims' (2003) 6 Yale Human Rights and Development Law Journal 1.
Bronen, Robin, 'Climate-Induced Community Relocations: Creating an Adaptive Governance Framework based in Human Rights Doctrine' (2011) 35 New York University Review of Law and Social Change 356.
Brown, Oli, 'Migration and Climate Change' (2008) IOM Migration Research Series No 31.
—— 'The Numbers Game' (2008) 31 Forced Migration Review 8.
—— 'The Rights of Peoples in Modern International Law' in James Crawford (ed), *The Rights of Peoples* (Clarendon Press, 1988).
Brownlie, Ian, 'The Relations of Nationality in Public International Law' (1963) 39 British Year Book of International Law 284.
Byravan, Sujatha and Sudhir Chella Rajan, 'The Ethical Implictions of Sea-Level Rise due to Climate Change' (2010) 24 Ethics and International Affairs 239.
—— 'Providing New Homes for Climate Change Exiles' (2006) 6 Climate Policy 247.
Cameron, James and Durwood Zaelke, 'Global Warming and Climate Change: An Overview of the International Legal Process' (1990) 5 American University Journal of International Law & Policy 249.

Campbell, John, 'Climate-Induced Community Relocation in the Pacific: The Meaning and Importance of Land' in Jane McAdam (ed), *Climate Change and Displacement: Multidisciplinary Perspectives* (Hart Publishing, 2010).

Campbell, John, Michael Goldsmith, and Kanyathu Koshy, *Community Relocation as an Option for Adaptation to the Effects of Climate Change and Climate Variability in Pacific Island Countries (PICs)* (Asia-Pacific Network for Global Change Research 2005).

Campbell, Kurt M and others, *The Age of Consequences: The Foreign Policy and National Security Implications of Global Climate Change* (Center for Strategic and International Studies and Center for a New American Security, November 2007).

Caney, Simon, 'Climate Change, Human Rights and Moral Thresholds' in Stephen Humphreys (ed), *Human Rights and Climate Change* (Cambridge University Press, 2010).

Carozza, Paolo G, 'Human Dignity and Judicial Interpretation of Human Rights: A Reply' (2008) 19 European Journal of International Law 931.

Cashman, Peter and Ross Abbs, 'Liability in Tort for Damage arising from Human-Induced Climate Change' in Rosemary Lyster (ed), *In the Wilds of Climate Law* (Australian Academic Press, 2010).

Cassese, Antonio, 'Can the Notion of Inhuman and Degrading Treatment Be Applied to Socio-Economic Conditions?' (1991) 2 European Journal of International Law 141.

Castles, Stephen, 'Afterword: What Now? Climate-Induced Displacement after Copenhagen' in Jane McAdam (ed), *Climate Change and Displacement: Multidisciplinary Perspectives* (Hart Publishing, 2010).

—— 'Environmental Change and Forced Migration: Making Sense of the Debate' (2002) UNHCR New Issues in Refugee Research, Working Paper No 70.

'Chairperson's Summary' (Nansen Conference on Climate Change and Displacement in the 21st Century, Oslo, 6–7 June 2011).

Chan, Johannes MM, 'The Right to a Nationality as a Human Right: The Current Trend Towards Recognition' (1991) 12 Human Rights Law Journal 1.

Christian Aid, *Human Tide: The Real Migration Crisis* (Christian Aid, 2007).

CNA Corporation, 'National Security and the Threat of Climate Change' (2007).

Cohen, Roberta and Megan Bradley, 'Disasters and Displacement: Gaps in Protection' (2010) 1 Journal of International Humanitarian Legal Studies 95.

—— 'Losing Ground? Tuvalu, the Greenhouse Effect and the Garbage Can' (2003) 44 Asia Pacific Viewpoint 89.

Connell, John, 'Environmental Change, Economic Development and Emigration in Tuvalu' (1999) 22 Pacific Studies 1.

—— 'Paradise Left? Pacific Island Voyagers in the Modern World' in James T Fawcett and Benjamin V Cariño (eds), *Pacific Bridges: The New Immigration from Asia and the Pacific Islands* (Center for Migration Studies, 1987).

Cooper, Jessica B, 'Environmental Refugees: Meeting the Requirements of the Refugee Definition' (1998) 6 New York University Environmental Law Journal 480.

Cordes-Holland, Owen, 'The Sinking of the Strait: The Implications of Climate Change for Torres Strait Islanders' Human Rights Protected by the *ICCPR*' (2008) 9 Melbourne Journal of International Law 405.

Corlett, David, 'Tuvalunacy, or the Real Thing?', *Inside Story* (27 November 2008).

Council of Europe, Presidency Note to Strategic Committee on Immigration Frontiers and Asylum 'Proposal for a Council Directive on Minimum Standards for the Qualification and Status of Third Country Nationals and Stateless Persons as Refugees or as Persons Who Oherwise Need International Protection' Doc 12148/02 LIMTE ASILE 43 (20 September 2002).

Council of Europe Parliamentary Assembly, Committee on Migration, Refugees and Population, 'Environmentally Induced Migration and Displacement: A 21st Century Challenge', Doc 11785 (23 December 2008).

—— 'Outcome of Proceedings of Working Party on Asylum', Doc 6128/C1 LIMITE ASILE 15 (16 February 2011).

Council of the European Union, 'Climate Change and International Security' (Report from the Commission and the Secretary-General/High Representative to the European Council) Doc No 7249/08 (3 March 2008).

—— 'Report from the Commission and the Secretary-General/High Representative to European Council: Climate Change and International Security', Doc 7249/08 LIMITE POLGEN 27, PESC 289, ENER 69, ENV 141, DEVGEN 28, RELEX 146 (3 March 2008).

—— 'Outcome of Proceedings', Doc 12199/02 LIMITE ASILE 45 (25 September 2002).

—— 'Presidency Note to Strategic Committee on Immigration, Frontiers and Asylum on 25 September 2002: Proposal for a Council Directive on Minimum Standards for the Qualification and Status of Third Country Nationals and Stateless Persons as Refugees or as Persons Who Otherwise Need International Protection', Doc 12148/02 LIMITE ASILE 43 (20 September 2002).

—— 'Note from President to Asylum Working Party: Discussion Paper on Subsidiary Protection' Doc 13167/99 LIMITE ASILE 41 (19 November 1999).

Craven, Matthew CR, 'The Problem of State Succession and the Identity of States under International Law' (1998) 9 European Journal of International Law 142.

CRIDEAU and CRDP, Faculty of Law and Economic Science, University of Limoges, 'Draft Convention on the International Status of Environmentally-Displaced Persons' (2008) 4 Revue Européene de Droit de l'Environnement 375 and updated version May 2010 <http://www.cidce.org/>.

Crisp, Jeff, '"Who Has Counted the Refugees?" UNHCR and the Politics of Numbers' (1999) UNHCR New Issues in Refugee Research, Working Paper No 12.

de Haan, Arjan, 'Livelihoods and Poverty: The Role of Migration—A Critical Review of the Migration Literature' (1999) 36 Journal of Development Studies 1.

de Haas, Hein, 'Turning the Tide? Why Development Will Not Stop Migration' (2007) 38 Development and Change 819.

Defence White Paper, *Defending Australia in the Asia Pacific Century: Force 2030* (Commonwealth of Australia, 2009).

den Heijer, Maarten, 'Whose Rights and Which Rights? The Continuing Story of *Non-Refoulement* under the European Convention on Human Rights' (2008) 10 European Journal of Migration and Law 277.

Department of Education, Employment and Workplace Relations (Australia), 'Interim Evaluation of the Pacific Seasonal Worker Pilot Scheme: Executive Summary' (2010).

Displacement Solutions, *Climate Change Displaced Persons and Housing, Land and Property Rights: Preliminary Strategies for Rights-Based Planning and Programming to Resolve Climate-Induced Displacement* (Displacement Solutions, 2009).

Docherty, Bonnie and Tyler Giannini, 'Confronting a Rising Tide: A Proposal for a Convention on Climate Change Refugees' (2009) 33 Harvard Environmental Law Review 349.

Doran, Peter T and Maggie Kendall Zimmerman, 'Examining the Scientific Consensus on Climate Change' (2009) 90 Eos, Transactions, American Geophysical Union 22.

Draft Report of the Working Group on the Universal Periodic Review: Republic of the Marshall Islands, UN Doc A/HRC/WG.6/9/L.10 (15 November 2010).

Drury, A Cooper, Richard S Olson and Douglas A Van Belle, 'The Politics of Humanitarian Aid: US Foreign Disaster Assistance 1964–1995' (2005) 67 Journal of Politics 454.

Dun, Olivia, François Gemenne and Robert Stojanov, 'Environmentally Displaced Persons: Working Definitions for the EACH-FOR Project' (2007) <http://www.each-for.eu/documents/Environmentally_Displaced_Persons_-_Working_Definitions.pdf> accessed 1 July 2011.

Dupuy, Pierre Marie, 'A Hard Look at Soft Law' (1990) 82 American Society of International Law Proceedings 371.

—— 'The Elusive "War Refugee": Violence and Discrimination in the Construct of "Refugee" as a Legal Category' (Refugee Studies Centre Lecture, University of Oxford, 10 November 2010).

Durieux, Jean-François, 'Salah Sheekh is a Refugee: New Insights into Primary and Subsidiary Forms of Protection' (2008) Refugee Studies Centre Working Paper Series No 49.

Durieux, Jean-François, and McAdam, Jane, '*Non-Refoulement* through Time: The Case for a Derogation Clause to the Refugee Convention in Mass Influx Emergencies' (2004) 16 International Journal of Refugee Law 4.

EACH-FOR, 'Synthesis Report' (14 May 2009) <http://www.each-for.eu/documents/EACH-FOR_Synthesis_Report_090515.pdf>.

Earthjustice, 'Global Warming and Human Rights: Testimony of Martin Wagner before the Inter-American Commission on Human Rights' (1 March 2007).

Edwards, Alice, 'Refugee Status Determination in Africa' (2006) 14 African Journal of International and Comparative Law 204.

Ehrhart, Charles and others, *Humanitarian Implications of Climate Change: Mapping Emerging Trends and Risk Hotspots* (2nd edn, OCHA and CARE, 2009).

Elahi, K Maudood, 'Impacts of Riverbank Erosion and Flood in Bangladesh: An Introduction' in K Maudood Elahi, K Saleh Ahmed and M Mafizuddin (eds), *Riverbank Erosion, Flood and Population Displacement in Bangladesh* (Riverbank Impact Erosion Study, Jahangirnagar University, 1991).

El-Hinnawi, Essam, *Environmental Refugees* (United Nations Environment Programme, 1985).

Elliott, Lorraine, 'Climate Migration and Climate Migrants: What Threat, Whose Security?' in Jane McAdam (ed), *Climate Change and Displacement: Multidisciplinary Perspectives* (Hart Publishing, 2010).

Engers, JF, 'The United Nations Travel and Identity Document for Namibians' (1971) 65 American Journal of International Law 571.

Equity BD, 'Climate Change Induced Forced Migrants: In Need of Dignified Recognition under a New Protocol' (December 2009).

European Council on Refugees and Exiles, 'Complementary Protection in Europe' (July 2009).

Falk, Richard, 'The Rights of Peoples (in Particular Indigenous Peoples)' in James Crawford (ed), *The Rights of Peoples* (Clarendon Press, 1988).

Farbotko, Carol, 'Representing Climate Change Space: Islographs of Tuvalu' (PhD thesis, University of Tasmania, School of Geography and Environmental Studies, 2008).

—— 'Tuvalu and Climate Change: Constructions of Environmental Displacement in the *Sydney Morning Herald*' (2005) 87 B Geografiska Annaler 279.

—— Assistant High Commissioner (Protection), 'The Refugee Convention at 60: Still Fit for Its Purpose?' (Workshop on Refugees and the Refugee Convention 60 Years On: Protection and Identity, Prato, 2 May 2011).

Feller, Erika, 'Statement by Ms Erika Feller, Director, Department of International Protection, UNHCR' (Strategic Committee on Immigration, Frontiers and Asylum, Brussels, 6 November 2002).

Ferris, Elizabeth, 'Humanitarian Perspectives on "Protection of Persons in the Event of Disasters"' (Roundtable on 'Canada and International Law', Department of Foreign Affairs and International Trade, 21 April 2009) <http://www.brockings.edu/speeches/2009/0421_natural_disasters_ferris.aspx>.

—— 'Natural Disaster- and Conflict-Induced Displacement: Similarities, Differences and Inter-Connections', prepared for the Annual Meeting of the Society for Applied Anthropology, Panel: 'Expanding the Borders of Disaster-Caused Displacement Research' (27 March 2008).

Fife, Rolf Einar, 'The Duty to Render Assistance at Sea: Some Reflections after *Tampa*' in Jarna Petman and Jan Klabbers (eds), *Nordic Cosmopolitanism: Essays in International Law for Martti Koskenniemi* (Martinus Nijhoff, 2003).

Findlay, Allan and Alistair Geddes, 'Critical Views on the Relationship between Climate Change and Migration: Some Insights from the Experience of Bangladesh' in Etienne Piguet, Antoine Pécoud, and Paul de Guchteneire (eds), *Migration and Climate Change* (Cambridge University Press and UNESCO Publishing, 2011).

Fischer, Horst, 'International Disaster Response Law Treaties: Trends, Patterns, and Lacunae' in IFRC, *International Disaster Response Laws, Principles and Practice: Reflections, Prospects and Challenges* (IFRC and Red Crescent Societies, 2003).

Fitzmaurice, Malgosia, 'Responsibility and Climate Change' (2010) 53 German Yearbook of International Law 89.

Fitzpatrick, Joan, 'Temporary Protection of Refugees: Elements of a Formalized Regime' (2000) 94 American Journal of International Law 279.

Foresight, *Migration and Global Envirnomental Change: Future Challenges and Opportunities: Final Project Report* (The Government Office for Science, London, 2011).

Foster, Michelle, '*Non-Refoulement* on the Basis of Socio-Economic Deprivation: The Scope of Complementary Protection in International Human Rights Law' [2009] New Zealand Law Review 257.

Framework on Durable Solutions for Internally Displaced Persons, UN Doc A/HRC/13/21/Add.4 (9 February 2010) <http://www2.ohchr.org/english/bodies/hrcouncil/docs/13session/A.HRC.13.21.Add.4_en.pdf>.

Franck, Thomas M and Paul L Hoffman, 'The Right of Self-Determination in Very Small Places' (1976) 8 New York University Journal of International Law and Politics 331.

Freestone, David and John Pethick, 'Sea Level Rise and Maritime Boundaries: International Implications of Impacts and Responses' in Gerald H Blake (ed), *Maritime Boundaries* (Routledge, 1994).

Friends of the Earth Australia, *A Citizen's Guide to Climate Refugees* (rev edn, Friends of the Earth Australia, 2007).

Fritze, Jessica G and others, 'Hope, Despair and Transformation: Climate Change and the Promotion of Mental Health and Wellbeing' (2008) 2 International Journal of Mental Health Systems 13.

Gebre, Yntiso, 'Resettlement and the Unnoticed Losers: Impoverishment Disasters among the Gumz Hosts in Ethiopia' (2003) 62 Human Organisation 50.

—— 'Climate-Induced Population Displacements in a 4°C+ World' (2011) 369 Philosophical Transactions of the Royal Society A 182.

—— 'How They Became the Human Face of Climate Change: Research and Policy Interactions in the Birth of the "Environmental Migration" Concept' in Etienne Piguet, Antoine Pécoud, and Paul de Guchteneire (eds), *Migration and Climate Change* (Cambridge University Press and UNESCO Publishing, 2011).

—— 'Environmental Changes and Migration Flows: Normative Frameworks and Policy Responses' (PhD thesis, Institut d'Etudes Politiques de Paris and University of Liège, 2009).

Gemenne, François, 'Climate Change and Forced Displacements: Towards a Global Environmental Responsibility? The Case of the Small Island Developing States (SIDS) in the South Pacific Ocean' (2006) Les Cahiers du CEDEM, Université de Liège <http://www.cedem.ulg.ac.be/m/cdc/12.pdf>.

German Advisory Council on Global Change, 'World in Transition: Climate Change as a Security Risk: Summary for Policy Makers' (WBGU Secretariat, 2007).

Gerrard, Nelson, ' "A Matter of Honour": The Constitution of New Iceland: Then and Now' (Building a New Relationship Conference, University of Manitoba, 27 October 2000).

Glahn, Benjamin, ' "Climate Refugees"? Addressing the International Legal Gaps—Part II' (*International Bar Association,* undated) <http://www.ibanet.org/Article/Detail.aspx?ArticleUid=3E9DB1B0-659E-432B-8EB9-C9AEEA53E4F6>.

Glazebrook, Susan, 'Human Rights and the Environment' (2009) 40 Victoria University of Wellington Law Review 293.

Gleditsch, Nils Petter, 'Armed Conflict and the Environment: A Critique of the Literature' (1998) 35 Journal of Peace Research 381.

Global Humanitarian Forum, *The Anatomy of a Silent Crisis: Human Impact Report: Climate Change* (Global Humanitarian Forum, 2009).

Global Protection Cluster Working Group, *Handbook for the Protection of Internally Displaced Persons* <http://www.unhcr.org/refworld/pdfid/4790cbc02.pdf>.

Goodwin-Gill, Guy S, 'Europe and the Right to be Granted Asylum' (Address at the Australian National University, Canberra, 29 September 2010) 10 (copy on file with author)

—— '*Non-Refoulement* and the New Asylum Seekers' (1986) 26 Virginia Journal of International Law 897.

Government of Bhutan, *Bhutan National Adaptation Programme of Action* (undated, submitted May 2006).

Government of Gambia, *National Adaptation Programme of Action (NAPA) on Climate Change* (November 2007).

Government of the Maldives, *National Adaptation Programme of Action* (2007).

Government of Mozambique, *National Adaptation Programme of Action (NAPA)* (December 2007).

Government of the People's Republic of Bangladesh, *Bangladesh Climate Change Strategy and Action Plan* (September 2009).

—— *Climate Change and Bangladesh* (September 2007).

Government of the Republic of Kiribati, *National Adaptation Program of Action (NAPA)* (January 2007).

Government of Samoa, *National Adaptation Programme of Action* (2005).

Government of Sao Tomé and Príncipe, *National Adaptation Programmes of Action on Climate Change* (December 2006).

Government of the Solomon Islands, *National Adaptation Programme of Action* (November 2008).

Government of Tuvalu, *Tuvalu's National Adaptation Programme of Action* (May 2007).

Government of Uganda, *Climate Change: Uganda National Adaptation Programmes of Action* (2007).

Government of the United Republic of Tanzania, *National Adaptation Programme of Action (NAPA)* (January 2007).

Grant, Tom D, 'Defining Statehood: The Montevideo Convention and its Discontents' (1999) 37 Columbia Journal of Transnational Law 403.

Green, Donna and Kirsty Ruddock, 'Could Litigation Help Torres Strait Islanders Deal with Climate Impacts?' (2009) 9 Sustainable Development Law and Policy 23.

Grote Stoutenburg, Jenny, 'When Do States Disappear? Thresholds of Effective Statehood and the Continued Recognition of "Deterritorialized" Island States' (Threatened Island Nations: Legal Implications of Rising Seas and a Changing Climate Conference, New York, 23–25 May 2011).

—— 'Statement' (Nansen Conference on Climate Change and Displacement in the 21st Century, Oslo, 6 June 2011).

Guterres, António, 'High Commissioner's Closing Remarks: 2010 Dialogue on Protection Gaps and Responses' (UNHCR, Geneva, 9 December 2010).

—— 'Keynote Address' (Ten Years of Guiding Principles on Internal Displacement: Achievements and Future Challenges, Oslo, 16 October 2008).

—— 'Opening Statement' (58th Session of the Executive Committee of the High Commissioner's Programme, Geneva, 1 October 2007).

Hall, Nina, 'Climate Change and Organizational Change in UNHCR' in Michelle Leighton, Xiomeng Shen and Koko Warner (eds), *Climate Change and Migration: Rethinking Policies for Adaptation and Disaster Risk Reduction* (UN University, 2011).

—— 'Rethinking Climate Refugees and Climate Conflict: Rhetoric, Reality and the Politics of Policy Discourse' (2010) 22 Journal of International Development 233.

Hartmann, Betsy, 'Population, Environment and Security: A New Trinity' (1998) 10 Environment and Urbanization 113.

Hathaway, James C, 'Is Refugee Status Really Elitist? An Answer to the Ethical Challenge' in Jean-Yves Carlier and Dirk Vanheule (eds), *Europe and Refugees: A Challenge?* (Kluwer Law International, 1997).

Hathaway, James C and R Alexander Neve, 'Making International Refugee Law Relevant Again: A Proposal for Collectivized and Solution-Oriented Production' (1997) 10 Haravard Human Rights Journal 115.

Haug, Ruth, 'Forced Migration, Processes of Return and Livelihood Construction among Pastoralists in Northern Sudan' (2002) 26 Disasters 70.

Henderson, John, 'Governance and Constitutional Issues: Reflections on Current Problems and Future Options' in Michael Powles (ed), *Pacific Futures* (Pandanus Books, 2006).

Henry, Sabine, Bruno Schoumaker, and Cris Beauchemin, 'The Impact of Rainfall on the First Out-Migration: A Multi-Level Event-History Analysis in Burkina Faso' (2004) 25 Population and Environment 423.

Hermann, Elfriede, 'Emotions, Agency and the Dis/placed Self of the Banabans in Fiji' in Toon van Meijl and Jelle Miedema (eds), *Shifting Images of Identity in the Pacific* (KITLV Press, 2004).

Hernández-Truyol, Berta E, and Matthew Hawk, 'Traveling the Boundaries of Statelessness: Global Passports and Citizenship' (2005) 52 Cleveland State Law Review 97.

Hodgkinson, David and others, '"The Hour When the Ship Comes In": A Convention for Persons Displaced by Climate Change' (2010) 36 Monash University Law Review 69.

Hogan, Claude E S, 'Seeking Refuge in the Mother Country: UK Maintained Strict Exclusion Policy Despite Erupting Volcano' (Montserrat Country Conference, Montserrat, 13–14 November 2002).

Homer-Dixon, Thomas F, 'On the Threshold: Environmental Changes as Causes of Acute Conflict' (1991) 16 International Security 76.

Huber, Daniel G and Jay Gulledge, 'Extreme Weather and Climate Change: Understanding the Link and Managing the Risk' (Center for Climate Energy Solution, December 2011).

—— 'Lessons from Past Forced Resettlement for Climate Change Migration' in Etienne Piguet, Antoine Pécoud, and Paul de Guchteneire (eds), *Migration and Climate Change* (Cambridge University Press and UNESCO Publishing, 2011).

—— 'Climate Change-Induced Mobility and the Existing Migration Regime in Asia and the Pacific' in Jane McAdam (ed), *Climate Change and Displacement: Multidisciplinary Perspectives* (Hart Publishing, 2010).

Hugo, Graeme, 'Environmental Concerns and International Migration' (1996) 30 International Migration Review 105.

Hulme, Mike, 'Commentary: Climate Refugees: Cause for a New Agreement?' (2008) 50(6) Environment 50.

Humphreys, Stephen, 'Competing Claims: Human Rights and Climate Harms' in Stephen Humphreys (ed), *Human Rights and Climate Change* (Cambridge University Press, 2010).

Hutton, David and C Emdad Haque, 'Human Vulnerability, Dislocation and Resettlement: Adaptation Processes of River-bank Erosion-induced Displacees in Bangladesh' (2004) 28 Disasters 41.

Immigration and New Zealand: A Statement of Current Immigration Policy (Immigration Division, Department of Labour, updated January 1986) (on file with author).

Inter-Agency Standing Committee Operational Guidelines on the Protection of Persons in Situations of Natural Disasters (Brookings–Bern Project on Internal Displacement 2011) <http://www2.ohchr.org/english/issues/idp/docs/OperationalGuidelines.pdf>.

Inter-Agency Standing Committee Task Force on Climate Change, 'Mid-Term Report for 2010' (9 July 2010).

—— 'Quick Guide to Climate Change Adaptation' (2010).

Inter-Agency Standing Committee, 'Disaster Risk Reduction Strategies and Risk Management Practices: Critical Elements for Adaptation to Climate Change' (Submission to the UNFCCC Ad Hoc Working Group on Long-Term Cooperative Action by the Informal Taskforce on Climate Change of the IASC and the International Strategy for Disaster Reduction, 11 November 2008).

——'Protection of Internally Displaced Persons: Inter-Agency Standing Committee Policy Paper' (December 1999).

Inter-Agency Standing Committee Working Group on Migration/Displacement and Climate Change, 'Displacement and Climate Change: Towards Defining Categories of Affected Persons' (20 September 2008) Working Paper First Draft.

Intergovernmental Panel on Climate Change, *Climate Change 2007: Impacts, Adaptation and Vulnerability: Contribution of Working Group II to the Fourth Assessment Report of the Intergovernmental Panel on Climate Change* (Cambridge University Press, 2008).

—— *Climate Change 2007: Synthesis Report: Summary for Policymakers* (Cambridge University Press, 2007).

—— *Climate Change: The IPCC Scientific Assessment: Final Report of Working Group I* (Cambridge University Press, 1990).

—— *IPCC Second Assessment: Climate Change 1995: A Report of the Intergovernmental Panel on Climate Change* (IPCC, 1995).

International Federation of the Red Cross and Red Crescent Societies, *World Disasters Report: Focus on Recovery* (IFRC and Red Crescent Societies, 2001).

—— 'Fourth Report on the Protection of Persons in the Event of Disasters by Mr Eduardo Valencia-Ospina, Special Rapporteur', 63rd Session, UN Doc A/CN.4/643 (11 May 2011).

——'Third Report on the Protection of Persons in the Event of Disasters by Mr Eduardo Valencia-Ospina, Special Rapporteur', 62nd Session, UN Doc A/CN.4/629 (31 March 2010).

—— 'Protection of Persons in the Event of Disasters, Statement of the Chairman of the Drafting Committee' (20 July 2010).

—— 'Report of the International Law Commission on the Work of its Sixty-Second Session (3 May–4 June and 5 July–6 August 2010)', UN Doc A/65/10.

—— 'Report on the Work of its Sixty-Second Session (3 May–4 June and 5 July–6 August 2010)', UN Doc A/65/10.

—— 'Fifth Report on the Expulsions of Aliens by Mr Maurice Kamto, Special Rapporteur', 61st Session, UN Doc A/CN.4/611 (27 March 2009).

—— 'Second Report on the Protection of Persons in the Event of Disasters by Mr Eduardo Valencia-Ospina, Special Rapporteur', 61st Session, UN Doc A/CN.4/615 (7 May 2009).

——'Protection of Persons in the Event of Disasters: Memorandum by the Secretariat: Addendum', 60th Session, UN Doc A/CN.4/590/Add.1 (26 February 2008).

—— 'Preliminary Report on the Protection of Persons in the Event of Disasters by Mr Eduardo Valencia-Ospina, Special Rapporteur', 60th Session, UN Doc A/CN.4/598 (5 May 2008).

—— 'Protection of Persons in the Event of Disasters: Memorandum by the Secretariat', 60th Session, UN Doc A/CN.4/590 (11 December 2007).

—— 'Report on the Work of its Fifty-Eighth Session (1 May–9 June and 3 July–11 August 2006)', UN Doc A/61/10.

International Law Commission, 'Prevention of Transboundary Harm from Hazardous Activities' in 'Report of the International Law Commission on the Work of its Fifty-Third Session', UN Doc A/56/10 (2001).

International Organization for Migration, *Assessing the Evidence: Environment, Climate Change and Migration in Bangladesh* (IOM, 2010).

—— *World Migration Report 2010: The Future of Migration: Building Capacities for Change* (IOM, 2010).

—— *Compendium of IOM's Activities in Migration, Climate Change and the Environment* (IOM, 2009).

—— Policy Brief, 'Migration, Climate Change and the Environment' (May 2009).

—— 'Discussion Note: Migration and the Environment', 94th session, Doc No MC/INF/288 (1 November 2007).

—— *Migration and the Environment* (IOM, 1992).

Islam, Waliul and M Ataur Rahman, 'Riverbank Erosion in Bangladesh: An Overview of Field Observation' (REIS Newsletter (3), Riverbank Impact Erosion Study, Jahangirnagar University, 1987).

Jacobs, Rebecca E, 'Treading Deep Waters: Substantive Law Issues in Tuvalu's Threat to Sue the United States in the International Court of Justice' (2005) 14 Pacific Rim Law and Policy Journal 103.

Jacobsen, Jodi L, 'Environmental Refugees: A Yardstick of Habitability' (1988) Worldwatch Paper 86.

Jäger, Jill and HL Ferguson (eds), *Climate Change: Science, Impacts and Policy*, Proceedings of the Second World Climate Conference (Cambridge University Press, 1991).

Jastram, Kate, 'Economic Harm as a Basis for Refugee Status and the Application of Human Rights Law to the Interpretation of Economic Persecution' in James C Simeon (ed), *Critical Issues in International Refugee Law: Strategies Towards Interpretative Harmony* (Cambridge University Press, 2010).

Julca, Alex, 'Multidimensional Re-creation of Vulnerabilities and Potential for Resilience in International Migration' (2011) 49 International Migration 30.

Kälin, Walter, 'Conceptualising Climate-Induced Displacement' in Jane McAdam (ed), *Climate Change and Displacement: Multidisciplinary Perspectives* (Hart Publishing, 2010).

—— 'Keynote Address: Climate Change, Migration Patterns and the Law' (International Association of Refugee Law Judges 8th World Conference, Cape Town, 28 January 2009).

—— 'The Climate Change–Displacement Nexus' (16 July 2008) <http://www.brookings.edu/speeches/2008/0716_climate_change_kalin.aspx?p=1> accessed 2 December 2008.

Kälin, Walter, and Claudine Haenni Dale, 'Disaster Risk Mitigation: Why Human Rights Matter' (2008) 31 Forced Migration Review 38.

Kälin, Walter, and Nina Schrepfer, 'Protecting People Crossing Borders in the Context of Climate Change: Normative Gaps and Possible Approaches' (Study for the Swiss Ministry of Foreign Affairs, April 2011) forthcoming as a paper in the UNHCR Legal and Protection Policy Research series.

Kemfert, Claudia and Richard SJ Tol, 'Equity, International Trade and Climate Policy' (2002) 2 International Environmental Agreements 23.

Kibreab, Gaim, 'Climate Change and Human Migration: A Tenuous Relationship?' (2009) 20 Fordham Environmental Law Review 357.

—— 'Environmental Causes and Impact of Refugee Movements: A Critique of the Current Debate' (1997) 21 Disasters 20.

Klein, Richard JT and Robert J Nicholls, 'Assessment of Coastal Vulnerability to Climate Change' (1999) 28 Ambio 182.

Kniveton, Dominic, 'Climate Change and Migration: Developing Methodologies' (Environment, Forced Migration and Social Vulnerability Conference presentation, Bonn, 9 October 2008).

Kniveton, Dominic and others, 'Climate Change and Migration: Improving Methodologies to Estimate Flows' (2008) IOM Migration Research Series No 33.

Knox, John H, 'Climate Change and Human Rights Law' (2009) 50 Virginia Journal of International Law 163.

—— 'Linking Human Rights and Climate Change at the United Nations' (2009) 33 Harvard Environmental Law Review 477.

Kolmannskog, Vikram and Finn Myrstad, 'Environmental Displacement in European Asylum Law' (2009) 11 European Journal of Migration and Law 313.

Kolmannskog, Vikram and Lisetta Trebbi, 'Climate Change, Displacement and Protection: A Multi-Track Approach' (2010) 92 International Review of the Red Cross 713.

Koser, Khalid, 'Climate Change and Internal Displacement: Challenges to the Normative Framework' in Etienne Piguet, Antoine Pécoud, and Paul de Guchteneire (eds), *Migration and Climate Change* (Cambridge University Press and UNESCO Publishing, 2011).

Koskenniemi, Martti, 'The Wonderful Artificiality of States' (1994) 88 Proceedings of the American Society of International Law 22.

Kothari, Uma, 'Migration and Chronic Poverty' (2002) Institute for Development Policy and Management, University of Manchester, Working Paper No 16.

Kozoll, Christopher M, 'Poisoning the Well: Persecution, the Environment, and Refugee Status' (2004) 15 Colorado Journal of International Environmental Law and Policy 271.

Laczko, Frank and Elizabeth Collett, 'Assessing the Tsunami's Effect on Migration',

Migration Information Source (Migration Information Source, April 2005) <http://www.migrationinformation.org/USfocus/display.cfm?id=299> accessed 13 August 2011.

Lambert, Hélène and Theo Farrell, 'The Changing Character of Armed Conflict and the Implications for Refugee Protection Jurisprudence' (2010) 22 International Journal of Refugee Law 237.

Legal Services, Immigration and Refugee Board of Canada, 'Consolidated Grounds in the *Immigration and Refugee Protection Act*: Persons in Need of Protection: *Risk to Life or Risk of Cruel and Unusual Treatment or Punishment*' (15 May 2002).

Leighton, Michelle, 'Desertification and Migration' in Pierre Marc Johnson, Karel Mayrand, and Marc Pacquin (eds), *Governing Global Desertification* (Ashgate, 2006).

Lein, Haakon, 'The Poorest and the Most Vulnerable? On Hazards, Livelihoods and Labelling of Riverine Communities in Bangladesh' (2009) 30 Singapore Journal of Tropical Geography 98.

—— 'Human Rights Obligations and Accountability in the Face of Climate Change' (2010) 38 Georgia Journal of International and Comparative Law 543.

Limon, Marc, 'Human Rights and Climate Change: Constructing a Case for Political Action' (2009) 33 Harvard Environmental Law Review 439.

Lonergan, Steve, 'The Role of Environmental Degradation in Population Displacement' (1998) 4 Environmental Change and Security Project Report 5.

Loper, Kelley, 'Human Rights, *Non-Refoulement* and the Protection of Refugees in Hong Kong' (2010) 22 International Journal of Refugee Law 404.

Loucaides, Loukis G, 'Environmental Protection through the Jurisprudence of the European Convention on Human Rights' in Loukis G Loucaides, *The European Convention on Human Rights: Collected Essays* (Martinus Nijhoff, 2007).

Loughry, Maryanne and Jane McAdam, 'Kiribati—Relocation and Adaptation' (2008) 31 Forced Migration Review 51.

Lowe, A Vaughan, 'Sustainable Development and Unsustainable Arguments' in Alan E Boyle and David Freestone (eds), *International Law and Sustainable Development: Past Achievements and Future Challenges* (Oxford University Press, 1999).

Mantouvalou, Virginia, 'Work and Private Life: *Sidabras and Dziautas v Lithuania*' (2005) 30 European Law Review 573.

Martin, Susan F, 'Managing Environmentally Induced Migration' in Frank Laczko and Christine Aghazarm (eds), *Migration, Environment and Climate Change: Assessing the Evidence* (IOM, 2009).

Mavroudi, Elizabeth, 'Palestinians and Pragmatic Citizenship: Negotiating Relationships between Citizenship and National Identity in Diaspora' (2008) 39 Geoforum 307.

McAdam, Jane, 'Individual Risk, Armed Conflict and the Standard of Proof in Complementary Protection Claims: The European Union and Canada Compared' in James C Simeon (ed), *Critical Issues in International Refugee Law: Strategies for Interpretative Harmony* (Cambridge University Press, 2010).

—— 'The Refugee Convention as a Rights Blueprint for People in Need of International Protection' in Jane McAdam (ed) *Forced Migration, Human Rights and Security* (Hart Publishing, 2008).

—— 'The European Union Qualification Directive: The Creation of a Subsidiary Protection Regime' (2005) 17 International Journal of Refugee Law 461.

McAdam, Jane and Maryanne Loughry, 'We Aren't Refugees', *Inside Story* (30 June 2009).

McAdam, Jane and Ben Saul, 'An Insecure Climate for Human Security? Climate-Induced Displacement and International Law' in Alice Edwards and Carla Ferstman (eds), *Human Security and Non-Citizens: Law, Policy and International Affairs* (Cambridge University Press, 2010).

—— 'Displacement with Dignity: International Law and Policy Responses to Climate Change Migration and Security in Bangladesh' (2010) 53 German Yearbook of International Law 233.

McConnell, Fiona, 'De facto, Displaced, Tacit: The Sovereign Articulations of the Tibetan Government-in-Exile' (2009) 28 Political Geography 343.

—— 'Governments-in-Exile: Statehood, Statelessness and the Reconfiguration of Territory and Sovereignty' (2009) Geography Compass 1901.

McCrudden, Christopher, 'Human Dignity and Judicial Interpretation of Human Rights' (2008) 19 European Journal of International Law 655.

McCue, Gregory S, 'Environmental Refugees: Applying International Environmental Law to Involuntary Migration' (1993) 6 Georgetown International Environmental Law Review 151.

McLeod, Deborah, 'Potential Impacts of Climate Change Migration on Pacific Families Living in New Zealand' in Bruce Burson (ed), *Climate Change and Migration: South Pacific Perspectives* (Institute of Policy Studies, Victoria University of Wellington, 2010).

Menghistu, F, 'The Satisfaction of Survival Requirements' in BG Ramcharan (ed), *The Right to Life in International Law* (Martinus Nijhoff, 1985).

Meze-Hausken, Elisabeth, 'Migration Caused by Climate Change: How Vulnerable are People in Dryland Areas?' (2000) 5 Mitigation and Adaptation Strategies for Global Change 379.

Millar, Ilona, 'There's No Place Like Home: Human Displacement and Climate Change' (2007) 14 Australian International Law Journal 71.

Millner, Felicity and Kirsty Ruddock, 'Climate Litigation: Lessons Learned and Future Opportunities' (2011) 36 Alternative Law Journal 27.

Morrissey, James, 'Environmental Change and Forced Migration: A State of the Art Review' (2009) Refugee Studies Centre, University of Oxford, Background Paper <http://www.rsc.ox.ac.uk/events/environmental-change-and-migration/EnvChangeandFmReviewWS.pdf> accessed 4 August 2011.

—— 'Rural–Urban Migration in Ethiopia' (2008) 31 Forced Migration Review 28.

Mortreux, Colette and Jon Barnett, 'Climate Change, Migration and Adaptation in Funafuti, Tuvalu' (2009) 19 Global Environmental Change 105.

Mushkat, Roda, 'Hong Kong and Succession of Treaties' (1997) 46 International and Comparative Law Quarterly 181.

—— 'Environmental Refugees: An Emergent Security Issue' (13th Economic Forum, Prague, 23–27 May 2005).

—— 'Environmental Refugees: A Growing Phenomenon of the 21st Century' (2002) 357 Philosophical Transactions of the Royal Society: Biological Sciences 609.

—— 'Environmental Refugees' (1997) 19 Population and Environment 167.

—— 'Environmentally-Induced Displacements: The State of the Art' in IOM/UNHCR, *Environmentally-Induced Population Displacements and Environmental Impacts Resulting from Mass Migration* (IOM/UNHCR International Symposium, Geneva, 1996).

Myers, Norman, 'Environmental Refugees in a Globally Warmed World' (1993) 43 BioScience 752.

Nicholls, Robert J and Richard SJ Tol, 'Impacts and Responses to Sea-Level Rise: A Global Analysis of the SRES Scenarios over the Twenty-First Century' (2006) 364 Philosophical Transactions of the Royal Society A 1073.

Norwegian Refugee Council, *Displacement due to Natural Hazard-Induced Disasters: Global Estimates for 2009 and 2010* (June 2011).

Oliver-Smith, Anthony, 'Communities after Catastrophe: Reconstructing the Material, Reconstituting the Social' in Stanley E Hyland (ed), *Community Building in the Twenty-First Century* (School of American Research Press, 2005).

Oraá, Jaime, 'The Protection of Human Rights in Emergency Situations under Customary International Law' in Guy S Goodwin-Gill and Stephan Talmon (eds), *The Reality of International Law: Essays in Honour of Ian Brownlie* (Oxford University Press, 1999).

Oreskes, Naomi, 'Beyond the Ivory Tower: The Scientific Consensus on Climate Change' (2004) 306 Science 1686.

Oxman, Bernard H, 'Human Rights and the United Nations Convention on the Law of the Sea' (1998) 36 Columbia Journal of Transnational Law 404.

Panayotou, Theodore, Jeffrey D Sachs, and Alix P Zwane, 'Compensation for "Meaningful Participation" in Climate Change Control: A Modest Proposal and Empirical Analysis' (2002) 43 Journal of Environmental Economics and Management 437.

Park, Susin, 'Climate Change and the Risk of Statelessness: The Situation of Low-Lying Island States', UNHCR Legal and Protection Policy Research Series, PPLA/2011/04 (May 2011).

Paskal, Cleo, 'How Climate Change is Pushing the Boundaries of Security and Foreign Policy' (2007) Chatham House Briefing Paper 4.

Paton, Kathryn L, 'At Home or Abroad: Tuvaluans Shaping a Tuvaluan Future' (Thesis submitted for Master of Development Studies, School of Geography, Environment and Earth Sciences, Victoria University of Wellington, 2009).

Paul, Bimal Kanti, 'Relief Assistance to 1998 Flood Victims: A Comparison of the Performance of the Government and NGOs' (2003) 169 The Geographical Journal 75.

Pelling, Mark and Juha I Uitto, 'Small Island Developing States: Natural Disaster Vulnerability' (2001) 3 Environmental Hazards 49.

Pender, James, 'Community-Led Adaptation in Bangladesh' (2008) 31 Forced Migration Review 54.

Penz, Peter, 'International Ethical Responsibilities to "Climate Change Refugees"' in Jane McAdam (ed), *Climate Change and Displacement: Multidisciplinary Perspectives* (Hart Publishing, 2010).

Perluss, Deborah and Joan F Hartman, 'Temporary Refuge: Emergence of a Customary Norm' (1986) 26 Virginia Journal of International Law 551.

'Petition to the Inter American Commission on Human Rights Seeking Relief from Violations Resulting from Global Warming Caused by Acts and Omissions of the United States' (7 December 2005) available at <http://www.earthjustice.org/library/legal_docs/petition-to-the-inter-american-commission-on-human-rights-on-behalf-of-the-inuit-circumpolar-conference.pdf> accessed 10 December 2009.

Poncelet, Alice, 'Bangladesh Case Study Report: "The Land of Mad Rivers"' (EACH-FOR, 2009).

Posner, Eric A, 'Climate Change and International Human Rights Litigation: A Critical Appraisal' (2007) 155 University of Pennsylvania Law Review 1925.

President Mohamed Nasheed (Maldives), 'Address to the United Nations General Assembly' (21 September 2009) <http://www.unmultimedia.org/tv/unifeed/d/13548.html>.

—— 'The Problems of Islands' (2000) 31 Victoria University of Wellington Law Review 427.

Quentin-Baxter, Alison, 'Niue's Relationship of Free Association with New Zealand' (1999) 30 Victoria University of Wellington Law Review 589.

Rahman, MZ Mizanur, 'Emigration and Development: The Case of a Bangladeshi Village' (2000) 38 International Migration 119.

Rahman, Reaz, 'Bangladesh's Fight against Terror: Lessons Learnt and Future Perspectives' in Mufley R Osmany and Mohammed Humayun Kabir (eds), *Global War on Terror: Bangladesh Perspective* (Bangladesh Institute of International and Strategic Studies, 2007).

Rahmstorf, Stefan and others, 'Recent Climate Observations Compared to Projections' (2007) 316 Science 709.

Raleigh, Clionadh, Lisa Jordan, and Idean Salehyan, 'Assessing the Impact of Climate Change on Migration and Conflict', Working Paper for Social Dimensions of Climate Change (World Bank, 2008).

Ramasamy, Sankar and others, 'The Recognised Seasonal Employer Policy: Seeking the Elusive Triple Wins for Development through International Migration' (2008) 23 Pacific Economic Bulletin 171.

Rashid, Syeda Rozana, 'Migration for Livelihood: Social Protection Issues of Rural Bangladeshis' in Chowdhury R Abrar and Janet Seeley (eds), *Social Protection and Livelihoods: Marginalised Migrant Workers of India and Bangladesh* (The University Press, 2009).

Rayfuse, Rosemary, 'W(h)ither Tuvalu? International Law and Disappearing States' (2009) University of New South Wales Faculty of Law Research Series Working Paper No 9 <http://law.bepress.com/unswwps/flrps09/art9/>.

Rayhan, Israt and Ulrike Grote, 'Coping with Floods: Does Rural–Urban Migration Play Any Role for Survival in Rural Bangladesh?' (2007) 1 Journal of Identity and Migration Studies 82.

Reed, RoseMary, 'Rising Seas and Disappearing Islands: Can Island Inhabitants Seek Redress under the Alien Tort Claims Act?' (2002) 11 Pacific Rim Law and Policy Journal 399.

Refugees International, 'Confronting Climate Displacement: Learning from Pakistan's Floods' (November 2010).

—— 'A Decision Framework for Environmentally Induced Migration' (2011) 49 International Migration 5.

Renaud, Fabrice and others, 'Control, Adapt or Flee: How to Face Environmental Migration?' (2007) 5 InterSecTions 19.

Report of the Committee to Review the Australian Overseas Aid Program (Commonwealth of Australia, 1984).

Robinson, Nehemiah, 'Convention relating to the Status of Stateless Persons: Its History and Interpretation' (Commentary, Institute of Jewish Affairs, World Jewish Congress, 1955, reprinted by UNHCR, 1997).

Röhl, Katharina, 'Fleeing Violence and Poverty: *Non-Refoulement* Obligations under the European Convention of Human Rights' (2005) UNHCR New Issues in Refugee Research, Working Paper No 111.

Sage, Yves-Louis, 'Self-Determination: "Shared Sovereignty": A Comparative Analysis' in Kennedy Graham (ed), *Models of Regional Governance for the Pacific: Sovereignty and the Future Architecture of Regionalism* (Canterbury University Press, 2008).

Schachter, Oscar, 'State Succession: The Once and Future Law' (1993) 33 Virginia Journal of International Law 253.

Scheinin, Martin, 'Economic and Social Rights as Legal Rights' in Allan Rosas and others (eds), *Economic, Social and Cultural Rights* (2nd edn, Kluwer, 2002).

Schmidt-Soltau, Kai, 'Protected Areas and Resettlement: What Scope for Voluntary Relocation?' (2007) 35 World Development 2182.

Schuck, Peter H, 'Refugee Burden-Sharing: A Modest Proposal' (1997) 22 Yale Journal of International Law 243.

Schwartz, Michelle, 'International Legal Protection for Victims of Environmental Abuse' (1993) 18 Yale Journal of International Law 355.

Schwartz, Peter and Doug Randall, 'An Abrupt Climate Change Scenario and its Implications for United States National Security' (October 2003).

Scott, Craig M, 'The Interdependence and Permeability of Human Rights Norms: Towards a Partial Fusion of the International Covenants on Human Rights' (1989) 27 Osgoode Hall Law Journal 769.

Seeley, Janet and others, '"The Family is Suffering": Challenges Faced by Migrants' Families Who Stay Behind in a Village in Rural Northwest Bangladesh' in Chowdhury R Abrar and Janet Seeley (eds), *Social Protection and Livelihoods: Marginalised Migrant Workers of India and Bangladesh* (The University Press, 2009).

—— *First Special Report: Government Response to the First Report from the International Development Committee, Session 1997–98: Montserrat* (HC 532) 16 February 1998.

—— *Sixth Report: Montserrat: Further Developments* (HC 726) 4 August 1998.

Select Committee on International Development (UK), *First Report, Montserrat* (HC 267) 18 November 1997.

—— *Economic Challenges facing Papua New Guinea and the Island States of the Southwest Pacific* (Commonwealth of Australia, 2009).

Senate Foreign Affairs, Defence and Trade Committee, *A Pacific Engaged: Australia's Relations with Papua New Guinea and the Island States of the South-West Pacific* (Commonwealth of Australia, 2003).

Shackley, Simon and Clair Gough, 'The Use of Integrated Assessment: An Institutional Analysis Perspective' (2002) Tyndall Centre for Climate Change Working Paper 14.

Shacknove, Andrew, 'Who is a Refugee?' (1985) 95 Ethics 274.

Shamshad, Rizwana, 'Politics and Origin of the India-Bangladesh Border Fence' (17th Biennial Conference of the Asian Studies Association of Australia, Melbourne, July 2008).

Shamsuddoha, Md and Rezaul Karim Chowdhury, 'Climate Refugee: Requires Dignified Recognition under a New Protocol' (EquityBD, April 2009).

—— 'The Environmental Jurisprudence of International Human Rights Tribunals' in Romina Picolotti and Jorge Daniel Taillant (eds), *Linking Human Rights and the Environment* (University of Arizona Press, 2003).

Shelton, Dinah, 'Environmental Rights' in Philip Alston (ed), *Peoples' Rights* (Oxford University Press, 2001).

Shen, Shawn, 'Noah's Ark to Save Drowning Tuvalu' (2007) 10 Just Change: Critical Thinking on Global Issues 18.

Shen, Shawn and François Gemenne, 'Contrasted Views on Environmental Change and Migration: The Case of Tuvaluan Migration to New Zealand' (2011) 49 International Migration e224.

Siddiqui, Tasneem, 'Introduction' in Tasneem Siddiqui (ed), *Migration and Development: Pro-Poor Policy Choice* (The University Press, 2005).

Siddiqui, Tasneem and Mohammad Jalal Uddin Sikder, 'Rural to Urban Migration for Domestic Work in Bangladesh' in Chowdhury R Abrar and Janet Seeley (eds), *Social Protection and Livelihoods: Marginalised Migrant Workers of India and Bangladesh* (The University Press, 2009).

Silverman, Martin G, 'Introduction: Locating Relocation in Oceania' in Michael D Lieber (ed), *Exiles and Migrants in Oceania* (The University Press of Hawaii, 1977).

—— 'Making Sense: A Study of a Banaban Meeting' in Michael D Lieber (ed), *Exiles and Migrants in Oceania* (The University Press of Hawaii, 1977).

Simma, Bruno, 'A Hard Look at Soft Law' (1990) 82 American Society of International Law Proceedings 371.

Sindico, Francesco, 'Climate Change: A Security (Council) Issue?' (2007) 1 Climate Change Law Review 29.

Smith, Dan and Janani Vivekananda, *A Climate of Conflict: The Links between Climate Change, Peace and War* (International Alert, 2007).

Sokolov, AP and others, 'Probalistic Forecast for Twenty-First-Century Climate Based on Uncertainties in Emissions (Without Policy) and Climate Parameters' (2009) 22 Journal of Climate 5175.

Soons, Alfred HA, 'The Effects of a Rising Sea Level on Maritime Limits and Boundaries' (1990) 37 Netherlands International Law Review 207.

Sopoaga, Enele S, 'Statement' (General Debate of the 56th Session of the United Nations General Assembly, New York, 16 November 2001).

Stahl, Charles W and Reginald T Appleyard, *Migration and Development in the Pacific Islands: Lessons from the New Zealand Experience* (AusAID, April 2007).

Storey, Hugo, 'The Internal Flight Alternative Test: The Jurisprudence Re-examined' (1998) 10 International Journal of Refugee Law 499.

Suhrke, Astri, 'Environmental Degradation and Population Flows' (1994) 47 Journal of International Affairs 473.

Suhrke, Astri and Visentin, Annamaria, 'The Environmental Refugee: A New Approach' [1991] Ecodecision 73.

Summary Conclusions of UNHCR's Expert Meeting on 'The Concept of Stateless Persons under International Law' (Prato, 27–28 May 2010).

Tabucanon, Gil Marvel and Brian Opeskin, 'The Resettlement of Nauruans in Australia: An Early Case of Failed Environmental Migration' (2011) 46 Journal of Pacific History (forthcoming).

Tacoli, Cecilia, 'Not Only Climate Change: Mobility, Vulnerability and Socio-Economic Transformations in Environmentally Fragile Areas of Bolivia, Senegal and Tanzania', Human Settlements Working Paper Series (International Institute for Environment and Development, 2011).

The Equal Rights Trust, *Unravelling Anomaly: Detention, Discrimination and the Protection Needs of Stateless Persons* (The Equal Rights Trust, 2010).

Thompson, Laura, Deputy Director General of IOM, 'Opening Remarks' (delivered by Peter Schatzer) (IOM International Dialogue on Migration 2011, Intersessional Workshop on Climate Change, Environmental Degradation and Migration, Geneva, 29 March 2011) 1.

Thompson, Martha and Izaskun Gaviria, 'Weathering the Storm: Lessons in Risk Reduction from Cuba' (Oxfam America, 2004).

Thürer, Daniel, 'Soft Law', in *Max Planck Encyclopedia of Public International Law* (Oxford University Press, online edition at <http://www.mpepil.com>).

—— 'The "Failed State" and International Law' (1999) 81 International Review of the Red Cross 731.

Tol, Richard SJ, 'Welfare Specification and Optimal Control of Climate Change: An Application of FUND' (2002) 24 Energy Economics 367.

Tol, Richard SJ and Roda Verheyen, 'State Responsibility and Compensation for Climate Change Damages: A Legal and Economic Assessment' (2004) 32 Energy Policy 1109.

Tong, Anote, President of Kiribati, 'Address' (Australian National University, Canberra, 19 June 2008) <http://news.anu.edu.au/?p=437>.

Türk, Volker, Director, Division of International Protection, UNHCR, 'Can Protection of Environmentally Displaced Persons Be Found in Existing Protection Regimes? What Are the Next Steps from a Protection Perspective?' (Nansen Conference on Climate Change and Displacement in the 21st Century, Oslo, 7 June 2011).

UN Committee on Economic, Social and Cultural Rights, General Comment No 2 'International Technical Assistance Measures (Art. 22 of the Covenant)', UN Doc E/1990/23 (2 February 1990).

—— General Comment No 3 'The Nature of States Parties Obligations (Art 2 Para 1 of the Covenant)', UN Doc E/1991/23 (14 December 1990).

—— General Comment No 7 'The Right to Adequate Housing (Art 11.1 of the Covenant): Forced Evictions', UN Doc E/1998/22 (20 May 1997).

—— General Comment No 11, 'Plans of Action for Primary Education (Article 14 of the International Covenant on Economic, Social and Cultural Rights)', UN Doc E/C.12/1999/4 (10 May 1999).

—— 'General Comment 12: The Right to Adequate Food (Art 11)', UN Doc E/C.12/1999/5 (12 May 1999).

—— 'General Comment No 13: The Right to Education (Art 13)', UN Doc E/C.12/1999/10 (8 December 1999).

—— 'General Comment No 14: The Right to the Highest Attainable Standard of Health (Art 12)', UN Doc E/C.12/2000/4 (11 August 2000).

—— 'General Comment No 15 (2002): The Right to Water (Arts 11 and 12)', UN Doc E/C.12/2002/11 (20 January 2003).

—— 'General Comment No 18 (2005): The Right to Work (Art 18)', UN Doc E/C.12/GC/18 (6 February 2006).

UN Committee on the Elimination of Racial Discrimination, 'Consideration of Reports Submitted by States Parties under Article 9 of the Convention: Concluding Observations of the Committee on the Elimination of Racial Discrimination: Japan', UN Doc CERD/C/JPN/CO/3-6 (16 March 2010).

UN Committee on the Rights of the Child, 'General Comment No 6 (2005): Treatment of Unaccompanied and Separated Children Outside Their Country of Origin', UN Doc CRC/GC/2005/6 (1 September 2005).

—— 'General Comment No 7 (2005): Implementing Child Rights in Early Childhood', UN Doc CRC/C/GC/7/Rev.1 (20 September 2006).

UN Development Programme, *Overcoming Barriers: Human Mobility and Development—Human Development Report 2009* (UNDP, 2009).

—— 'Report of the Secretary General, International Cooperation on Humanitarian Assistance in the Field of Natural Disasters, from Relief to Development', UN Doc A/60/227 (12 August 2005).

UN General Assembly, 'International Co-operation to Avert New Flows of Refugees: Note by the Secretary-General', UN Doc A/41/324 (13 May 1986).

—— 'General Comment No 6: The Right to Life (Art 6)' (30 April 1982).

—— 'General Comment No 14: Nuclear Weapons and the Right to Life (Art 6)' (9 November 1984).

UN Human Rights Committee, 'General Comment 15: The Position of Aliens under the Covenant' (11 April 1986).

——'General Comment 18: Non-Discrimination' (10 November 1989).

—— 'General Comment No 20: Replaces General Comment 7 concerning Prohibition of Torture and Cruel Treatment or Punishment (Art 7)' (10 March 1992).

—— 'General Comment No 31: The Nature of the General Legal Obligation Imposed on States Parties to the Covenant', UN Doc CCPR/C/21/Rev.1/Add.13 (26 May 2004).

UN Human Rights Council Advisory Committee, 'Report to the Human Rights Council on the First Session of the Advisory Committee', UN Doc A/HRC/AC/2008/1/L.11 (15 August 2008).

—— 'Report of the Office of the United Nations High Commissioner for Human Rights on the Relationship between Climate Change and Human Rights', UN Doc A/HRC/10/61 (15 January 2009).

UN Human Rights Council, 'Report of the Advisory Committee on its First Session (Geneva, 4–15 August 2008)', UN Doc A/HRC/10/2, A/HRC/AC/2008/1/2 (3 November 2008).

UN Office of the High Commissioner for Human Rights, 'Discussion Paper: Protecting the Human Rights of Internally Displaced Persons in Natural Disasters: Challenges in the Pacific' (Davui Printery, 2011).

—— 'Report of the Office of the United Nations High Commissioner for Human Rights on the Relationship between Climate Change and Human Rights', UN Doc A/HRC/10/61 (15 January 2009).

—— 'OHCHR Study on the Relationship between Climate Change and Human Rights: Submissions and Reference Documents Received' (2008).

UN Resident and Humanitarian Coordinator's Office, 'Longer-Term Disaster Displaced: A Forgotten Group' (Field Bulletin, Nepal, 11 July 2011).

UN Secretary-General, 'Climate Change and Its Possible Security Implications: Report of the Secretary-General', UN Doc A/64/350 (11 September 2009).

UN Special Rapporteur on the Right to Food, 'Implementation of General Assembly Resolution 60/251 of 15 March 2006 Entitled "Human Rights Council"', UN Doc A/HRC/4/30 (19 January 2007).

—— 'Prevention of Discrimination: Prevention of Discrimination and Protection of Indigenous Peoples: Expanded Working Paper by Françoise Hampson on the Human Rights Situation of Indigenous Peoples in States and Other Territories Threatened with Extinction for Environmental Reasons', UN Doc E/CN.4/Sub.2/2005/28 (16 June 2005).

—— 'The Human Rights Situation of Indigenous Peoples in States and Territories Threatened with Extinction for Environmental Reasons: Working Paper prepared by Mrs Françoise Hampson, Member of the Working Group on Indigenous Populations', UN Doc E/CN.4/Sub.2/AC.4/2004/CRP.1 (13 July 2005).

UN Sub-Commission on the Promotion and Protection of Human Rights, 'Prevention of Discrimination: Prevention of Discrimination and Protection of Indigenous Peoples: Report of the Working Group on Indigenous Populations on its Twenty-First Session', UN Doc E/CN.4/Sub.2/2003/22 (11 August 2003).

UN Under-Secretary-General for Humanitarian Affairs and Emergency Relief Coordinator, 'Opening Remarks at the Dubai International Humanitarian Aid and Development Conference and Exhibition "DIHAD 2008 Conference"' (Dubai, 8 April 2008).

UN-Habitat, *State of the World's Cities 2008/2009* (Earthscan, 2008).

UNHCR Executive Committee Conclusion No 103 (LVI) 'The Provision of International Protection including through Complementary Forms of Protection' (2005).

—— No 22 (XXXII) 'Protection of Asylum-Seekers in Situations of Large-Scale Influx' (1981)

UNHCR Executive Committee of the High Commissioner's Programme, 'UNHCR's Role in Support of an Enhanced Humanitarian Response for the Protection of Persons afffected by Natural Disasters', UN Doc EC/62/SC/CRP.19 (6 June 2011).

——'Draft Report of the 51st Meeting of the Standing Committee (21–23 Jnne 2011)' UN Doc EC/62/SC/CRP. 25 (16 September 2011).

—— 'Progress Report on Resettlement', UN Doc EC/61/SC/CRP.11 (31 May 2010).

—— 'Statelessness: Prevention and Reduction of Statelessness and Protection of Stateless Persons', UN Doc EC/57/SC/CRP.6 (14 February 2006).

UNHCR Expert Roundtable on Climate Change and Displacement, 'Summary of Deliberations on Climate Change and Displacement' (Bellagio, 22–25 February 2011).

UNHCR High Commissioner's Forum, 'Convention Plus Core Group on Addressing Irregular Secondary Movements of Refugees and Asylum-Seekers', UN Doc FORUM/2005/7 (8 November 2005).

—— 'Making Comprehensive Approaches to Resolving Refugee Problems More Systematic', UN Doc FORUM/2004/7 (16 September 2004).

UNHCR Regional Office for the Pacific, 'Roundtable Discussion on Climate Change and Human Security in the Pacific' (Suva, Fiji, 16 September 2008).

—— 'International Cooperation to Share Burden and Responsibilities', Discussion Paper (Expert Meeting in Amman, Jordan, 27–28 June 2011).

—— *Global Trends 2010* (UNHCR, 2010).

—— 'High Commissioner's Dialogue on Protection Challenges Breakout Session 1: Gaps in the International Protection Framework and in its Implementation, Report by the Co-Chairs' (8–9 December 2010).

—— *A Thematic Compilation of Executive Committee Conclusions* (4th edn, UNHCR, April 2009).

—— 'Forced Displacement in the Context of Climate Change: Challenges for States under International Law', Submission to the 6th Session of the Ad Hoc Working Group on Long-Term Cooperative Action under the Convention (20 May 2009).

UNHCR, 'Climate Change, Natural Disasters and Human Displacement: A UNHCR Perspective' (23 October 2008, revised 14 August 2009).

—— *Asylum in the European Union: A Study of the Implementation of the Qualification Directive* (UNHCR, 2007).

—— 'Note on International Protection', UN Doc A/AC.96/1008 (4 July 2005).

—— 'Guidelines on International Protection: "Internal Flight or Relocation Alternative" within the Context of Article 1A(2) of the 1951 Refugee Convention and/or 1967 Protocol relating to the Status of Refugees', UN Doc HCR/GIP/03/04 (23 July 2003).

—— *Handbook on Procedures and Criteria for Determining Refugee Status under the 1951 Convention and the 1967 Protocol relating to the Status of Refugees*, UN Doc HCR/IP/4/Eng/Rev.1 (2nd edn, 1992).

—— 'Protection of Persons of Concern to UNHCR Who Fall Outside the 1951 Convention: A Discussion Note', UN Doc EC/1992/SCP.CRP.5 (2 April 1992).

UNHCR supported by the International Organization for Migration and the Norwegian Refugee Council, 'Climate Change and Statelessness: An Overview' (Submission to the 6th Session of the Ad Hoc Working Group on Long-Term Cooperative Action (AWG-LCA 6) under the UN Framework Convention on Climate Change (UNFCCC), 1–12 June 2009).

US Citizenship and Immigration Services, 'Temporary Protected Status—Haiti' (last updated 1 February 2010).

Vanheule, Dirk, 'The Qualification Directive: A Milestone in Belgian Asylum Law' in Karin Zwaan (ed), *The Qualification Directive: Central Themes, Problem Issues, and Implementation in Selected Member States* (Wolf Legal Publishers, 2007).

Verheyen, Roda, 'The Legal Framework of Adaptation and Adaptive Capacity' in Richard JT Klein, Saleemal Huq, and Joel B Smith (eds), *Climate Change, Adaptive Capacity and Development* (Imperial College Press, 2003).

Voigt, Christina, 'State Responsibility for Climate Change Damages' (2008) 77 Nordic Journal of International Law 1.

Waldock, Humphrey, 'General Course on Public International Law' (1962) 106 Recueil des Cours de l'Académie de Droit International 5.

Waldron, Jeremy, 'Liberal Rights: Two Sides of the Coin' in Jeremy Waldron, *Liberal Rights: Collected Papers 1981–1991* (Cambridge University Press, 1993).

Warbrick, Colin, 'Recognition of States: Recent European Practice' in Malcolm D Evans (ed), *Aspects of Statehood and Institutionalism in Contemporary Europe* (Dartmouth Press, 1997).

—— 'Climate Change Induced Displacement: Adaptation Policy in the Context of the UNFCCC Climate Negotiations' (2011) UNHCR Legal and Protection Policy Research Series, PPLA/2011/02.

Warner, Koko, 'Assessing Institutional and Governance Needs Related to Environmental Change and Human Migration' (Study Team on Climate-Induced Migration, German Marshall Fund of the United States, June 2010).

Warner, Koko and others, *In Search of Shelter: Mapping the Effects of Climate Change on Human Migration and Displacement* (CARE International and others, 2009).

Wasem, Ruth E and Karma Ester, 'Temporary Protected Status: Current Immigration Policy and Issues' (Congressional Research Service, 9 September 2010).

Williams, Angela, 'Turning the Tide: Recognizing Climate Change Refugees in International Law' (2008) 30 Law & Policy 502.

Wood, William B, 'Ecomigration: Linkages between Environmental Change and Migration', in Aristide R Zolberg and Peter M Benda (eds), *Global Migrants, Global Refugees: Problems and Solutions* (Berghahn Books, 2001).

World Bank Office Dhaka, 'Dhaka: Improving Living Conditions for the Urban Poor', Bangladesh Development Series Paper No 17 (World Bank, 2007).

—— *World Development Report: Development and Climate Change* (International Bank for Reconstruction and Development/World Bank, 2010).

World Bank, *Migration and Remittances Factbook* (International Bank for Reconstruction and Development/World Bank, 2008).

World Food Programme, International Federation of Red Cross and Red Crescent Societies (IFRC), and UN Office for the Coordination of Humanitarian Affairs, 'Addressing the Humanitarian Challenges of Climate Change: Regional and National Perspectives: Preliminary Findings from the IASC Regional and National Level Consultations' (2009).

Yasmin, Lailufar, 'Bangladesh–India Relations: Neighbourliness Beyond Terrorism' in Imtiaz Ahmed (ed), *Terrorism in the 21st Century: Perspectives from Bangladesh* (The University Press, 2009).

Zagor, Matthew, 'Elementary Considerations of Humanity' in Karine Bannelier, Theodore Christakis, and Sarah Heathcote (eds), *The ICJ and the Evolution of International Law: The Enduring Impact of the Corfu Channel Case* (Routledge, 2012).

Zartman, I William, 'Introduction: Posing the Problem of State Collapse' in I William Zartman (ed), *Collapsed States: The Disintegration and Restoration of Legitimate Authority* (Lynne Rienner Publishers, 1995).

—— 'Protecting Environmentally Displaced People: Developing the Capacity of Legal and Normative Frameworks' (2011) Refugee Studies Centre Research Report.

Zetter, Roger, 'Legal and Normative Frameworks' (2008) 31 Forced Migration Review 62.

Zukerman, Wendy, 'Shape-Shifting Islands Defy Sea-Level Rise', *New Scientist* (London, 2 June 2010).

Index